VOLUME FORTY ONE

THE ENZYMES

RNA Modification

VOLUME FORTY ONE

THE ENZYMES

RNA Modification

Edited by

GUILLAUME F. CHANFREAU
University of California, Los Angeles, CA, United States

ELSEVIER

ACADEMIC PRESS
An imprint of Elsevier

Academic Press is an imprint of Elsevier
50 Hampshire Street, 5th Floor, Cambridge, MA 02139, United States
525 B Street, Suite 1800, San Diego, CA 92101-4495, United States
The Boulevard, Langford Lane, Kidlington, Oxford OX5 1GB, United Kingdom
125 London Wall, London, EC2Y 5AS, United Kingdom

First edition 2017

ISBN: 978-0-12-811777-4
ISSN: 1874-6047

For information on all Academic Press publications
visit our website at https://www.elsevier.com/books-and-journals

Publisher: Zoe Kruze
Acquisition Editor: Kirsten Shankland
Editorial Project Manager: Naomi Robertson
Production Project Manager: Surya Narayanan Jayachandran
Cover Designer: Vicky Pearson

Typeset by SPi Global, India

CONTENTS

CONTRIBUTORS

Paul F. Agris
The RNA Institute, State University of New York, Albany, NY, United States

Juan D. Alfonzo
Department of Microbiology, Ohio State Biochemistry Program, The Center for RNA Biology, The Ohio State University, Columbus, OH, United States

Peter A. Beal
University of California, Davis, CA, United States

Guillaume F. Chanfreau
University of California, Los Angeles, CA, United States

Meemanage D. De Zoysa
University of Rochester Medical Center, Center for RNA Biology, Rochester, NY, United States

Emily Eruysal
The RNA Institute, State University of New York, Albany, NY, United States

Brian D. Gregory
University of Pennsylvania; Cell and Molecular Biology Graduate Program; Genomics and Computational Biology Graduate Program, University of Pennsylvania, Philadelphia, PA, United States

Anthony K. Henras
Laboratoire de Biologie Moléculaire Eucaryote, Centre de Biologie Intégrative, Université de Toulouse, CNRS, UPS, Toulouse, France

Yves Henry
Laboratoire de Biologie Moléculaire Eucaryote, Centre de Biologie Intégrative, Université de Toulouse, CNRS, UPS, Toulouse, France

Ya-Ming Hou
Thomas Jefferson University, Philadelphia, PA, United States

Odile Humbert
Laboratoire de Biologie Moléculaire Eucaryote, Centre de Biologie Intégrative, Université de Toulouse, CNRS, UPS, Toulouse, France

Olga Kolaj-Robin
Institut de Génétique et de Biologie Moléculaire et Cellulaire (IGBMC); Centre National de Recherche Scientifique (CNRS) UMR 7104; Institut National de Santé et de Recherche Médicale (INSERM) U964; Université de Strasbourg, Illkirch, France

Isao Masuda
Thomas Jefferson University, Philadelphia, PA, United States

Ryuma Matsubara
Thomas Jefferson University, Philadelphia, PA, United States

Katherine M. McKenney
Department of Microbiology, Ohio State Biochemistry Program, The Center for RNA Biology, The Ohio State University, Columbus, OH, United States

Amithi Narendran
The RNA Institute, State University of New York, Albany, NY, United States

Célia Plisson-Chastang
Laboratoire de Biologie Moléculaire Eucaryote, Centre de Biologie Intégrative, Université de Toulouse, CNRS, UPS, Toulouse, France

Yves Romeo
Laboratoire de Biologie Moléculaire Eucaryote, Centre de Biologie Intégrative, Université de Toulouse, CNRS, UPS, Toulouse, France

Mary Anne T. Rubio
Department of Microbiology, Ohio State Biochemistry Program, The Center for RNA Biology, The Ohio State University, Columbus, OH, United States

Kathryn Sarachan
The RNA Institute, State University of New York, Albany, NY, United States

Bertrand Séraphin
Institut de Génétique et de Biologie Moléculaire et Cellulaire (IGBMC); Centre National de Recherche Scientifique (CNRS) UMR 7104; Institut National de Santé et de Recherche Médicale (INSERM) U964; Université de Strasbourg, Illkirch, France

Joanna I. Sulkowska
Center of New Technologies, University of Warsaw, Warsaw, Poland

Ryuichi Takase
Thomas Jefferson University, Philadelphia, PA, United States

Lee E. Vandivier
University of Pennsylvania; Cell and Molecular Biology Graduate Program, University of Pennsylvania, Philadelphia, PA, United States

Ville Y.P. Väre
The RNA Institute, State University of New York, Albany, NY, United States

Yuru Wang
University of California, Davis, CA, United States

Yi-Tao Yu
University of Rochester Medical Center, Center for RNA Biology, Rochester, NY, United States

Yuxuan Zheng
University of California, Davis, CA, United States

PREFACE

The field of RNA modifications has been experiencing a renaissance in recent years. Decades ago, some of the earliest studies focusing on RNA modifications revealed their role in influencing the decoding of genetic messenger RNAs by transfer RNAs (tRNAs). While RNA modifications were initially identified in these tRNAs and in ribosomal RNAs (rRNAs) because of their high abundance, the advent of modern sequencing technologies has enabled researchers to investigate the global extent of RNA modifications on the transcriptome. These recent studies have revealed a large variety of modifications and of modification sites in all major classes of cellular RNAs. The enzymes that catalyze the addition of chemical modifications to RNA are as diverse as the chemical groups that are added to RNAs. These enzymes range from single protein chains to large ribonucleoprotein complexes, in which the presence of an RNA guide targets the modification enzyme to its RNA substrates. The specificity of RNA-modifying enzymes is essential, as misguided RNA modifications can negatively impact the process of genetic decoding during translation. Moreover, the presence of RNA modifications has a major influence on the metabolic fate of these modified RNAs, as these modifications can promote their correct folding, nucleocytoplasmic trafficking, translation, or turnover. Finally, RNA modifications can provide recognition sites for proteins that interact with modified RNAs. This volume of *The Enzymes* highlights some of the most recent advances in the field of RNA modifications and attempts to provide a comprehensive review of the structure, function, and specificity of RNA-modifying enzymes as well as a survey of the role of these modifications on RNA metabolism. Throughout these chapters, it will become apparent that these modifications provide key functional groups, as disruption of RNA modification reactions due to mutations in RNA-modifying enzymes results in a variety of inherited diseases, highlighting the importance of these modifications in cellular metabolism.

I thank all the authors who generously contributed to this volume of *The Enzymes*. I would also like to express my gratitude to Naomi Robertson at Elsevier for handling and copyediting this volume.

Guillaume F. Chanfreau

CHAPTER ONE

The Importance of Being Modified: The Role of RNA Modifications in Translational Fidelity

Paul F. Agris[1], Amithi Narendran, Kathryn Sarachan, Ville Y.P. Väre, Emily Eruysal
The RNA Institute, State University of New York, Albany, NY, United States
[1]Corresponding author: e-mail address: pagris@albany.edu

Contents

The Enzymes, Volume 41
ISSN 1874-6047
http://dx.doi.org/10.1016/bs.enz.2017.03.005

Abstract

The posttranscriptional modifications of tRNA's anticodon stem and loop (ASL) domain represent a third level, a third code, to the accuracy and efficiency of translating mRNA codons into the correct amino acid sequence of proteins. Modifications of tRNA's ASL domain are enzymatically synthesized and site specifically located at the anticodon wobble position-34 and 3′-adjacent to the anticodon at position-37. Degeneracy of the 64 Universal Genetic Codes and the limitation in the number of tRNA species require some tRNAs to decode more than one codon. The specific modification chemistries and their impact on the tRNA's ASL structure and dynamics enable one tRNA to decode cognate and "wobble codons" or to expand recognition to synonymous codons, all the while maintaining the translational reading frame. Some modified nucleosides' chemistries prestructure tRNA to read the two codons of a specific amino acid that shares a twofold degenerate codon box, and other chemistries allow a different tRNA to respond to all four codons of a fourfold degenerate codon box. Thus, tRNA ASL modifications are critical and mutations in genes for the modification enzymes and tRNA, the consequences of which is a lack of modification, lead to mistranslation and human disease. By optimizing tRNA anticodon chemistries, structure, and dynamics in all organisms, modifications ensure translational fidelity of mRNA transcripts.

1. INTRODUCTION

1.1 Universal Genetic Code

The Universal Genetic Code of 64 triplet codons encoded in DNA and transcribed into each messenger RNA (mRNA) is degenerate (Fig. 1). As a general rule, 61 codons represent the 20 amino acids during translation on the ribosome. Three codons are read by protein factors as translation stops. Transfer RNAs (tRNAs) decode the mRNA codons while bringing the cognate amino acid to the ribosome for incorporation into the growing peptide chain. The ribosome is one of the most complex of enzymes. With the bacterial and eukaryote decoding or aminoacyl site (A-site) and the peptidyl transferase center (PTC) consisting almost completely of RNA [1,2], the ribosome is an RNA-based enzyme with large RNA substrates. Even the mitochondrial ribosome, which is more highly proportioned of proteins, has an RNA-rich decoding and PTC [3]. In humans, the ribosome's substrates include some 25,000 mRNAs for proteins of less than 100 amino

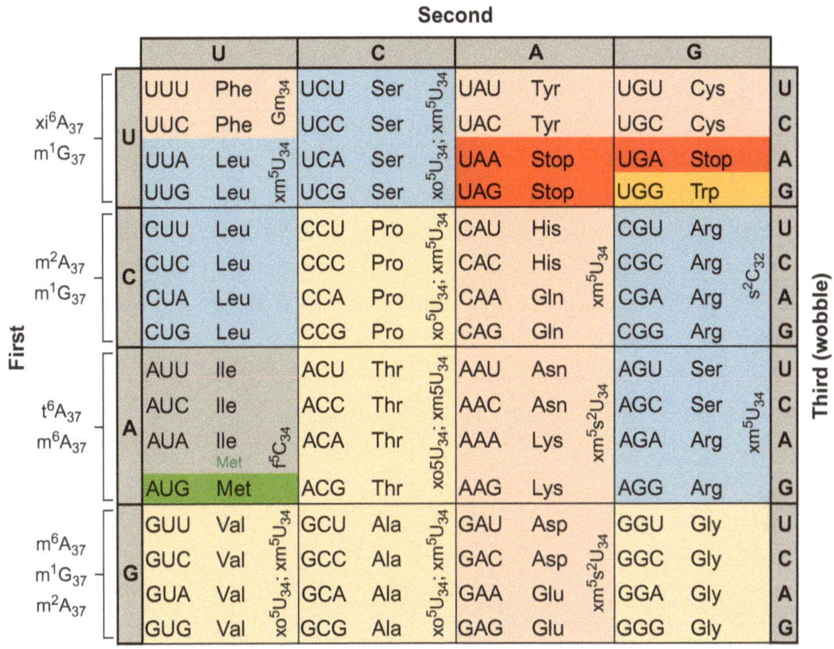

Fig. 1 The Universal Genetic Code. Twofold degenerate codons are highlighted in *tan*; threefold (Ile) in *gray*, fourfold in *yellow*, and sixfold in *blue*. The single codons of Met and Trp are highlighted in *green* and *orange*, respectively, and the three stop codons are highlighted in *red*. The figure is annotated with the abbreviations for those modified nucleosides found in the anticodon domain of tRNAs responding to the codons and discussed in this review. The chemical structures and full names of the modifications are found in Fig. 3.

acids in length to more than 20,000 amino acids [4]; GTP; a number of protein factors; and some 40 transfer RNAs (tRNA) each aminoacylated with one specific amino acid. The ribosome's enzymatic activity in translating mRNA codons into the amino acid sequences is a "processive" mechanism of action. The ribosome sequentially and covalently links the growing peptide chain to the amino terminus of the incoming amino acid that is bound to its specific tRNA through an ester bond. tRNAs are aminoacylated at their universal 3′-terminal adenosine by amino acid-specific, aminoacyl-tRNA synthetases. The aminoacylated tRNA is brought to the ribosome by a protein, elongation factor, that recognizes that the tRNA is aminoacylated, but is not engaged in codon recognition and specificity.

Codon recognition by aminoacylated tRNA, acceptance and accommodation of the tRNA by the ribosome, peptide bond formation, translocation of the peptidyl-tRNA from the decoding or aminoacyl site (A-site) to the peptidyl-site (P-site), and the movement of unacylated tRNA from the P-site to the exit (E-site) are processive. Without leaving the mRNA template, the ribosome progresses from codon to codon from the mRNA's translational start sequence and first codon AUG for methionine to the 3′- termination codons UAG, UAA, and UGA, and release sequences. The three nucleosides of each aminoacylated tRNA's anticodon recognize a three-nucleotide mRNA codon and bind to the codon in a sequence- and frame-specific manner to deliver the 20 common amino acids for protein synthesis. The aminoacylated tRNA is accepted at the A-site in response to the complementary mRNA codon, and a proofreading mechanism consisting of some nine hydrogen bonds formed between the ribosomal RNA (rRNA), tRNA, and codon ensures a high fidelity in translation. Accurate interaction of the anticodon with codon is very much dependent on the initial Watson–Crick (A●U; G●C) complementarity of the hydrogen bonding of the first two base pairs. Other enzymes having processive mechanisms of action include the DNA- and RNA polymerases. While protein synthesis is accurate to 1 in 10,000 or 20,000 amino acids at a rate of 10–20 peptide bonds formed per second [5,6], DNA- and RNA polymerases have errors of less than 1 in 10^9 at 50–100 s of base pairs per second including proofreading steps [7,8] and 1 in 10^5 at a rate of 8–85 bases per second [9,10], respectively, in vivo.

tRNA has to be considered one of the most unique of enzyme substrates. The ribosome structure and its processive catalytic mechanism [2] require the tRNA substrates to have a uniformity in chemistry and structure. Nevertheless, some 40 tRNA molecules, 22 within the human mitochondria, possess enough unique chemical diversity and structural malleability to be recognized by 20 aminoacyl-tRNA synthetases and to decode 61 mRNA codons on the ribosome. How do tRNAs achieve uniformity of chemistry and structure and yet a necessary distinctiveness for protein synthesis? Although seemingly contradictory, tRNAs use posttranscriptional modification chemistries to achieve both a stable structure mandated by the ribosome and the originality required for protein recognition determinants and codon reading.

1.2 Posttranscriptional RNA Modification

All RNAs are involved in activities that contribute directly or indirectly to the regulation, accuracy, and efficiency of gene expression that is fulfilled in translating the mRNA codons into the amino acid sequences of proteins. As in translating one language into another, the accuracy of that translation is dependent not only on a literal word-to-word equivalency but also on context, frame, and punctuation. The enzymatically synthesized, site specifically positioned, posttranscriptional modifications of RNA (Fig. 2) could be considered the punctuation marks in the translation of the genetic code. It is difficult to find among RNA species (mRNA; tRNA; ribosomal, rRNA; small nuclear, snRNA; micro, miRNA; long noncoding, lncRNA) one that is not posttranscriptionally modified [11].

The Universal Genetic Code for the 20 amino acids, the first code, and the operational RNA code for the recognition of tRNA by its cognate aminoacyl-tRNA synthetase, the second code [12], did not take into account the posttranscriptional modifications of RNA, for little was known about them at the time, except that they existed. The ubiquitously occurring, conserved modified nucleosides constitute a yet to be fully appreciated third code in regulating translation, its accuracy, and efficiency [13]. Even

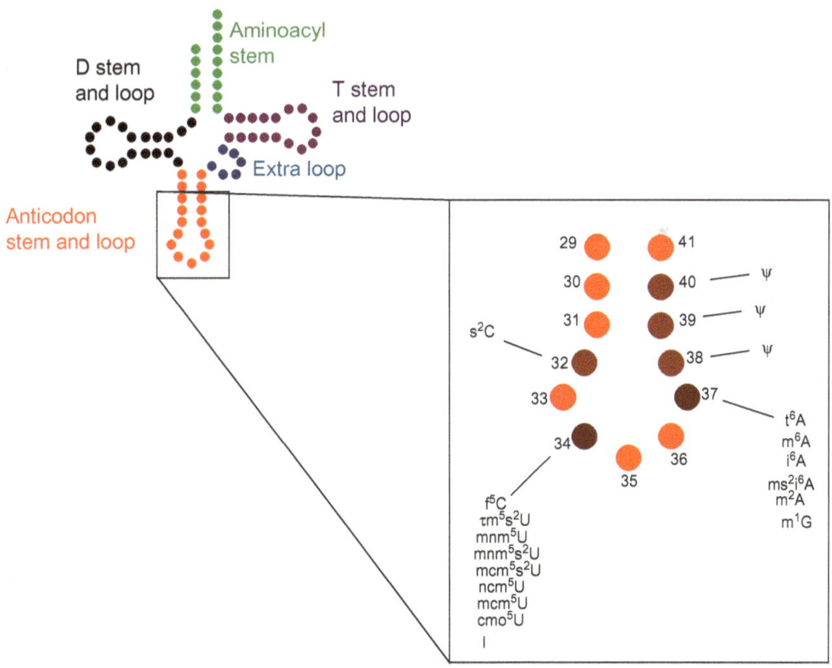

Fig. 2 tRNA and its anticodon stem and loop (ASL) domain. *Left*: General cloverleaf structure of a 76-nucleotide tRNA with the aminoacyl-accepting stem in *green*, dihydrouridine (D) stem and loop in *black*, ASL in *red*, extra loop in *blue*, and ribothymidine (T) stem and loop in *plum*. *Right*: An enlargement of the ASL domain, with the abbreviations of the important modifications discussed in this review listed. A *darker color hue* in *plum* represents more important modification sites.

with more than half a century of research, more than 100 different posttranscriptional modifications of RNAs remain mostly an undetermined regulating factor to translation [14,15]. The posttranscriptional modification of RNA, designated as the epitranscriptome [16], now usually refers exclusively to mRNAs. mRNAs have a half-dozen different types of modifications identified to date [17]. These represent some of the most common of all modifications, the 2′-O methylations of the four major nucleosides adenosine (Am), guanosine (Gm), cytosine (Cm), and uridine (Um); the isomer of uridine with a carbon–carbon glycosidic bond pseudouridine (Ψ); and the base methylations N^6-methyladenosine (m^6A), 1-methyladenosine (m^1A), 7-methylguanosine (m^7G), 5-methylcytosine (m^5C), and 5-hydroxymethylcytosine (hm^5C) [18]. Though these are simple chemical alterations, their influence in regulating translation appears to be truly significant [19–21]. We are only just beginning to understand the relevance of mRNA modifications to decoding [22].

In recent years, the coding regions of mRNA have been found punctuated with posttranscriptional modifications [17]. Early in the study of mRNA biochemistry it was recognized that mRNA had modified nucleosides in noncoding regions, such as the 5′-CAP [23]. Some of the mRNA modifications are transient in that methyltransferases "write" a methylation and demethylases "erase" the methyl group while other proteins, as "readers," recognize the "mark" and bind the RNA, affecting its structure and function [24]. Some have been proven to affect translation and gene expression [25–28]. The biochemical mechanism by which translation is affected is at present speculative. However, nucleobase modifications of nucleosides within the mRNA coding region will alter recognition of codons by tRNA. Any modification of the Watson–Crick hydrogen bonding face of a nucleoside will change tRNA's anticodon base pairing to the codon. Thus, the methyl of an m^6A in an mRNA coding region negates canonical Watson–Crick tRNA base pairing, possibly changing the accuracy and rate of decoding or affecting premature termination, and even the eventual acceptance of the mRNA for translation. The erase mechanism can then functionalize that codon for tRNA reading.

1.3 RNA Modification and the Accuracy and Efficiency of Translation

Accuracy of decoding mRNA at the ribosome's aminoacyl- or A-site depends on tRNA recognition of a complementary mRNA codon. tRNA as the decoder of mRNA is also posttranscriptionally modified for the accuracy and efficiency of reading mRNA codons [11,29–33]. More than

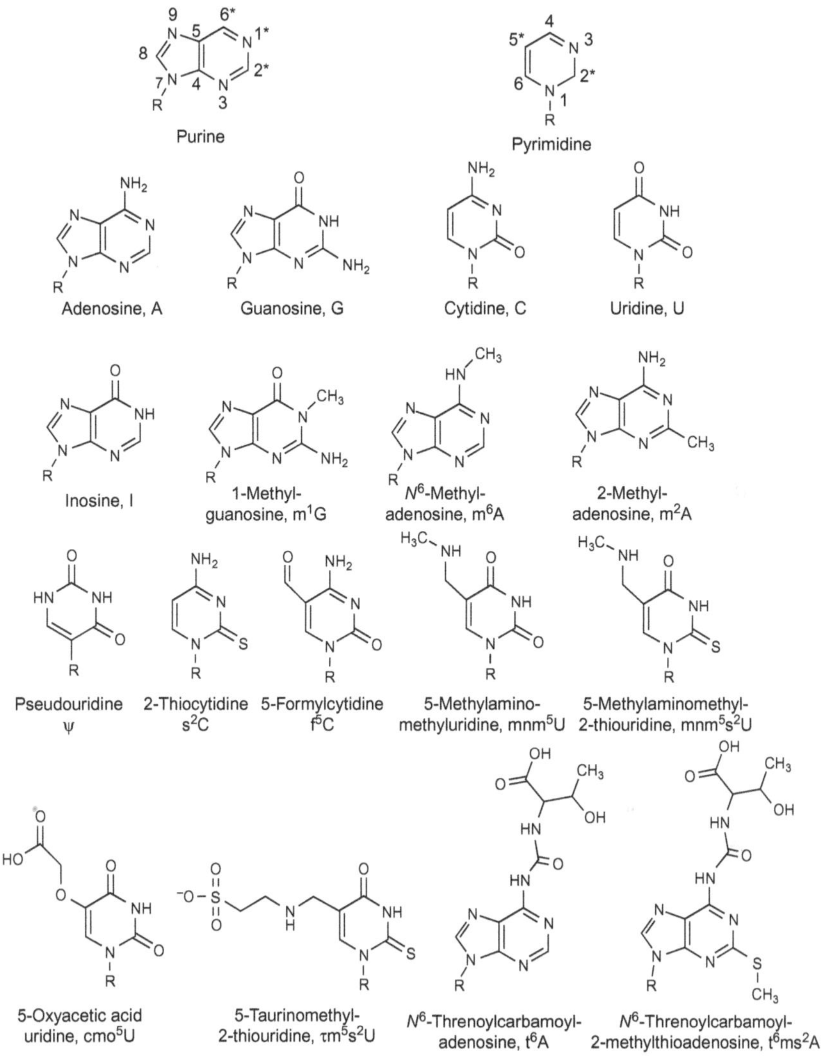

Fig. 3 Chemical structures of the major and modified nucleosides of tRNA. *Top row*: General structures of purine and pyrimidine nucleosides with numbering of the atoms. R represents ribose and the asterisk (*) marks the most important and/or frequent sites of modification. *Second row*: Conventional RNA nucleosides and their letter abbreviations. *Third through fifth rows*: The modified nucleosides discussed in this review with their shorthand notions [14,15].

90 different modifications found in tRNAs are site specifically and enzymatically synthesized (Fig. 3). Fully one-quarter of a eukaryotic, cytoplasmic tRNA's nucleosides can be found in a modified form [34,35]. The locations of modifications within tRNA's conserved cloverleaf secondary and "L"-shaped tertiary structures have been known for decades because those

sites of modification have been found to be conserved for specific types of modified nucleosides. Modifications outside of the anticodon region are usually of the most common type found in all RNAs, such as 2′-O-methyl, Ψ, m^7G, and m^1A, but also include modifications that have become synonymous with tRNA structural domains, the dihydrouridine, D, stem and loop and the ribothymidine, T, stem and loop. For the most part, their individual contributions to directing the folding of tRNA into its functioning form have only of late become clearly apparent in physicochemical terms.

tRNAs are also modified in their anticodons and surrounding sequences. In fact, modifications in the anticodon stem and loop domain of tRNAs (ASL) (Figs. 2 and 3) are the most diverse and the most complex chemically and structurally of all modifications found in RNAs [36]. The ASL domain of tRNAs is critical to accurate and efficient translation, and thus, the modifications found in this domain are equally important. ASL modifications contribute directly to the correct and operationally effective decoding of mRNA codons. After the Universal Genetic Code and the operational recognition of tRNA, they constitute a third regulatory level of translational control no less, and probably more important, than those modified nucleosides found in the coding regions of mRNAs.

The 40 some tRNA species outnumber the 20 amino acids, except for instance in the human mitochondria where tRNAs and their genes number 22. Thus, there are fewer tRNA species than there are codons by a considerable amount in the mitochondrion but also in the cytoplasm of any organism. tRNA recognition of a codon through Watson–Crick canonical base pairing is irrespective of the amino acid attached to the tRNA, depending almost completely on the tRNA's anticodon sequence [37,38]. Naturally occurring mutations that give rise to premature stop codons only to be suppressed by tRNAs prove this point. Suppressor tRNAs in responding to premature stop codons in mRNAs are a naturally occurring example of the introduction of an amino acid in which the tRNA anticodon complements a stop codon, not an amino acid codon [39,40]. tRNA complementation of a stop codon has allowed for natural expansion of the genetic code [41] as well as the introduction of nonstandard amino acids via stop codon recognition by tRNA [42].

The efficiency and accuracy of translation in reading mRNA codons is shaped by the base modifications of tRNA's anticodon, the anticodon loop, and the stem [29]. With the codes being degenerate and with fewer tRNA species than the 61 amino acid codes, some tRNAs need to respond to multiple codons. In reading multiple codons effectively, tRNAs recognize exact

Watson–Crick complements of their anticodon, G●C, C●G, A●U, and U●A at the first two base pairs. Degeneracy of the codes and limitation in the number of tRNA species predicates recognition of a "wobble" codon at the third base pair [43]. Wobble codon reading occurs when the nucleoside of the codon is recognized by a nucleoside of the tRNA anticodon other than the complementary base. Such a "wobble" base recognition has been realized for 50 years ever since Crick proposed the wobble hypothesis [43] for the third position of the codes, for instance U●G and G●U, and I●A/U/C [30]. For accurate insertion of the correct amino acid, tRNA discrimination of the third codon base is critically important to "mixed codon boxes" that have more than one amino acid represented. In contrast, some tRNA anticodon nucleoside modifications expand the ability of tRNA species to read as many as all four codons in a fourfold degenerate codon box [44–46]. In this review, we present examples of tRNA anticodon and loop modifications that make possible unique solutions for individual tRNA recognition of a single codon, two synonymous codons, all four codons of a fully degenerate codon box, and the six codons of a single amino acid.

2. HUMAN MITOCHONDRIAL $tRNA^{Met}$ DECODES THE 1:3 DEGENERATE CODON BOX

2.1 Human Mitochondrial and Cytoplasmic $tRNA^{Met}$

The mammalian mitochondrial DNA encodes all of the 22 functional tRNAs, one for each of the 18 amino acids and two each for serine and leucine [47] that are required for the synthesis of 13 critical proteins necessary for ATP synthesis and energy production [48,49]. In contrast to cytoplasmic protein synthesis, human mitochondrial tRNA for the amino acid methionine (mt $tRNA^{Met}{}_{CAU}$ where CAU is the anticodon) is unique in that a single gene codes for the tRNA used for initiation of the translation initiation start codon and for elongation. The universal AUG codon and what is nominally the cytoplasmic isoleucine codon, AUA, are both read as methionine by this one mitochondrial $tRNA^{Met}{}_{CAU}$. In the cytoplasm of all cells of all living organisms, two classes of methionine tRNAs decode the only AUG methionine codon: (i) initiator $tRNA^{Met}{}_{CAU}$ used for initiation of protein synthesis and (ii) elongator $tRNA^{Met}{}_{CAU}$ that functions to insert methionine into the growing peptide chain. In the cytoplasm, AUA codes for isoleucine and is not used as a methionine codon. However, the AUA codon is present as 20% of the initiator codon in mitochondrial mRNAs and as 80% of internal elongator methionine codons [50,51]. The prevalence of the AUA codon in the mammalian mitochondrial genome indicates that a unique modification exists that allows the recognition of the AUA codon by the single $tRNA^{Met}{}_{CAU}$ (mt $tRNA^{Met}{}_{CAU}$). Although the Crick wobble hypothesis allows for one tRNA to read more than one codon [43], it does not predict the unfavorable base pairing of CAU to AUA due to the C∘A mismatch. This is overcome by the presence of a unique posttranscriptional modification of $tRNA^{Met}{}_{CAU}$ at wobble position-34 transforming cytidine to 5-formylcytidine (f^5C_{34}). This modification is critical for the single mt $tRNA^{Met}$ to recognize both the nonuniversal AUA and the universal AUG codons [52,53]. N-terminal formylation of methionine is important for proper initiation of translation by bacterial and eukaryotic organelle $tRNA^{Met}{}_{CAU}$ [54,55]. In mitochondria, it is catalyzed by the mitochondrial methionyl-$tRNA^{Met}$ transformylase (MTFMT) [56,57]. The formylated fMet-$tRNA^{Met}$ is used as an initiator tRNA, while the nonformylated methionyl-$tRNA^{Met}$ incorporates methionine into the elongating peptide chain.

Although the basic cloverleaf structure of mammalian mitochondrial tRNAs resembles the canonical tRNAs, they have weaker tertiary interactions in their three dimensional structure due to the lack of several conserved nucleotides that facilitate and contribute to the L-shaped tertiary structure. These features in the human mt $tRNA^{Met}{}_{CAU}$ include (i) a smaller D-loop that lacks G_{18} and G_{19} residues which enables interactions with the T-loop in the tertiary structure; (ii) a Mg^{2+}-binding site critical to folding of the full-length tRNA; (iii) a short variable loop lacking G_{47}; (iv) two adjacent pyrimidine:pyrimidine base pairs, U∘U and U∘Ψ in the T-stem; (v) the presence of only six nucleotides instead of seven in the T-stem; and (vi) the absence of TΨC in the T-loop [58,59]. The loss of these conserved residues indicates that the tertiary interactions occurring in the shortened D- and T-loops of $hmtRNA^{Met}$ are weaker and different from canonical cytoplasmic tRNAs [58] with the highly conserved (Type 0) cloverleaf structure (Fig. 4).

2.2 Posttranscriptional Modification of Mitochondrial $tRNA^{Met}$

Posttranscriptionally modified cytidines have been shown to control switching between the Universal Genetic Code and a deviant code in the isoleucine/methionine codon box. The unique posttranscriptional modification of the wobble position-34 C to f^5C_{34} allows mt $tRNA^{Met}{}_{CAU}$ to engage in unfavorable base pairing of the CAU anticodon to the AUA codon [52,53]. This facilitates the single human mt $tRNA^{Met}{}_{CAU}$ to be

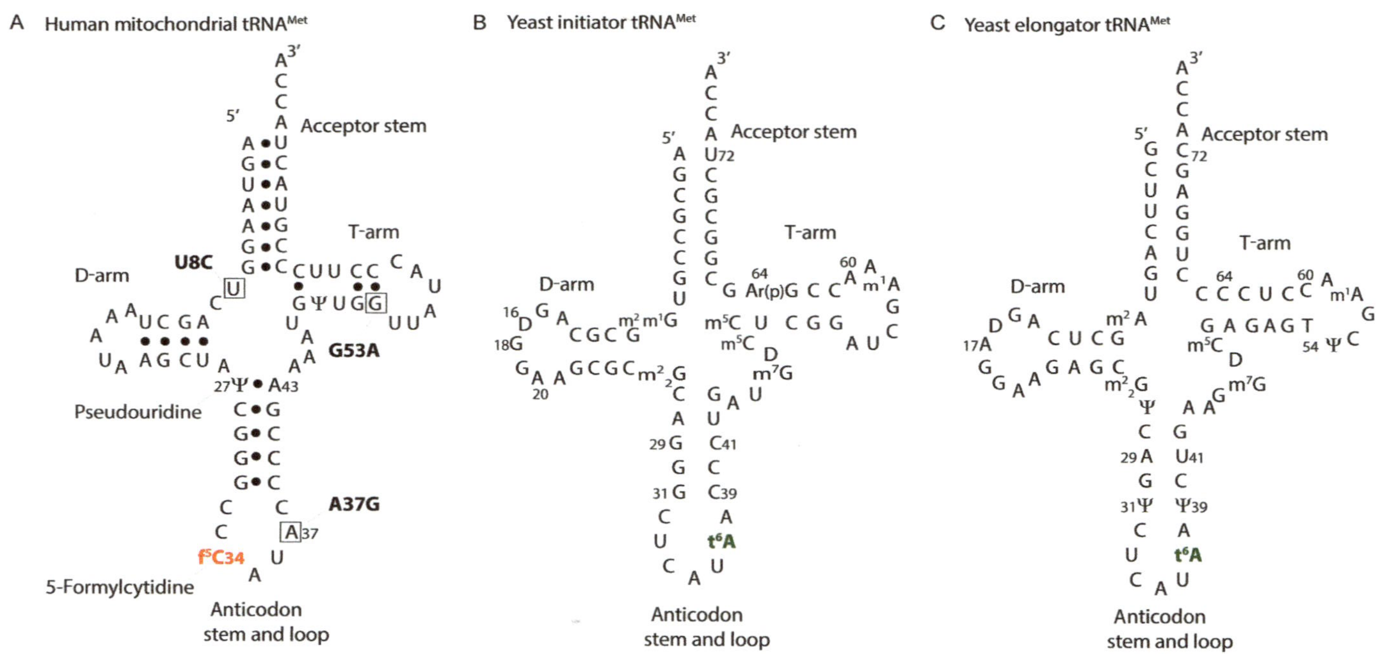

Fig 4 See figure legend on next page.

D Modified nucleoside structures in tRNA^{Met}

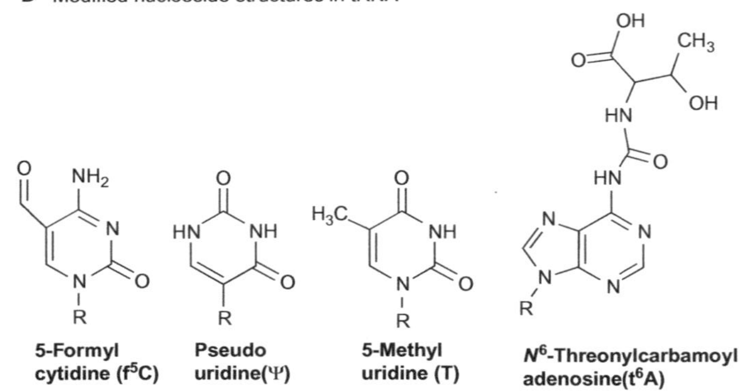

Fig. 4 Nucleoside sequence and cloverleaf structures of tRNA^{Met}. (A) Human mitochondrial tRNA^{Met} displaying the condensed variable loop and the shortened D- and T-arms. Nucleotides affected by disease-causing point mutations are *boxed*. (B) *Saccharomyces cerevisiae* initiator tRNA^{Met}. Ar(p) at position-64 refers to the 2′-O-ribosyl phosphate modification, which is an identity element of initiator tRNA^{Met} in yeast [60]. (C) *S. cerevisiae* elongator tRNA^{Met}; 5-methyluridine at position-54, indicated by T, is an important determinant of elongator tRNA^{Met}. (D) Chemical structures of the modifications f^{5}C$_{34}$, Ψ at various noted positions, and t^{6}A$_{37}$ found in the three methionyl tRNA [61].

brought to the P-site of the ribosome for translation initiation as well as to interact with mitochondrial elongation factor Tu (EF-Tu_{mt}) and bind to the aminoacyl A-site of the ribosome during peptide elongation. The single modification of f^5C_{34} alters the binding kinetics enabling the ASL to bind the Ile codons AUU and AUC, as well as AUA, in the P-site and to initiate translation in mitochondria [59]. An in vitro reconstituted mitochondrial translation system demonstrated that the modified f^5C_{34} nucleoside of $tRNA^{Met}{}_{CAU}$ was critical for the recognition of the nonuniversal AUA codon as methionine [62]. The f^5C modification is not only found in the mt $tRNA^{Met}{}_{CAU}$ of bovine and rodents but is thought to be universal in all mammalian mt $tRNA^{Met}{}_{CAU}$ [53,63,64]. In addition, it is also present in the mt $tRNA^{Met}{}_{CAU}$ of several other species including nematodes, squids, frogs, chicken, and fruit flies [65–67].

The first reported synthesis and codon binding analysis of the human mt $tRNA^{Met}$ ASL (hmtASL^{Met}-f^5C_{34}) compared to the unmodified ASL showed that the f^5C_{34} modification enhanced the motional dynamics of the loop, destabilizing the ASL [68]. The modification decreased the thermodynamic stability of the RNA, observed as a diminished melting temperature. It also reduced base stacking interactions as seen in the decreased ellipticity of the circular dichroism (CD) spectrum at 270 nm. NMR and CD studies made apparent that the modification directed the conformational dynamics of the anticodon bases, facilitating binding to the P- and A-sites. A five-base-paired stem consisting of three G●C/C●G base pairs proximal to the anticodon loop also contributes to the thermodynamic stability of RNA stacking in $tRNA^{Met}$ species, increasing the affinity of initiator $tRNA^{Met}$ to binding to the P-site [59,68,69]. The wobble position f^5C_{34} was important for the binding of the hmt$ASL^{Met}{}_{CAU}$ to the AUA codon at the A-site in canonical Watson–Crick geometry, and the rare imino-oxo tautomer of cytidine stabilized the f^5C_{34} base pair with two hydrogen bonds [70] (Fig. 5). This is in contrast to the restrictive modification of 4-acetylcytidine (ac^4C_{34}) that prevents the cytoplasmic $tRNA^{Met}{}_{CAU}$ from reading the AUA codon [71], and of lysidine (k^2C) [72] and agmatidine (agm^2C) [73,74] from preventing $tRNA^{Ile}{}_{CAU}$ from reading AUG in bacteria and archaea.

A recent report demonstrated the synthesis of the 5-formyl group of f^5C through consecutive reactions involving methylation, hydroxylation, and oxidation [75]. The previously uncharacterized methyltransferase, NOL1/NOP2/Sun domain-containing protein 3 (NSUN-3), catalyzed the methylation of C_{34} of mt $tRNA^{Met}$ in the presence of *S*-adenosylmethionine

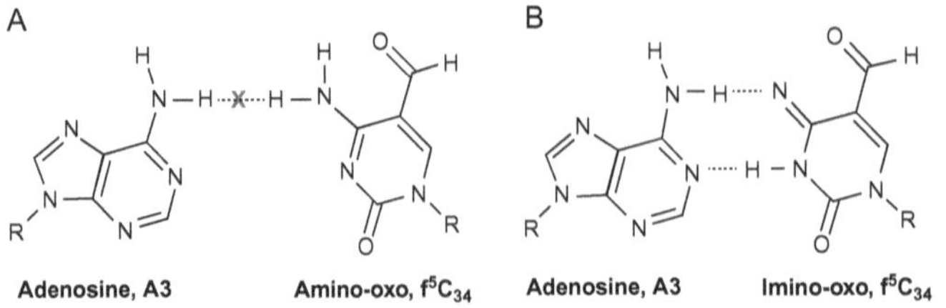

Fig. 5 The tautomeric forms of 5-formylcytidine, f^5C_{34}. Tautomerism of f^5C_{34} allows for the stereochemistry of the Watson–Crick base pairing between tRNA's f^5C_{34} and A3 of the AUA codon to approximate that of a canonical U●A pair. (A) Steric repulsion between the common amino-oxo form of f^5C_{34} N4H and A_3N^6H marked by an X. (B) Favorable interactions between the imino-oxo form of f^5C_{34} and A_3 [70]. *Dotted lines* indicate hydrogen bonding between nucleoside residues [70].

(AdoMet) to form 5-methylcytidine (m^5C) in the initial step of f^5C generation in human mitochondria. Human embryonic kidney cells lacking NSUN-3 showed reduced mitochondrial protein synthesis and oxygen consumption, causing deficient mitochondrial activity [75]. Mass spectrometric analysis indicated that f^5C_{34} was absent in the NSUN3-KO cells and was replaced by m^5C in these cells.

2.3 Other Human Mitochondrial $tRNA^{Met}$ Properties

Base pairing between f^5C_{34} and A3 of the AUA codon was observed through crystallographic studies of the f^5C_{34}-containing ASL bound to the 30s subunit of *Thermus thermophilus* [70]. The authors concluded that the most likely base pairing occurs due to an exceptional shift in the tautomeric equilibrium of f^5C_{34} from the more common amino-oxo form to the less commonly formed imino-oxo tautomer that enables interaction with A_3 (Fig. 5) [70]. In addition, this study included a structural investigation of the substitution of cytidine for "invariant" uridine at position-33 (C_{33}) as it occurs in mammalian initiator $tRNA^{Met}$. The invariant U_{33} is responsible for the sharp U-turn in the backbone between U_{33} and the nucleoside in the wobble position, allowing anticodon to codon binding. One characteristic of the U-turn is the hydrogen bonding between U_{33} and the phosphate of nucleoside 36 (CAU). A weak interaction was seen between C_{33} and U_{36}, stabilizing the sharp turn of the backbone and base pairing of f^5C_{34}. These studies using an induced-fit model not only resulted in the A-site binding conformation but also suggested a favorable canonical U-turn allowing interaction between C_{33} and U_{36}, permitting it to fit into the P-site in a slightly different conformation [70].

The $hmtASL^{Met}$-f^5C lacks the universal t^6A_{37} (*N*^{6}-threonylcarbamoyladenosine) modification present in most cytoplasmic methionyl tRNAs and found in almost all tRNAs responding to codons beginning with A (ANN codons) [76]. The A_{37} modification in elongator tRNAs promotes stacking of the bases that stabilize the first anticodon–codon base pairing. The absence of t^6A has been implicated in an increase in +1 and −1 frameshift in cytoplasmic translation, as well as an increase in non-AUG start codons [76]. Since unmodified, $hmtASL^{Met}$-f^5C forms a stable U_{36}●A1 base pair [70], it is possible that the lack of this modification enables efficient binding to the AUA nonstart codon along with the f^5C_{34} modification that enhances the motional dynamics of the ASL [68]. The unusual intraloop base pairing between C_{32} and the unmodified A_{37} [70] can also promote a more elongator-like conformation; this would expand the codon recognition of f^5C_{34}, allowing a single tRNA to perform both as an initiator and an elongator. The human mt $tRNA^{Met}{}_{CAU}$ is similar to the cytoplasmic elongator $tRNA^{Met}{}_{CAU}$ in *Escherichia coli*, whereby they both have a wobble position-34 modification. In *E. coli*, it is N^4-acetylcytidine (ac^4C), whereas in humans this modification is 2′-*O*-methylcytidine (Cm) [59,77].

2.4 Human Mitochondrial tRNA Disease

Mutations of mitochondrial DNA have been associated with numerous mitochondrial diseases caused by pathogenic mutations in mitochondrial tRNA. Mutations have been found in every mitochondrial gene and are associated with several maternally inherited human genetic diseases. More than 250 mt-tRNA mutations [78] represent a large proportion of all reported mutations leading to disease. Mitochondrial dysfunction caused by point mutations in mitochondrial tRNA is associated with most maternally inherited human genetic diseases. Other mutations affecting tRNA modifications are nuclear-encoded point mutations in the genes encoding mitochondrial tRNA modification. They have been linked to mitochondrial dysfunction and contribute to molecular pathogenesis [79,80]. Three diseases caused by mutations in the gene for human mt $tRNA^{Met}$ arise from a single point mutation: (i) the T4409C mutation leads to a change in the conserved U_8 to C in the corner of the acceptor and D-stems, causing mitochondrial myopathy that results in dystrophic muscles and intolerance to exercise [81]; (ii) the A4435G mutation resulting in an A_{37} to G_{37} change in the anticodon loop acts as a modulator of Leber's hereditary optic neuropathy (LHON), a maternally inherited disorder leading to rapid bilateral loss of central vision [82]; and (iii) the G4450A mutation causes the loss of the third G●C base pair in the T-stem, leading to splenic lymphoma, mainly in lymphocytes resulting in abnormal mitochondria, causing severe disruption of the respiratory chain (Fig. 4A; mt tRNA^{Met}) [83].

The single point mutation of T4409C, resulting in a U_8-to-C_8 replacement, in the gene for hmtRNA$^{Met}{}_{CAU}$ causes mitochondrial myopathy [58,81]. Structural probing and molecular reconstitution experiments revealed that the C_8 substitution leads to a drastic disruption of the tRNA structure, and the loss of one or more critical Mg^{2+} binding sites on the tRNA [58]. The U_8-to-C_8 substitution reduced aminoacylation and resulted in the lack of a formylated mt methionyl-tRNA$^{Met}{}_{CAU}$. Thus, correct initiation and the degree of initiation were reduced, which led to the absence of a stable ternary complex formation with eEF1 to participate in chain elongation [58].

The maternally transmitted LHON disease impaired vision of an Asian family of three generations. They had the rare A4435G mutation which is associated with the ND4 G11778A mutation in families of Chinese origin [82]. The A4435G mutation is located in the 3′-end adjacent to the anticodon (A_{37}) of mt tRNA$^{Met}{}_{CAU}$. A_{37} in tRNA^{Met} species is highly conserved from bacteria to human mitochondria, and contributes to high fidelity of codon recognition and structural stabilization of tRNAs. Molecular and genetic analysis of mt DNA identified this novel A_{37} mutation as contributing to a significant loss of steady-state levels of tRNA$^{Met}{}_{CAU}$. The presence of the A4435G mutation increased the expression of the mitochondrial dysfunction associated with the G11778A mutation [82]. In addition, the A4435G mutant has been shown to have reduced m^5C_{34} activity, the first step in f^5C_{34} biosynthesis, indicating a possible lack of NSUN3, and hypomodified f^5C_{34} in these patients [75].

2.5 Summary

Human mitochondria have a unique mechanism to utilize the single hmtRNA$^{Met}{}_{CAU}$ between initiation of protein synthesis and elongation of the protein chain. The modification 5-formylcytidine at wobble position-34 permits this single tRNA^{Met} to recognize and decode the universal AUG codon as well as the AUA codon as methionine. Codon binding and structural and thermodynamic studies have indicated stable base pairing between the noncanonical CAU and AUA bases at the wobble position, facilitated by the presence of the unique f^5C_{34} modification found in the single tRNA. The rare imino-oxo tautomer of the modified cytidine stabilizes this base pairing and expands the use of the AUA codon in human mitochondria. Substitution of C_{33} for U_{33} in hmtASL$^{Met}{}_{CAU}$ supported a C-turn conformation similar to the U-turn conformation seen with the invariant uridine-33. Importantly, the f^5C_{34} modification contributes a third element of translational control by enabling wobble base pairing to isoleucine codons.

3. tRNA DECODING OF THE TWOFOLD, 2:2, DEGENERATE CODON BOX

3.1 $tRNA^{Lys}$ Twofold Degenerate Codons AAA and AAG

Asparagine and lysine share a twofold degenerate codon box in the Universal Genetic Code (Fig. 1) [29]. Their codons differ only in the last base being a pyrimidine, U or C for asparagine, and an A or G for lysine. Aspartic and glutamic acids, and histidine and glutamine share twofold degenerate codon boxes and similarly have codons differing only in the third base, pyrimidine or purine. Though phenylalanine, tyrosine, and cysteine are encoded by two codons each, phenylalanine shares a twofold degenerate codon box with leucine, which is one of the three amino acids encoded by six codons. Tyrosine's two codons differ from two translational stop codons only in the third base, and cysteine's two codons differ only in the third base from a stop codon and that for tryptophan. Thus, the accurate and efficient reading of mRNA codons by the tRNAs for these amino acids requires discrimination at the third base of the codon. The ability of tRNA's first or wobble anticodon base at position-34 becomes all that more important in distinguishing between codons. The number of tRNAs used for the recognition of these codons varies from amino acid to amino acid and from organism to organism [29,84]. Here, we discuss a twofold codon degeneracy using the AAA and AAG lysine codons as examples. The importance of correctly modified $tRNA^{Lys}_{UUU}$ is not limited to accurate translation, because it is also important in maintaining the translational reading frame and translocation on the ribosome. Hypomodified or incorrectly modified mt $tRNA^{Lys}_{UUU}$ has been linked to mitochondrial diseases such as MELAS and MERRF in humans [85], and cytoplasmic $tRNA^{Lys}_{UUU}$ is the primer for reverse transcriptase of all lentiviruses including HIV. Mutations in modification of human $tRNA^{Lys}_{UUU}$ have been associated with a possible mistake in proinsulin translation and are a risk factor in Type 2 Diabetes (T2D) [86,87]. The low enthalpy of the three U●A base pairs in the interaction of $tRNA^{Lys}_{UUU}$ with its cognate codon must be compensated with stabilizing modifications [88]. The modifications present in the ASL are essential for not only codon recognition but additional functions as well [88–90].

E. coli, unlike eukaryotic cells, contains only one tRNA for lysine and it has the anticodon UUU. $tRNA^{Lys}{}_{UUU}$ recognizes both AAA and AAG codons [88,90]. Archaea, like prokaryotes, for the most part contain but a single $tRNA^{Lys}{}_{UUU}$ which recognizes both AAA and AAG codons [91]. The number of eukaryotic $tRNA^{Lys}{}_{UUU}$ species varies among organisms. Mammals such as humans and rabbits contain three species of $tRNA^{Lys}$ that differ in their anticodons, as well as in nucleoside sequence and modifications. Both $tRNA^{Lys1}{}_{CUU}$ and $tRNA^{Lys2}{}_{CUU}$ contain the CUU anticodon which recognizes AAG. $tRNA^{Lys3}{}_{UUU}$ contains the UUU anticodon capable of recognizing both AAA and AAG with a preference for the former (Fig. 6). In developing tissue, $tRNA^{Lys2}{}_{CUU}$ may appear in a hypomodified form, sometimes referred to as $tRNA^{Lys4}{}_{CUU}$, whereas $tRNA^{Lys5}{}_{CUU}$ and $tRNA^{Lys6}{}_{CUU}$ refer to the various stages in between [92].

The smaller genomes of organelles encode fewer tRNAs than the nuclear genome in eukaryotes, approximately 22 in mitochondria and 30 in chloroplasts [94]. Indeed, much like in *E. coli*, mitochondria utilize a single mt $tRNA^{Lys}{}_{UUU}$ for the recognition of both AAA and AAG codons [95,96]. Additionally, it has been shown that it is feasible to design *E. coli* strains with reduced codon degeneracy, where degenerate codons are replaced with allowed codons [97]. There are some exceptions to the Universal Genetic Code, however. In some species, such as echinoderms and platyhelminthes, the lysine AAA codon in the mitochondria is translated as asparagine, and in starfish mitochondria, mt $tRNA^{Lys}{}_{CUU}$ is the sole lysine tRNA [98]. One of the most common coding errors for lysine tRNA is mistranslation as asparagine [92,99]. Similar to mammalian mitochondria, chloroplasts utilize a single $tRNA^{Lys}{}_{UUU}$ for translation [100]. Therefore, $tRNA^{Lys}{}_{UUU}$ must be able to read both AAA and AAG codons in all organisms where it is utilized. This is made possible by extensive modifications of the ASL ($ASL^{Lys}{}_{UUU}$) at positions-34 and -37, in particular [29,88,101].

3.2 Wobble Position Uridine-34 Modifications

In humans and other mammals $tRNA^{Lys1}{}_{CUU}$ and $tRNA^{Lys2}{}_{CUU}$, position-34 contains an unmodified cytosine; these $tRNA^{Lys}{}_{CUU}$ species read only the AAG codon. $tRNA^{Lys3}{}_{UUU}$ as well as the bacterial and organelle $tRNA^{Lys}{}_{UUU}$ contains the UUU anticodon that is capable of decoding both AAA and AAG codons [92,93]. In order for codons ending in a purine to be efficiently read, cognate tRNAs often contain a modified uridine nucleoside

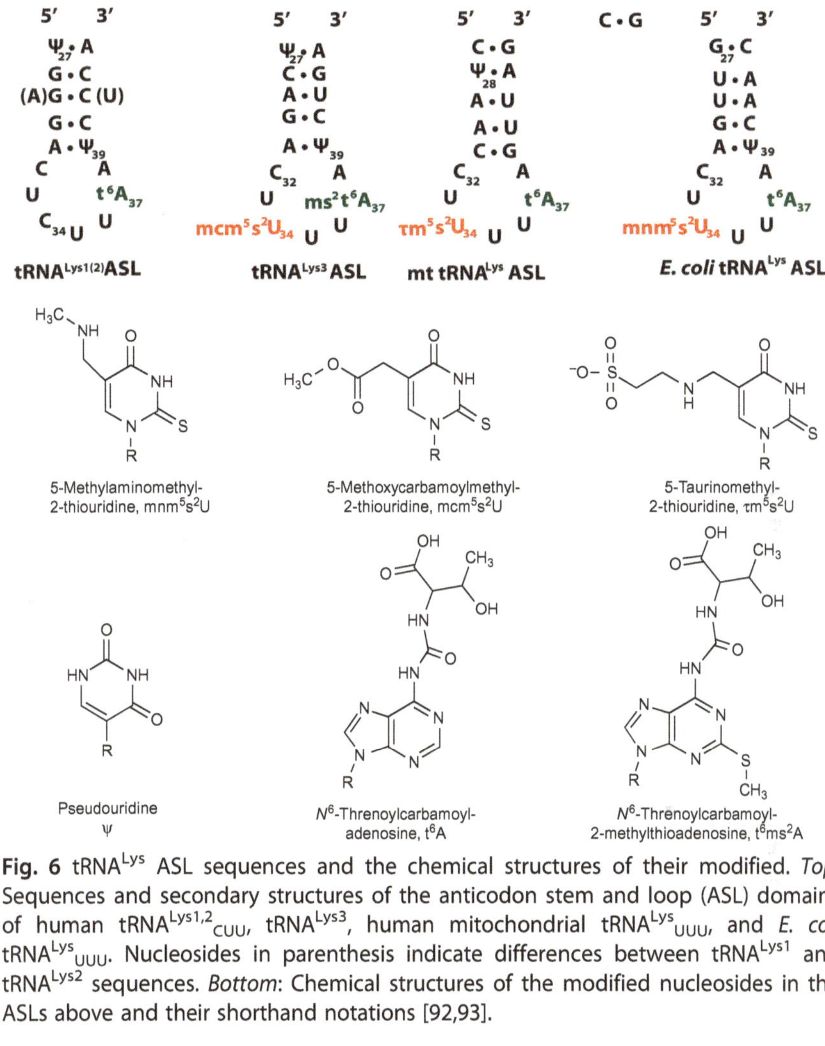

Fig. 6 tRNA^{Lys} ASL sequences and the chemical structures of their modified. *Top*: Sequences and secondary structures of the anticodon stem and loop (ASL) domains of human tRNA^{Lys1,2}$_{CUU}$, tRNA^{Lys3}, human mitochondrial tRNA^{Lys}$_{UUU}$, and *E. coli* tRNA^{Lys}$_{UUU}$. Nucleosides in parenthesis indicate differences between tRNA^{Lys1} and tRNA^{Lys2} sequences. *Bottom*: Chemical structures of the modified nucleosides in the ASLs above and their shorthand notations [92,93].

at the wobble position, where 5-methyl-2-thiouridine derivatives (xm^{5}s^{2}U) are commonly found (Fig. 6) [99,102]. In *E. coli*, the one tRNA^{Lys}$_{UUU}$ contains 5-methylaminomethyl-2-thiouridine at position-34 (mnm^{5}s^{2}U$_{34}$). Similarly, human tRNA^{Lys3}$_{UUU}$ contains 5-methoxycarbonylmethyl-2-thiouridine at position-34 (mcm^{5}s^{2}U$_{34}$; Fig. 6) [90], and mammalian mt tRNA^{Lys}$_{UUU}$ has 5-taurinomethyl-2-thiouridine (τm^{5}s^{2}U$_{34}$; Fig. 6) in the wobble position [96,103,104]. In several bacteria as well as in a murine leukemia cell line, the sulfur in carbon position-2 of uridine-34 can be replaced by selenium in an ATP-dependent process, resulting in a mixture of thiolated and selenated tRNAs for the amino acids glutamate, glutamine, and lysine. The s^2U is a required precursor for the reaction to occur, but the mnm^5 group is not [92].

3.2.1 Functional Significance of tRNA U_{34} Modifications

The functional significance of U_{34} modifications is substantial. Many of these modifications are 5-position derivatives of 5-methyl-2-thiouridine, or xm^5s^2U (Fig. 6). The xm^5s^2 modifications restrict the conformational flexibility of the anticodon, thus providing translational fidelity [45,96,105]. A complete knockout of U_{34} modifications is lethal in *Saccharomyces cerevisiae* [106]. Hypomodified $tRNA^{Lys}{}_{UUU}$ cannot efficiently induce the expected conformational changes in the 30S ribosomal subunit during translation as it cannot form favorable enough interactions with its cognate codon, causing the translation to stall [88,107]. The xm^5U modification in bacterial and mt $tRNA^{Lys}{}_{UUU}$ (mnm^5U, mcm^5U, and τm^5U) at the wobble position stabilizes the invariant U-turn that occurs between U_{33} and U_{34}. This in turn stabilizes the anticodon–codon duplex [88,108,109]. In prokaryotes, this is achieved through a hydrogen bond between 2′-OH of U_{33} and the amino group of $mnm^5s^2U_{34}$ [109,110]. The xm^5U modification is essential for a sure and stable recognition of the AAG wobble codon in addition to the AAA codon. For instance, the $mcm^5s^2U_{34}$ shifts from the keto to enol tautomeric form in binding to G3, achieving a stable wobble base pair in the Watson–Crick geometry [111]. Due to its contribution to enhanced structural rigidity and preordering of the anticodon loop, $mnm^5s^2U_{34}$ enables wobble base pairing with G, and thus allows recognition of both codons [88,112].

The mnm^5U modification in the absence of the s^2 group does not appear to restrict a U_{34}●pyrimidine interaction. However, the poor base stacking of the UUU anticodon makes such restriction unnecessary as a U_{34}∘U3 interaction would abrogate any base stacking between U_{34} and U_{35}, leading to a free energy penalty that would make codon recognition unfeasible. Thus, the role of the xm^5U modification is to increase translational fidelity by enhancing purine recognition, but not to restrict pyrimidine recognition; this is perhaps predictable from thermodynamic stabilities [88,113]. NMR studies have shown that a modified wobble base is important for the stacking between anticodon positions-34 and -35, since in the absence of modifications noncanonical and closed conformations are highly populated for the UUU anticodon. However, in human $tRNA^{Lys3}{}_{UUU}$ the mcm^5U

modification does not seem to enhance base stacking between U_{34} and U_{35} in a significant manner [110]. Therefore, the thio modification is required to compensate for the poor stacking of uridine [110]. The 2-thio group allows the ribose of U_{34} to adopt C3′-*endo* puckering augmenting the U-turn and thus facilitates binding with purines and restricting recognition of NNY codons [105,110,114–117]. The thio group stabilizes the U-A base pair, but it is mildly destabilizing for U-G base pairs [29,116].

Hypomodified tRNAs, such as the mammalian $tRNA^{Lys3}_{UUU}$ lacking the fully modified mcm^5s^2U at the wobble position, cannot efficiently decode their cognate codons, resulting in codon-specific ribosomal pausing, which in turn effects negative consequences on protein homeostasis [107]. Hypomodified $tRNA^{Lys}_{UUU}$ is defective in translocation from the A- to P-sites of the ribosome [89]. Interestingly though, the hypomodified bacterial $tRNA^{Lys}_{UUU}$ is less frequently misread as asparagine than fully modified $tRNA^{Lys}_{UUU}$ [99,108]. The increased misreading may simply be due to increased efficiency induced by the modifications. The fact that misreading increases for the mnm^5U-modified $tRNA^{Lys}_{UUU}$ compared to unmodified tRNA, but has the opposite effect for $tRNA^{Glu}_{UUC}$, implies that the effect of the modification depends on the structural context as well as the chemical nature of the modification [108]. Thus, hypomodification can lead to mistranslation of alternate codons in twofold degenerate codon boxes of the Universal Genetic Code, or to changes in the translational reading frame leading to aberrant protein synthesis.

3.2.2 Modified Nucleoside-Dependent Synonymous Codon Bias at Uridine-34 and Stress

A mechanism of modified nucleoside-dependent synonymous codon bias regulates gene expression through wobble base pairing at position-34, and is most affected by the uridine-34 modifications. tRNA reading of synonymous codons, those degenerate codons for which the identical amino acid is represented, is "silent" in that the protein is not altered (Fig. 1). Yet, the reading of synonymous codons can lead to different functional outcomes [118–120]. Because tRNA modifications alter tRNA's codon usage or bias, translation can be affected by changes in modifications that support synonymous codon bias in the mRNAs of specific proteins. Variously modified aminoacyl-tRNAs are not necessarily prevalent in equal concentration to codons present, nor is the hypomodified state of the wobble-34 modification most favorable for anticodon–codon binding [119]. Also, optimal reading of codons by fully modified tRNA resulting at what appears to be the most advantageous translation rates is not always the most suitable for a particular cell condition. Thus, particular codons are read by hypomodified tRNA with the result that specific proteins harboring these synonymous codons are expressed. This regulation of translation can occur in response to factors external to cells, such as signal transduction, extracellular RNA, environmental factors and contaminants, illegal and legal drug use. The wobble position-34 uridine modifications have been of particular interest in that their codon biases are altered in response to cellular stress, such as oxidative stress [121–123]. These changes in U_{34} modification are associated with the translation of specific stress response proteins, the mRNAs of which have a particular codon bias, including the recoding of stop codons [124,125]. In this way, tRNA modifications are "reprogrammed" to support a cell's efficient translation of specialized proteins. Selenated or thiolated uridine-34 may help sense damage caused by irradiation [92]. A lack of the U_{34} thiolation has also been shown to increase sensitivity to oxidative stress in *S. cerevisiae* [126], whereas incomplete U_{34} modifications did not affect a drug-induced stress response in *S. cerevisiae* compared to controls [107].

3.3 Modifications of Adenosine-37, 3′-Adjacent to the Anticodon

Modifications at position-37 of tRNA are commonly found in a variety of tRNAs, with the type of modification depending on the base found at position-36 [127]. While not essential in all bacteria, the N^6-threonylcarbamoyladenosine (t^6A) modification at position-37 is found in tRNAs reading ANN codons in all domains of life [88,110,128–130]. It is required for efficient aminoacylation of several tRNAs, tRNA binding to the A-site of the ribosome and codon recognition in ANN codons, efficient translocation, reading frame maintenance, and preventing leaky scanning of initiation codons, and proper recognition of stop codons [88–90,131,132]. The lysine tRNAs of prokaryotes, mammalian mitochondria as well as mammalian $tRNA^{Lys3}$ contain the t^6A modification [88,110,128]. Adenosine at position-37 can also contain the 2-methylthio-N^6-threonylcarbamoyladenosine (ms^2t^6A) or the cyclic-N^6-threonylcarbamoyladenosine (ct^6A) [131,133,134].

The t^6A modification improves the stability of the codon–anticodon duplex through cross-strand stacking with the first base of the codon [88,110]. Furthermore, neither t^6A nor ms^2t^6A modification results in an increase in stability of free tRNA in molecular dynamics simulations, which would be expected for a modification responsible for stabilizing the

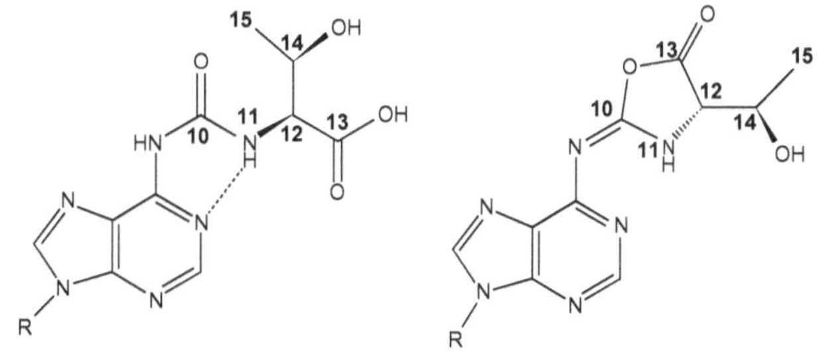

Fig. 7 *N*6-Threonylcarbamoyladenosine and its cyclic derivative, t6A and ct6A. *Left:* Chemical structure of t6A with the atoms of the modification numbered. The hydrogen bonding that produces a pseudocyclic conformation of the modified nucleoside is shown as a *dashed line. Right:* The cyclic form of t6A, ct6A, with the corresponding atoms numbered [88,133].

codon–anticodon interaction in the ribosome [135]. In fact, the unmodified $tRNA^{Lys3}{}_{UUU}$ ASL elicits a higher melting temperature and increased stability because of intraloop hydrogen bonding compared to the open loop, t^6A_{37}-modified $ASL^{Lys3}{}_{UUU}$ [90]. Pseudouridine (Ψ) at position-39 enhances the stability of both unmodified and t^6A_{37}-modified $ASL^{Lys3}{}_{UUU}$, yet the unmodified ASL with or without Ψ_{39} is incapable of binding poly-A programmed ribosomes [90,132].

A model of $tRNA^{Lys}{}_{UUU}$ with the circular t^6 modification, ct^6A_{37}, at the ribosomal A-site [133] compared to the structure of $tRNA^{Lys3}{}_{UUU}$ containing the t^6A modification at the ribosomal A-site [88] predicts the effects of the ct^6A modification on codon recognition and ribosomal function. The ct^6A37 modification has been shown to enhance the decoding efficiency much like the native t^6A_{37} modification [133]. In both cases, the modification allows the nucleoside to stack with the first adenosine (A1) of the codon and across the anticodon–codon base pair, thus stabilizing the interaction (Fig. 7). A hydrogen bond between N1 and N11 is found in both cases [88,133], enhancing the base stacking ability of t^6A_{37} with A1 as this causes the formation of a third ring that expands the area of the base [88,90]. The C14-OH group in ct^6A_{37} can form a hydrogen bond with N7 of A1, further stabilizing the interaction [133]. The oxazolone ring is also thought to be planar with respect to the adenine [133]. The threonyl group of t^6A_{37} has rotational freedom about the N11-C12 bond [88]. This, due to steric hindrance, prohibits the incorporation of t^6A into the helix, which in turn together with its tricyclic structure allows the stacking interaction with the first adenine of the codon [88].

3.3.1 Functional Significance of tRNA A_{37} Modifications

In a number of organisms, there is but one $tRNA^{Lys}{}_{UUU}$ that reads both AAA and AAG codons [88,90,91,94]. In humans and other mammals, $tRNA^{Lys3}{}_{UUU}$ decodes AAA and AAG, and is expressed at a ratio of 1:2 with $tRNA^{Lys1,2}{}_{CUU}$ that reads only AAG [94]. To enhance the affinity of what is an energetically weak UUU-AAA anticodon–codon interaction, the tRNA is heavily modified with U_{34} and A_{37} contributing to binding affinity, translocation ability, and maintenance of the translational reading frame [88,109]. Modifications are necessary in order for $tRNA^{Lys}{}_{UUU}$ to read the AAG codon [88,89,109]. The t^6A_{37} modification restores cognate codon binding to unmodified $tRNA^{Lys}{}_{UUU}$, but does not enable translocation of the $ASL^{Lys}{}_{UUU}$ from the A-site to the P-site of the ribosome [89]. The modification mnm^5U_{34} also restores codon binding but not translocation. Interestingly, the modification s^2U_{34} enhanced translocation, but not to the degree of t^6A_{37} and mnm^5U_{34} together.

Together the modifications enabled cognate and wobble codon binding and resulted in a 25-fold increase in translocation in comparison to the unmodified $ASL^{Lys}{}_{UUU}$ [89]. Besides its role in translocation, the modification t^6A_{37} is an essential, strong positive specificity determinant of many but not all prokaryote isoleucyl-tRNA synthetases [129]. The modification may also be a recognition determinant of eukaryote and viral proteins. The viral nucleocapsid protein of HIV, NCp7, recognizes t^6A_{37} [136]. Peptides, composed of 15 amino acids and selected to mimic NCp7, not only recognize the modification but are also equally capable of denaturing the modified $ASL^{Lys3}{}_{UUU}$ as does the protein [136,137]. Modification of already modified N^6-adenosine-37 (t^6A_{37}, i^6A_{37}) with the 2-methylthio moiety (ms^2) is found in specific tRNA species in almost all organisms [14]. Mutations in the gene for the enzyme, *Cdkal1*, responsible for ms^2-group of $ms^2t^6A_{37}$ in human $tRNA^{Lys3}{}_{UUU}$ have been shown to be a reliable risk assessment gene for T2D [86,87]. The homozygous recessive mutation of *cdkal1* has a 1.50 risk of T2D, comparable to the *brca 1* and *2* risks of breast cancer. With an incompletely modified $tRNA^{Lys3}{}_{UUU}$, responsible for inserting Lys88 positioned adjacent to Arg89 at which a crucial protease cleavage separates the insulin A-chain from the C-peptide, the Lys88 codon may not be read, resulting in a proinsulin processing issue, or aberrant protein.

3.4 Summary

The twofold degeneracy within the Universal Genetic Code requires a distinction in decoding between NNA/G and NNU/C. Nature's solution to the twofold degeneracy is to ensure that tRNA anticodons read the appropriate, amino acid-specific codons and not those of the amino acid or stop codons sharing the first and second codon bases. Here again, modifications contribute a third code to controlling translation of mRNAs. In the decoding of twofold degenerate codons, modifications of tRNA's wobble position-34 add precision to the reading of the third codon base. Keto-enol tautomerism enables strong wobble codon reading in a canonical base pair geometry. Modification of the invariant purine at position-37 prestructures and restricts dynamics of the ASL domain to maintain the translational reading frame. The purine-37 modification is all that more important when the first anticodon–codon base pair is a U●A or A●U. The inherent weakness of these base pairs is counterbalanced by the van der Waals forces of the 3′-adjacent modification, creating a hydrophobic platform that is evident for such tRNAs as $tRNA^{Lys}_{UUU}$.

4. tRNA DECODING OF CODON FOURFOLD DEGENERACY

4.1 Single tRNA Reads All Four Codons of the Prokaryote Fourfold Degenerate Codon Box

The amino acids alanine, leucine, proline, serine, threonine, and valine each claim, at least, an entire set of four codons, a completely degenerate codon box within the Universal Genetic Code (Fig. 1). Within each set of fourfold degenerate codons, the first two nucleosides of the codons are conserved, while the divergence occurs at the third nucleoside. For each of these amino acids within prokaryotic organisms, there is a single corresponding tRNA that can recognize all four of these codons. In prokaryotes, these tRNAs contain the 5-oxyacetic acid modification at the position-34 uridine (cmo^5U_{34}) (Fig. 8) [138]. Although pyrimidine:pyrimidine base pairs were originally thought to be disallowed under the original wobble hypothesis, tRNAs containing cmo^5U_{34} can recognize codons ending in U and even C in addition to codons ending with the canonical A or the traditional wobble G [30,139–142]. The interaction between the cmo^5U_{34} modification and the phosphate backbone has been found to promote the C2′-*endo* conformation of the ribose ring, which subsequently allows for the U∘U base pair that would otherwise be too short (Fig. 9) [142].

There are multiple tRNAs for the aforementioned fourfold degenerate amino acid codons, but the tRNA species containing the cmo^5U_{34} modification may be sufficient for viability of prokaryotic organisms. Three tRNAs are responsible for decoding four proline codons in *Salmonella*

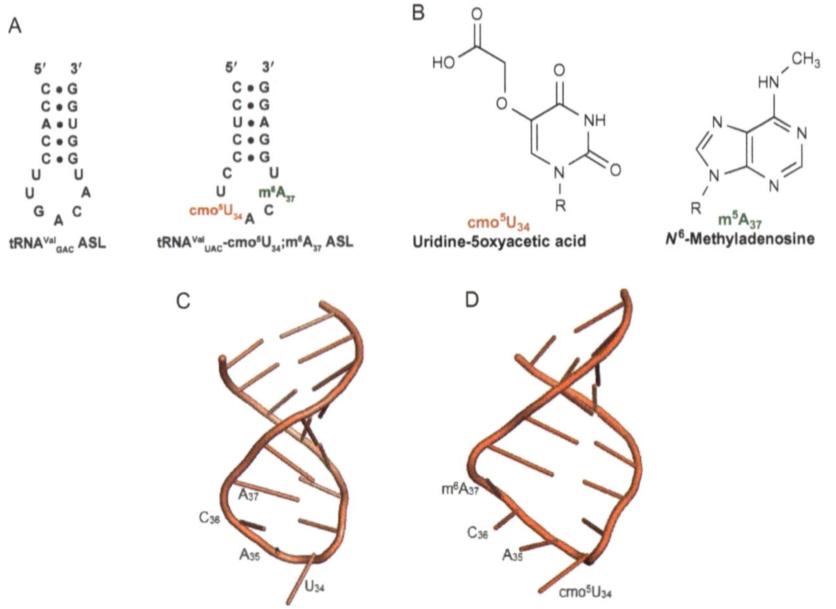

Fig. 8 The primary sequences and nucleoside modifications of the *E. coli* tRNA^{Val} isoacceptors and the structures of ASL^{Val}$_{UAC}$ and ASL^{Val}$_{UAC}$-cmo^{5}U$_{34}$;m^{6}A$_{37}$ compared. (A) The anticodon domain primary sequences of the three tRNA^{Val} isoacceptors from *E. coli* and their modifications: uridine-5-oxyacetic acid at position-34, *red* and N^6-methyladenosine at position-37, *green*. The two tRNA^{Val} species containing the GAC anticodon have identical ASLs with variations in sequence in other regions of the tRNA molecule. (B) The chemical structures, names, and abbreviations of the modifications that appear in the *E. coli* tRNA^{Val}$_{UAC}$-cmo^{5}U$_{34}$;m^{6}A$_{37}$. (C) Structure of unmodified ASL^{Val}$_{UAC}$, *pink*. (D) Fully modified ASL^{Val}$_{UAC}$ containing the modifications cmo^{5}U$_{34}$ and m^{6}A$_{37}$, ASL^{Val}$_{UAC}$-cmo^{5}U$_{34}$;m^{6}A$_{37}$, *red*, showing the base stacking of cmo^{5}U$_{34}$ and an open loop structure. The structures of the unmodified and modified ASL^{Val}$_{UAC}$ were determined by NMR [45] (Protein Data Bank 2JR4 for unmodified ASL^{Val3}$_{UAC}$ and 2JRG for ASL^{Val3}$_{UAC}$-cmo^{5}U$_{34}$;m^{6}A$_{37}$).

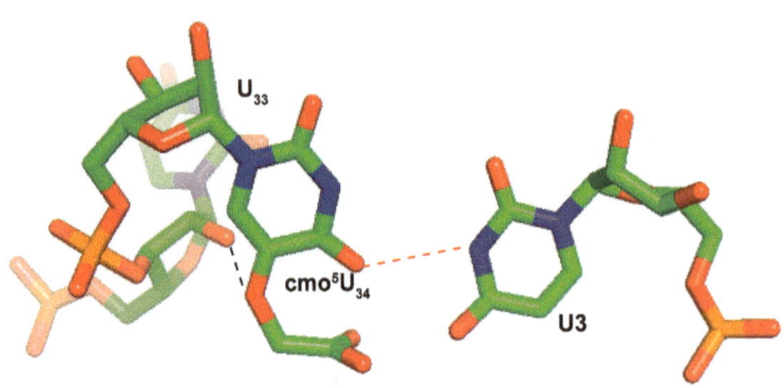

Fig. 9 cmo^{5}U$_{34}$∘U3 wobble pair. Base pairing between cmo^{5}U$_{34}$ of the ASL^{Val}$_{UAC}$ and U3 of the GUU codon [46] (Protein Data Bank 2UUB). A hydrogen bond occurs between the 2′-OH of the invariant U$_{33}$ (*background*) of the ASL^{Val}$_{UAC}$ and O5 of cmo^{5}U$_{34}$ (*black dashed line*), structuring cmo^{5}U$_{34}$ for base pairing with U3. The hydrogen bond that allows for the base pair between O2 of cmo^{5}U$_{34}$ and N2 of U3 (*red dashed line*).

enterica: $tRNA^{Pro}{}_{CGG}$, $tRNA^{Pro}{}_{GGG}$, and $tRNA^{Pro}{}_{UGG}$-cmo^5U_{34}. When the genes for $tRNA^{Pro}{}_{CGG}$ and $tRNA^{Pro}{}_{GGG}$ are deleted, the organism is still viable, indicating the ability of $tRNA^{Pro}{}_{UGG}$-cmo^5U_{34} to read all four proline codons [143]. Furthermore, knockout of *cmoA* and *cmoB*, two genes implicated in the synthetic pathway of the cmo^5U_{34} modification, in a strain of *S. enterica* containing only $tRNA^{Pro}{}_{UGG}$-cmo^5U_{34} resulted in reduced decoding efficiency of the four proline codons by the hypomodified tRNA; the cmo^5U_{34} modification, as opposed to the 5-hydroxyuridine 34 (ho^5U_{34}) and 5-methoxyuridine 34 (mo^5U_{34}) modifications, must be vital for recognition of $tRNA^{Pro}{}_{UGG}$-cmo^5U_{34} by near-cognate codons within the *Salmonella* system [143].

In *E. coli*, the cmo^5U_{34}-modified $tRNA^{Ala}{}_{UGC}$ can bind to the GCC codon within the ribosomal A-site, but this binding was found to be seven times less stable than binding to the GCA codon. The rate of GTP hydrolysis during ribosomal translocation with the GCC codon differed from that of the cognate codon only by a factor of 2.5 [144]. In fact, the cmo^5U_{34} modification may be necessary in order to prevent the ribosome from stalling in the presence of an anticodon–codon pair in which a U∘U or C∘U base pair occurs at the wobble position. Translocation from the A-site to the P-site of the ribosome did not occur when unmodified *E. coli* $tRNA^{Val}{}_{UAC}$ was bound to the GUU valine codon [89].

4.2 Eukaryote tRNAs That Wobble in Reading Fourfold Degenerate Codons

While the cmo^5U_{34} modification is exclusive to prokaryotic tRNA, eukaryotic tRNA species that decode for the amino acids with fourfold degenerate codons, among others, contain similar modifications at the 5-position carbon of uridine-34 (Fig. 1) [145]. These eukaryotic modifications lack the methoxy group connectivity of cmo^5U_{34} yet include a carboxylic acid derivative. Perhaps this resemblance is enough to enhance recognition at the wobble position in more complex organisms, expanding the wobble capacity of eukaryotic tRNAs as well. In *S. cerevisiae*, the 5-carbamoylmethyluridine-34 modification (ncm^5U_{34}) present in $tRNA^{Val}{}_{UAC}$, $tRNA^{Thr}{}_{UGU}$, and $tRNA^{Ser}{}_{UGA}$ enhances the efficiency of wobble pairing to the GUG, ACG, and UCG codons, respectively [112]. In fact, fully 25% of the *S. cerevisiae* tRNA wobble position-34 uridines are modified to be 5-carbamoylmethyluridine, ncm^5U_{34} or mcm^5U_{34} for translational accuracy and efficiency. Synthesis of these modifications requires the Elongator complex associated with elongating RNA polymerase II transcription (see chapter "Structures and activities of the

Elongator complex and its co-factors" by Kolaj-Robin and Séraphin). The Elongator complex is composed of six elongator protein subunits and appears conserved among eukaryotes. In addition, three protein subunits of the yeast *Kluyveromyces lactis* killer toxin (zymocin) insensitivity genes are involved in this modification [146]. tRNAs with the modifications mcm^5s^2, $tRNA^{Glu}{}_{UUC}$, $tRNA^{Lys3}{}_{UUU}$, and $tRNA^{Gln}{}_{UUG}$ are cleaved by zymocin toxin (γ-toxin) at the 3′-side of the modified nucleoside. Resistance to γ-toxin is observed in the modification-deficient mutants [146]. Lack of Elongator activity results in hypomodified tRNA which, in turn, causes defects [147]. ELPC-1 and ELPC-3, the *Caenorhabditis elegans* homologues to the *S. cerevisiae* Elongator complex subunits Elp1 and Elp3p, are essential to the synthesis of the ncm^5U_{34} and mcm^5U_{34} modifications within nematode tRNAs [148]. The xm^5- and s^2 modifications are primary to maintaining the translational reading frame and thus accuracy of protein synthesis. As ELPC-1 and ELPC-3 are expressed in ASE neurons, *elpc-1* and *elpc-3* mutants demonstrate defective neuronal function, as exhibited through the inability of the mutants to adapt the expected experience-dependent behavioral changes in response to having previously been exposed to 100 mM sodium chloride in nutrient-deficient conditions [148].

4.3 Roles of Both Wobble Positions-34 and -37 Modifications in Reading Fourfold Degenerate Codons

In addition to the cmo^5U_{34} modification, the *E. coli* isoaccepting $tRNA^{Val}{}_{UAC}$ contains another modification on the adenosine adjacent to the anticodon. The N^6-methyladenosine modification (m^6A_{37}) (Fig. 8), which naturally occurs at position-37 of the $tRNA^{Val}{}_{UAC}$ species with cmo^5U_{34}, plays an important role in structurally ordering the ASL for binding to the codon. Modifications at this conserved purine have been proposed to negate intraloop hydrogen bonding, thereby allowing the ASL to attain an open loop structure for binding to the codon, and to simultaneously enhance base stacking (Fig. 8). Base stacking effectively orders the ASL structure [30]. Determination of thermodynamic stability by UV melting of the ASL of *E. coli* $tRNA^{Phe}$ containing the N^6-isopentenyladenosine (i^6A_{37}) modification at nucleoside-37 supported the hypothesis that an intraloop hydrogen bond was disrupted [149]. Similar results were determined for the ASL of human $tRNA^{Lys}{}_{UUU}$ containing the N^6-threonylcarbamoyladenosine (t^6A_{37}) modification at position-37 [90].

Differences in thermal stabilities were observed for the fully modified $tRNA^{Val}{}_{UAC}$ with cmo^5U_{34} and m^6A_{37} in comparison to that of the unmodified $tRNA^{Val}_{UAC}$ [45]. While the melting temperatures, T_m, did not differ substantially, the hyperchromicity of the modified $tRNA^{Val}_{UAC}$ was more than twice that of the unmodified $tRNA^{Val}_{UAC}$ [45]. Coupled with the entropic contributions of the modifications, this increased hyperchromicity indicated the increased base stacking, resulting in a more ordered ASL that arose from the presence of the cmo^5U_{34} and m^6A_{37} modifications [45]. The increased ellipticity of the fully modified $tRNA^{Val}_{UAC}$ hairpin in comparison to that of the unmodified $tRNA^{Val}_{UAC}$ hairpin, obtained by CD spectroscopy, also confirms the increased base stacking within the ASL of the fully modified $tRNA^{Val}_{UAC}$ [45].

The fully modified $tRNA^{Val}_{UAC}$ with cmo^5U_{34} and m^6A_{37} decodes the GUA, GUG, GUU, and GUC codons [140]. Within the A-site of the 30S ribosomal subunit, the binding affinities of both the fully modified $tRNA^{Val}_{UAC}$ and the unmodified $tRNA^{Val}_{UAC}$ to each of the four codons were measured [45]. There was a little difference between the dissociation constants for fully modified $tRNA^{Val}_{UAC}$ and the unmodified tRNA when binding the cognate GUA codon. Whereas the fully modified $tRNA^{Val}_{UAC}$ bound to the GUG and GUU codons ($K_d = 1.96$ and $1.93\,\mu M$, respectively), the unmodified $tRNA^{Val}_{UAC}$ did not have measurable dissociation constants for these three valine codons [45]. The K_d for $tRNA^{Val}_{UAC}$-cmo^5U_{34};m^6A_{37} binding to the GUC codon remains undetermined [45]. Crystal structures have been elucidated for the binding of the fully modified $tRNA^{Val}_{UAC}$ on the small subunit of the ribosome in response to mRNA containing each of the valine codons, including the GUC codon [46]. The previously unexpected U∘C wobble pairing in the presence of the cmo^5U_{34} modification must occur, despite being a weak interaction.

The cmo^5U_{34} and m^6A_{37} modifications play important roles in structuring the ASL domain of the fully modified $tRNA^{Val}_{UAC}$. The comparatively low root-mean square deviation values obtained from NOESY spectra suggest that the fully modified ASL^{Val}_{UAC} is more ordered than unmodified ASL^{Val}_{UAC} due to the steric hindrance encountered by the N^6-methyl of the m^6A_{37} modification and to the geometry of the cmo^5U_{34} modification brought about by hydrogen bonding between the base and the 5-oxy group [45]. Additionally, both the fully modified and modified ASLs contained some, but not all, defining characteristics of a U-turn at U_{33}, a type of π-turn that confers stability onto the ASL through base-phosphate stacking and hydrogen bonding [45]. With respect to this feature, the ASL^{Val}_{UAC} with cmo^5U_{34} and m^6A_{37} contains the shorter hydrogen bond between U_{33} O2′ and A_{35} N7 [45].

4.4 Summary

The existence of a single species of tRNA that can decode each codon of a fourfold degenerate set is undoubtedly advantageous. Referring back to *S. enterica*, the presence of the modified $tRNA^{Pro}{}_{UGG}$ with cmo^5U_{34} in the absence of $tRNA^{Pro}{}_{CGG}$ and $tRNA^{Pro}{}_{GGG}$ allows for survival of the organism, albeit at a reduced growth rate [143]. In some cases, the cmo^5U_{34} modification may prove absolutely essential, even for reading of cognate codons. For instance, the unmodified $tRNA^{Ala}{}_{UGC}$ cannot recognize the cognate GCA codon or the near-cognate GCG in accordance with traditional base pairing and wobble pairing, respectively, in the ribosomal P-site [150]. In decoding of the wobble codons ending in G, cmo^5U_{34} is crucial [44]. The cmo^5U_{34} is apparently in the enol form and facilitates formation of a Watson–Crick geometry when binding G3. An intramolecular hydrogen bonding stabilizes the conformation of the anticodon for codon binding [46].

5. THREE AMINO ACIDS ARE ENCODED BY SIX SYNONYMOUS CODONS EACH

5.1 Sixfold Degeneracy and Reading of the Arginine Codons

Within the Universal Genetic Code, three amino acids are coded by six different codons: arginine, leucine, and serine. These six synonymous codons always comprise one whole and an additional half of a twofold degenerate codon box, such that ambiguity exists not only in the identity of the base at the third codon position, but in that of the bases at the first and second positions as well (Fig. 1). This poses a chemical challenge for their set of tRNA isoacceptors, which must together recognize and discriminate among all six degenerate codons while sharing the specific structure and chemistry required for their common recognition by a sole cognate aminoacyl-tRNA synthetase and by the ribosomal A-site. In the case of the *E. coli* arginine tRNA isoacceptors, a compelling case may be made that a suite of posttranscriptional nucleoside modifications, combined with the chemistry imparted by the identity of the anticodon nucleosides themselves, provides the necessary identity determinants for these whole-and-split codon box tRNA isoacceptors to achieve their function.

In *E. coli*, the six codons for arginine are decoded by five tRNA isoacceptors [77,151,152]. Usage of these six codons varies widely, from >2% of all codons for the most common CGU and CGC to <0.2% for the very rare AGA and AGG [51,153–155]. Of the four less common codons, three (CGG, AGA, and AGG) are decoded by individual isoacceptors tRNA^{Arg3}, tRNA^{Arg4}, and tRNA^{Arg5}, respectively, while the fourth, CGA, is decoded by the same isoacceptors (tRNA^{Arg1} and tRNA^{Arg2}) that also decode the two most common codons CGU and CGC [77]. The primary sequences, and therefore chemical identities, of the *E. coli* arginine tRNA isoacceptors differ between members of the family; this is particularly true in the region of the anticodon, which must vary in its sequence among isoacceptors to account for the variation in the first and third nucleosides in the six synonymous arginine codons (Fig. 10). Despite these chemical differences, sufficient chemical similarity must exist to permit the single *E. coli* arginyl-tRNA synthetase (ArgRS) to charge all five isoacceptors [156]. Furthermore, three of the isoacceptors contain bases in the first position of the anticodon that theoretically permit wobble recognition of near-cognate codons, an expansion tRNA decoding capacity that appears to

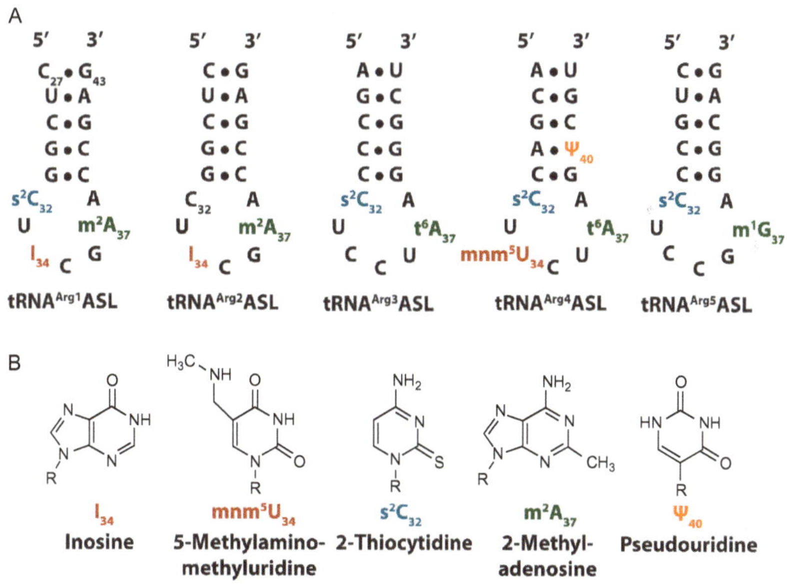

Fig. 10 The primary sequences and nucleoside modifications of the *E. coli* tRNA^{Arg} isoacceptors. (A) The anticodon domain primary sequences of the five tRNA^{Arg} isoacceptors from *E. coli* and their modifications: 2-thiocytidine at position-32 (*blue*); inosine and 5-methylaminomethyluridine at position-34 (*red*); 2-methyladenosine, N^6-threonylcarbamoyladenosine, and 1-methylguanosine at position-37 (*green*); and pseudouridine at position-40 (*orange*). (B) The chemical structures, names, and abbreviations of the modifications that appear in the five *E. coli* tRNA^{Arg} isoacceptors.

be either regulated or disallowed in vivo [77]. These conflicting chemical requirements—the recognition of a variety of mRNA codons, sufficient similarity to enable charging by ArgRS, and functional regulation of wobble decoding—are facilitated by a suite of isoacceptor-specific posttranscriptional nucleoside modifications.

5.2 Modification-Dependent tRNA Reading of the Arginine Codons

A number of chemical signatures and naturally occurring posttranscriptional modifications are in fact common to all or nearly all of the *E. coli* tRNA arginine isoacceptors. A highly conserved adenosine occupies position-38 at the 3′-terminus of the tRNA's ASL domain in all five isoacceptors, and thus, a mismatch base pair between A_{38} and C_{32}/s^2C_{32} can be formed [77,138]. These mismatches typically feature a stable hydrogen bond between the O2 of C_{32} and the N6 of A_{38} and may, together with the invariant uridine at position-33, be important for permitting the ASL to adopt the "U-turn" structural motif for correct tRNA recognition in the ribosomal A-site [110,138,157,158].

A conserved cytidine at position-35, the second position of the anticodon, is also common to all five *E. coli* tRNA isoacceptors for arginine [138]. C_{35}, positioned to form a Watson–Crick base pair with the invariant guanine that occupies position two of all six arginine codons, is known to provide at least a portion of the common chemical signature that serves as a strong identity determinant for aminoacylation; it is recognized through backbone interactions with ArgRS His22 and His23 and a stacking interaction with Trp569 [159]. Structurally, because C_{35} must be exposed to the solvent in order to make specific contacts with the ArgRS backbone and/or side chains, but must also accommodate ribosomal binding and mRNA decoding, the ASL domain of arginine tRNA isoacceptors must be significantly deformable. A significant conformational change must occur between binding to the aminoacyl-tRNA synthetase and adopting the canonical U-turn. This deformability is, in fact, visible in a comparison between the ribosome-bound and synthetase-bound structures of the tRNA, which differ considerably in the ASL region [159,160]. Elements of their respective modification schemes are also common among the arginine tRNA isoacceptors. Each has a modified purine at position-37, either a 2-methyladenosine (m^2A; $tRNA^{Arg1}$ and $tRNA^{Arg2}$), an N^6-threonylcarbamoyladenosine (t^6A_{37}; $tRNA^{Arg3}$ and $tRNA^{Arg4}$), or a 1-methylguanosine (m^1G; $tRNA^{Arg5}$) [77,138].

The two *E. coli* isoacceptors responsible for decoding not only the common CGU and CGC codons but also the seldom-used CGA sequence are $tRNA^{Arg1}{}_{ICG}$ and $tRNA^{Arg2}{}_{ICG}$. These isoacceptors, which have identical primary sequences and differ only in that $tRNA^{Arg1}{}_{ICG}$ lacks the adenosine nucleotide at position-20 (A_{20}) in its dihydrouridine loop and contains the rare modification 2-thiocytidine at position-32 (s^2C_{32}), are sometimes referred to jointly as $tRNA^{Arg1,2}{}_{ICG}$ [161,162]. Interestingly, both A_{20} and s^2C_{32} are present in the other three arginine isoacceptors, $tRNA^{Arg3,4,5}$ [77,161,162]. Both $tRNA^{Arg1}{}_{ICG}$ and $tRNA^{Arg2}{}_{ICG}$ also contain posttranscriptional nucleoside modifications at position-34 (inosine, I_{34}) and position-37 (2-methyladenosine, m^2A_{37}), which have also been shown to play significant roles in allowing $tRNA^{Arg1,2}$ to achieve its functionality [77,138]. In $tRNA^{Arg1}$, the absence of A_{20} from the molecule's dihydrouridine loop has been shown to negatively affect aminoacylation efficiency [163]. Mutant $tRNA^{Arg2}$ transcripts in which the expected A_{20} is either substituted or deleted result in a 370-fold decrease in aminoacylation activity by ArgRS, suggesting that $tRNA^{Arg1}{}_{ICG}$ is constitutively aminoacylated and available to participate in translation at a lower rate of efficiency than $tRNA^{Arg2}{}_{ICG}$ [164].

5.3 The Modification 2-Thiocytidine Modulates Inosine Reading of A, U, and C

The unusual 2-thiocytidine modification also plays a role in differentiating between the properties of the $tRNA^{Arg1}$ and $tRNA^{Arg2}$ isoacceptors. s^2C_{32} is a relatively uncommon modification, present only in a subset of bacterial and archaeal species; in *E. coli*, it can be found in four of the five arginine tRNA isoacceptors ($tRNA^{Arg1,3,4,5}$) and $tRNA^{Ser}{}_{GCU}$ at position-32 in the ASL [14,77]. Functionally, the presence of s^2C_{32} was first demonstrated to promote reading frame maintenance and to reduce the rate of tRNA selection into the ribosomal A-site and was suggested to play a role in improving translational accuracy by reducing the speed of translation, particularly during suboptimal growth conditions [165]. A series of ribosomal A-site binding assays using differentially modified $tRNA^{Arg1,2}$ ASLs later definitively demonstrated that the $ASL^{Arg1,2}{}_{ICG}$ containing either a single s^2C_{32} or m^2A_{37} modification cannot efficiently wobble decode CGA, while the unmodified $ASL^{Arg1,2}{}_{ICG}$ recognizes, per the expectations of the wobble hypothesis, CGC, CGU, and CGA [43,166]. This experiment definitively established the role of s^2C_{32} in $tRNA^{Arg1,2}$ as restricting the ability of the

$tRNA^{Arg1}{}_{ICG}$ isoacceptor to decode the rarest of its three wobble codons [166].

The functional difference imparted by the s^2C_{32} modification in $tRNA^{Arg1}{}_{ICG}$ cannot be explained by structural differences in solution; NMR structural analysis of the differentially modified $tRNA^{Arg1,2}{}_{ICG}$ ASL constructs employed in the ribosomal binding assay showed that neither the presence of s^2C_{32} nor of m^2A_{37} altered the solution structure of the $ASL^{Arg1,2}{}_{ICG}$ significantly from the highly stable 5′-$U_{33}NCG_{36}$-3′ motif observed for the unmodified ASL and found as a tetraloop [166–168]. When a subset of the $ASL^{Arg1,2}{}_{ICG}$ constructs were crystallized bound to their cognate mRNA in the context of the 30S bacterial ribosomal subunit, they adopted the characteristic U-turn structural motif but failed to indicate any significant source of structural difference among them (W. Cantara, personal communication). However, biophysical characterization suggested that s^2C_{32} contributions to stability and thermodynamics properties, reducing the melting temperature of the $ASL^{Arg1,2}{}_{ICG}$ hairpin, increasing the free energy of folding, and decreasing base stacking interactions compared to the unmodified ASL [166]. A computational analysis of the possibly dynamic effect of s^2C_{32} on the properties of $tRNA^{Arg1,2}{}_{ICG}$ in its mRNA-bound context on the ribosome led to the suggestion that the presence of a purine:purine I_{34}-A_3 basepair in the A-site wobble position of the unmodified $tRNA^{Arg1,2}{}_{ICG}$ distorts the U-turn from its canonical structural motif and precipitates the formation of a network of water molecules that stabilize the distorted U-turn. Subsequent substitution of the larger, less electronegative sulfur atom for oxygen in the position-32 cytosine disrupts this hydrogen bonding network, such that the unconventional distorted U-turn cannot be simultaneously maintained within the geometric constraints of a large purine:purine I_{34}-A_3 base pair (S. Ranganathan and S. Vangaveti, personal communication).

The posttranscriptional modification of some $tRNA^{Arg1,2}{}_{ICG}$ transcripts with s^2C_{32} provides the cell with a regulatory strategy to address the allotment of cellular resources to the translation of the rare codon CGA. Of the two isoacceptors $tRNA^{Arg1}$ and $tRNA^{Arg2}$, only $tRNA^{Arg2}$, bearing an unmodified C_{32}, is capable of decoding CGA [166]. Under typical circumstances, when CGA-containing mRNA transcripts are rare, the $tRNA^{Arg1,2}$ population may be regulated at the level of posttranscriptional modification to favor s^2C_{32}-modified $tRNA^{Arg1}$ and the decoding of CGC and CGU only; under stress responses or in other circumstances where a more robust response to CGA is required, the modification of $tRNA^{Arg1,2}$ transcripts can be reduced, favoring $tRNA^{Arg2}$.

Although the $tRNA^{Arg1,2}{}_{ICG}$ isoacceptors have the ability, depending upon modification profile, to decode three codons, the remaining three isoacceptors recognize only one codon each [169,170]. Two of these, $tRNA^{Arg3}$ and $tRNA^{Arg5}$, contain the anticodons CCU and CCG, respectively, and, consistent with the expectations of the wobble hypothesis, do not form non-Watson–Crick wobble base pairs with any nucleosides other than guanosine at the third position of their cognate codons AGG and CGG [43,77]. Both isoacceptors contain a modified purine at position-37 and a 2-thiocytidine modification at position-32 [138]. While the function of s^2C_{32} in these arginine tRNA isoacceptors is not known and cannot be accounted for by a need for the restriction of wobble decoding, the position-37 purine modifications t^6A_{37} and m^1G_{37} have been, respectively, shown to order the ASL domain in an open conformation for the formation of the U-turn motif and to prevent translational frameshifting while enhancing codon–anticodon affinity [90,111,171,172].

5.4 $tRNA^{Arg}$ Species Decoding a Twofold Degenerate Codon Box

$tRNA^{Arg4}{}_{UCU}$, however, provides another example in which nucleoside identity and modification scheme combine to achieve specific functional results. A wide variety of bacterial and archaeal species contain verified $tRNA^{Arg}$ species with UCU anticodons for the decoding of the codon AGA; candidate sequences for an equivalent tRNA species in several eukaryotic genomes, while not yet confirmed experimentally, have also been identified [77,173]. The *E. coli* $tRNA^{Arg4}{}_{UCU,}$ known to be solely responsible for the decoding of the codon AGA in that species, contains four endogenous posttranslational modifications: 2-thiocytidine at position-32 (s^2C_{32}); 5-methylaminomethyluridine at wobble position-34 (mnm^5U_{34}); t^6A_{37}; and Ψ_{40} [77,174]. $tRNA^{Arg4}{}_{UCU}$ shares with the bacterial $tRNA^{Lys}{}_{UUU}$, which decodes both AAA and AAG, a nearly identical ASL sequence and both the mnm^5U_{34} and t^6A_{37} modifications [77]. Although the well-studied human $tRNA^{Lys}{}_{UUU}$ does not share s^2C_{32} or Ψ_{40}, its biophysical properties are otherwise so similar to those of *E. coli* $tRNA^{Arg4}{}_{UCU}$ that it was expected that functional properties of two tRNA species would likewise resemble each other [111]. However, it was found that in $tRNA^{Arg4}{}_{UCU}$, the mnm^5U_{34} is required for this isoacceptor to recognize the rare codon AGA, but does not allow wobble pairing to AGG.

In contrast, mnm^5U_{34} in *E. coli* $tRNA^{Lys}{}_{UUU}$ demonstrably confers the ability to read both AAA and AAG codons [132,175]. The only differences in the loop sequences of *E. coli* $tRNA^{Lys}{}_{UUU}$ and $tRNA^{Arg4}{}_{UCU}$ lie in the identity of the second anticodon nucleoside (C_{35} for $tRNA^{Arg4}$; A_{35} for $tRNA^{Lys}$) and the s^2C_{32} modification in $tRNA^{Arg4}{}_{UCU}$ [138]; like $tRNA^{Arg1,2}{}_{ICG}$, it was initially thought that the presence of s^2C_{32} would serve an inhibitory role in AGG recognition [166]. However, A-site codon-specific ribosome binding assays performed on three synthesized $tRNA^{Arg4}{}_{UCU}$ ASL constructs (unmodified $ASL^{Arg4}{}_{UCU}$, singly modified $ASL^{Arg4}{}_{UCU}$–s^2C_{32}, and triply modified $ASL^{Arg4}{}_{UCU}$–mnm^5U_{34};t^6A_{37};Ψ_{40}) showed that only the triply modified ASL bound AGA with a physiological dissociation constant (188 ± 21 nM), while no construct, with or without modifications, bound AGG (W. Cantara, personal communication).

While the presence of s^2C_{32} surprisingly appears not to restrict the wobble interaction of $tRNA^{Arg4}$ with the AGG codon as it does in $tRNA^{Arg1,2}$, biophysical analysis shows that its modification suite contributes to a reduction in base stacking and overall molecular order, similar to effects observed in the fully modified ASL of $ASL^{Lys}{}_{UUU}$. Preliminary NMR analysis further indicated that modifications modulate the pK_a of A_{38} to stabilize the noncanonical $C_{32}{\circ}A_{38}$ base pair and enhance cognate decoding by enhancing the base stacking interactions between G_{31} and s^2C_{32}. In this instance, although modifications clearly affect the thermodynamic properties and overall function of $tRNA^{Arg4}{}_{UCU}$, the chemical determinant for the ability of $tRNA^{Arg4}{}_{UCU}$ to discriminate between A- and G-ending codons lies in the presence of modifications together with the sequence identity of the nucleotide, C_{35}, occupying the second position of the anticodon (W. Cantara, personal communication).

The *E. coli* tRNA arginine isoacceptors provide an excellent example of a system in which the chemical identity of the nucleotides comprising the RNA, together with their posttranscriptional modifications, operates in tandem to achieve several functional goals for the family of isoacceptors: discrimination between six chemically distinct arginine codon from whole- and split-codon boxes and differing from each other at their first and third nucleotide positions; universal recognition for amino acid charging by a single arginyl aminoacyl-tRNA synthetase; and regulation of wobble decoding to influence the ability of the cell to decode the rare codons CGA, AGA, and AGG. For example, the modification state of $tRNA^{Arg1,2}{}_{ICG}$ regulates whether the isoacceptor decodes all three codons to which it theoretically wobble pairs or only to the two common codons (CGU and CGC) [166]; this regulatory role for modifications counteracts the need for multiple isoacceptors to decode the common and rare codons and be recognizable to the aminoacyl-tRNA synthetase, while also allowing the translation of CGA-containing genes to be regulated by a single modification enzyme under appropriate cellular conditions. It is also clear from the $tRNA^{Arg}$ family of isoacceptors that modifications perform different roles depending upon their chemical environment; the rare s^2C_{32} modification, for example, has a clear inhibitory effect on wobble decoding of $tRNA^{Arg1,2}$ but permits and enhances cognate codon binding for $tRNA^{Arg4}{}_{UCU}$ [166]. Such dual functionality may also exist for mnm^5U_{34} in $tRNA^{Arg4}{}_{UCU}$ and $tRNA^{Lys}{}_{UUU}$ to explain their respective abilities to inhibit or allow wobble decoding.

5.5 Summary

Sixfold degenerate codons across two codon boxes differ not only in the nucleoside occupying their wobble positions but also in the nucleoside present in their first (and, for serine, second) positions. The tRNA isoacceptors decoding these degenerate codons must simultaneously recognize and discriminate between these divergent codons while remaining universally recognizable to a single aminoacyl-tRNA synthetase and able to bind the ribosomal A-site. In *E. coli*, posttranscriptional modifications assist the five arginine tRNA isoacceptors in accomplishing distinguishing their cognate and wobble codons. In the $tRNA^{Arg1,2}{}_{ICG}$ isoacceptors, the variable presence of a 2-thiocytidine at position-32 inhibits the ability of the tRNA to wobble decode the rare arginine codon CGA due to steric and hydrogen bonding effects resulting from the replacement of an oxygen with a larger and less electronegative sulfur atom. $tRNA^{Arg4}{}_{UCU}$, by contrast, is also restricted in its wobble recognition of the near-cognate AGG codon, but depends for this effect not on its s^2C_{32} modification but on the chemical identity of the codon nucleosides together with the presence of four endogenous posttranscriptional modifications. The *E. coli* arginine tRNA isoacceptors provide an excellent example of the functional specificity conferred by posttranscriptional modifications acting in unique local chemical environments, as well as pointing out how RNA modification-based strategies of cellular regulation might influence the cell's response to codon bias and the translation of rare codons.

6. POSTTRANSCRIPTIONAL MODIFICATION OF tRNA IS REQUIRED FOR ACCURATE AND EFFICIENT DECODING

Posttranscriptional modifications evolved and were conserved at tRNA nucleoside positions-34 and -37 to both expand and restrict the capability of isoaccepting tRNAs to read cognate and wobble codons. Though the effects are opposite in function, in both cases the modifications ensure translational fidelity of mRNA transcripts. These site-specific, context-dependent roles for tRNA modifications emphasize their ability to fine-tune tRNA molecular function through chemistry and conformational dynamics in control of gene expression [29,31,45]. This occurs even to the extent that a single modification may perform different functions depending on the sequence and chemical context of the ASL [108].

In their use of modification chemistries, tRNA molecules achieve the simultaneous goals of structural uniformity and ASL sequence variety through the use of a conformational dynamics and chemistries. tRNA structural uniformity is prescribed by the constraints of the ribosomal translational machinery [2,176,177]. Every tRNA species must have ASL domain structures that conform to the spatial limitations and chemistries of the A- and P-sites of the ribosome. To be productively accepted into the decoding site, the backbone chemistries of tRNA's ASL must be structured to engage in stable hydrogen bonding to the rRNAs, bacterial 16 and 23S RNAs, and eukaryote 18 and 28S RNAs. Yet, within the variety of their individual sequence chemistries, each of approximately 40 tRNAs needs to read one or more of the 61 amino acid codons. The individual chemistries, structures, and conformational dynamics required for accurate and efficient decoding of mRNA are provided to the tRNA in the form of prestructuring the ASL domain with modified nucleosides [111].

Instances in which a codon box is split between multiple amino acids (Fig. 1) exemplify how modifications prestructure a tRNA to distinguish the correct codon(s) from the incorrect codon(s). As described, mt $tRNA^{Met}$ modifications allow this one tRNA to read AUG and AUA codons as the initial and internal methionine codes. The $tRNA^{Val}{}_{UAC}$ cmo^5U_{34} and m^6A_{37} modifications are crucial for accurate codon recognition of fourfold degenerate codons. Prokaryote and mammalian $tRNA^{Lys}{}_{UUU}$ are fully functional in accurate codon recognition, translocation, and maintenance of the reading frame only when fully modified. When the first and second codon bases are shared as they are for lysine and asparagine codons, the introduction of a modification at wobble position-34, such as the 2-thiouridine-34, enables the $tRNA^{Lys}{}_{UUU}$ to bind the AAA and AAG lysine codons while hindering recognition of the asparagine AAC and AAU codons [150]. This level of translational regulation through the contribution of modified nucleosides leads to the accurate binding of the anticodon to the cognate and wobble codons within the ribosomal A-site and peptide bond formation. There is still opportunity for the anticodon–codon mismatch within the P-site to trigger early termination of translation [6].

Modified nucleosides are required of $tRNA^{Arg}$ species for accurate distinction of sixfold degenerate codons and modulation of wobble position inosine-34 recognition of A, U, and C. The sixfold degenerate codons for leucine claim the entire CUN codon box and share the UUN (where N is A, C, G, or U) mixed codon box with phenylalanine. MiaA, a tRNA isopentenyltransferase implicated in the biosynthesis of the N^6-isopentenyl-2-methylthioadenosine-37 ($ms^2i^6A_{37}$) modification found in $tRNA^{Leu}{}_{UAA}$ and $tRNA^{Leu}{}_{CAA}$ [178], is required for the accurate translation of the leucine-containing *rpoS* and *iraP* gene products in *E. coli*, whose mRNA transcripts code for leucine using the complementary UUA and UUG codons [179].

Nature rarely discards chemistries that prove successful for one purpose only to be advanced and duplicated for another purpose. In doing so with RNA-modified nucleosides, the value has been conserved. Through site-specific placement and structure/activity relationships, modifications evolved to optimize translational accuracy and efficiency. The ASL was optimized with modifications that prestructures conformation, lowers energy barriers, and negates the need to refashion the chemistry, structure, and dynamics for codon recognition and ribosome binding. The first reportings of modified nucleosides in RNA in general, and in tRNA (s-RNA) in particular, occurred over 50 years ago [180]. The modifications of tRNA's ASL domain are the most studied and understood. Yet, scientists worldwide continue to gain insight into how this encoded third level of translational control within tRNA and other RNAs provide accurate, efficient, and responsive decoding of the Universal Genetic Codes.

ACKNOWLEDGMENT

Our review emphasizes and cites some of the most recent findings regarding the chemistry, structure, and functions of posttranscriptional RNA modifications, focusing on tRNA's anticodon domain. As in any such digest of work, we apologize for inadvertently missing the contributions of some, particularly since interest in RNA modifications has exponentially increased and publications are in a large variety of journals. The summation presented here was only possible through half a century of collaborations that we have had with undergraduate and graduate students, postdoctoral fellows, and colleagues during a long and continuing career supported by grants from NIH, NSF, and DOD, including at present to PFA (NSF, CHE-1407042; DOD, W81XWH-15-PRMRP-DA) and to Manal Swairjo, PI (San Diego State University) and PFA co-PI (NIH, 1R01GM110588).

REFERENCES

[1] Z. Liu, C. Gutierrez-Vargas, J. Wei, R.A. Grassucci, M. Sun, N. Espina, S. Madison-Antenucci, L. Tong, J. Frank, Determination of the ribosome structure to a resolution of 2.5 A by single-particle cryo-EM, Protein Sci. 26 (1) (2017) 82–92.

[2] E. Zimmerman, A. Yonath, Biological implications of the ribosome's stunning stereochemistry, ChemBioChem 10 (1) (2009) 63–72.

[3] B.J. Greber, N. Ban, Structure and function of the mitochondrial ribosome, Annu. Rev. Biochem. 85 (2016) 103–132.

[4] L. Brocchieri, S. Karlin, Protein length in eukaryotic and prokaryotic proteomes, Nucleic Acids Res. 33 (10) (2005) 3390–3400.

[5] C. Kurland, D. Hughes, M. Ehrenberg, Limitations of translational accuracy, in: F.C. Neidhardt, R. Curtis III, J.L. Ingraham, E.C.C. Lin, J.K.B. Low, B. Magasarák, W.S. Reznikoff, M. Riley, M. Schaachler, H.E. Umbarger (Eds.), *Escherichia coli* and *Salmonella*: Cellular and Molecular Biology, ASM Press, Washington, DC, 1996, pp. 979–1004.

[6] H.S. Zaher, R. Green, Quality control by the ribosome following peptide bond formation, Nature 457 (7226) (2009) 161–166.

[7] S.D. McCulloch, T.A. Kunkel, The fidelity of DNA synthesis by eukaryotic replicative and translesion synthesis polymerases, Cell Res. 18 (1) (2008) 148–161.

[8] M. Mechali, Eukaryotic DNA replication origins: many choices for appropriate answers, Nat. Rev. Mol. Cell Biol. 11 (10) (2010) 728–738.

[9] H. Kwak, J.T. Lis, Control of transcriptional elongation, Annu. Rev. Genet. 47 (2013) 483–508.

[10] P. Maiuri, A. Knezevich, A. De Marco, D. Mazza, A. Kula, J.G. McNally, A. Marcello, Fast transcription rates of RNA polymerase II in human cells, EMBO Rep. 12 (12) (2011) 1280–1285.

[11] M. Duechler, G. Leszczynska, E. Sochacka, B. Nawrot, Nucleoside modifications in the regulation of gene expression: focus on tRNA, Cell. Mol. Life Sci. 73 (16) (2016) 3075–3095.

[12] P. Schimmel, R. Giegé, D. Moras, S. Yokoyama, An operational RNA code for amino acids and possible relationship to genetic code, Proc. Natl. Acad. Sci. U.S.A. 90 (19) (1993) 8763–8768.

[13] P.F. Agris, The importance of being modified: an unrealized code to RNA structure and function, RNA 21 (4) (2015) 552–554.

[14] W.A. Cantara, P.F. Crain, J. Rozenski, J.A. McCloskey, K.A. Harris, X. Zhang, F.A. Vendeix, D. Fabris, P.F. Agris, The RNA modification database, RNAMDB: 2011 update, Nucleic Acids Res. 39 (2011) D195–D201 (Database issue).

[15] M.A. Machnicka, K. Milanowska, O. Osman Oglou, E. Purta, M. Kurkowska, A. Olchowik, W. Januszewski, S. Kalinowski, S. Dunin-Horkawicz, K.M. Rother, et al., MODOMICS: a database of RNA modification pathways—2013 update, Nucleic Acids Res. 41 (2013) D262–D267 (Database issue).

[16] Y. Saletore, K. Meyer, J. Korlach, I.D. Vilfan, S. Jaffrey, C.E. Mason, The birth of the Epitranscriptome: deciphering the function of RNA modifications, Genome Biol. 13 (10) (2012) 175.

[17] W.V. Gilbert, T.A. Bell, C. Schaening, Messenger RNA modifications: form, distribution, and function, Science 352 (6292) (2016) 1408–1412.
[18] B. Delatte, F. Wang, L.V. Ngoc, E. Collignon, E. Bonvin, R. Deplus, E. Calonne, B. Hassabi, P. Putmans, S. Awe, et al., Transcriptome-wide distribution and function of RNA hydroxymethylcytosine, Science 351 (6270) (2016) 282–285.
[19] S. Geula, S. Moshitch-Moshkovitz, D. Dominissini, A.A. Mansour, N. Kol, M. Salmon-Divon, V. Hershkovitz, E. Peer, N. Mor, Y.S. Manor, et al., m^6A mRNA methylation facilitates resolution of naïve pluripotency toward differentiation, Science 347 (6225) (2015) 1002–1006.
[20] A. Maity, B. Das, *N6*-methyladenosine modification in mRNA: machinery, function and implications for health and diseases, FEBS J. 283 (9) (2016) 1607–1630.
[21] B.S. Zhao, C. He, Pseudouridine in a new era of RNA modifications, Cell Res. 25 (2) (2015) 153–154.
[22] J. Choi, K.W. Ieong, H. Demirci, J. Chen, A. Petrov, A. Prabhakar, S.E. O'Leary, D. Dominissini, G. Rechavi, S.M. Soltis, et al., N^6-methyladenosine in mRNA disrupts tRNA selection and translation-elongation dynamics, Nat. Struct. Mol. Biol. 23 (2) (2016) 110–115.
[23] F. Rottman, A.J. Shatkin, R.P. Perry, Sequences containing methylated nucleotides at the 5′ termini of messenger RNAs: possible implications for processing, Cell 3 (3) (1974) 197–199.
[24] G. Cao, H.-B. Li, Z. Yin, R.A. Flavell, Recent advances in dynamic m^6A RNA modification, Open Biol. 6 (4) (2016) 160003.
[25] P.J. Batista, B. Molinie, J. Wang, K. Qu, J. Zhang, L. Li, D.M. Bouley, E. Lujan, B. Haddad, K. Daneshvar, et al., m^6A RNA modification controls cell fate transition in mammalian embryonic stem cells, Cell Stem Cell 15 (6) (2014) 707–719.
[26] E.M. Kennedy, H.P. Bogerd, A.V. Kornepati, D. Kang, D. Ghoshal, J.B. Marshall, B.C. Poling, K. Tsai, N.S. Gokhale, S.M. Horner, et al., Posttranscriptional m^6A editing of HIV-1 mRNAs enhances viral gene expression, Cell Host Microbe 19 (5) (2016) 675–685.
[27] S. Schwartz, Cracking the epitranscriptome, RNA 22 (2) (2016) 169–174.
[28] J. Zhou, J. Wan, X. Gao, X. Zhang, S.R. Jaffrey, S.B. Qian, Dynamic m^6A mRNA methylation directs translational control of heat shock response, Nature 526 (7574) (2015) 591–594.
[29] P.F. Agris, Decoding the genome: a modified view, Nucleic Acids Res. 32 (1) (2004) 223–238.
[30] P.F. Agris, F.A. Vendeix, W.D. Graham, tRNA's wobble decoding of the genome: 40 years of modification, J. Mol. Biol. 366 (1) (2007) 1–13.
[31] E.M. Gustilo, F.A. Vendeix, P.F. Agris, tRNA's modifications bring order to gene expression, Curr. Opin. Microbiol. 11 (2) (2008) 134–140.
[32] N. Ranjan, M.V. Rodnina, tRNA wobble modifications and protein homeostasis, Translation 4 (1) (2016) e1143076.
[33] R. Wang, Z. Luo, K. He, M.O. Delaney, D. Chen, J. Sheng, Base pairing and structural insights into the 5-formylcytosine in RNA duplex, Nucleic Acids Res. 44 (10) (2016) 4968–4977.
[34] J.E. Jackman, J.D. Alfonzo, Transfer RNA modifications: nature's combinatorial chemistry playground, Wiley Interdiscip. Rev. RNA 4 (1) (2013) 35–48.
[35] M.A. Machnicka, A. Olchowik, H. Grosjean, J.M. Bujnicki, Distribution and frequencies of post-transcriptional modifications in tRNAs, RNA Biol. 11 (12) (2014) 1619–1629.
[36] P.F. Agris, The importance of being modified: roles of modified nucleosides and Mg^{2+} in RNA structure and function, Prog. Nucleic Acid Res. Mol. Biol. 53 (1996) 79–129.

[37] F. Chapeville, F. Lipmann, G.V. Ehrenstein, B. Weisblum, W.J. Ray, S. Benzer, On the role of soluble ribonucleic acid in coding for amino acids, Proc. Natl. Acad. Sci. U.S.A. 48 (6) (1962) 1086–1092.
[38] P.R. Effraim, J. Wang, M.T. Englander, J. Avins, T.S. Leyh, R.L. Gonzalez Jr., V.W. Cornish, Natural amino acids do not require their native tRNAs for efficient selection by the ribosome, Nat. Chem. Biol. 5 (12) (2009) 947–953.
[39] M.R. Capecchi, G.N. Gussin, Suppression in vitro: identification of a serine-sRNA as a "nonsense" suppressor, Science 149 (3682) (1965) 417–422.
[40] D.L. Engelhardt, R.E. Webster, R.C. Wilhelm, N. Zinder, *In vitro* studies on the mechanism of suppression of a nonsense mutation, Proc. Natl. Acad. Sci. U.S.A. 54 (6) (1965) 1791–1797.
[41] A. Ambrogelly, S. Palioura, D. Soll, Natural expansion of the genetic code, Nat. Chem. Biol. 3 (1) (2007) 29–35.
[42] T.H. Wright, M.R. Vallee, B.G. Davis, From chemical mutagenesis to post-expression mutagenesis: a 50 year odyssey, Angew. Chem. Int. Ed. Engl. 55 (20) (2016) 5896–5903.
[43] F.H. Crick, Codon–anticodon pairing: the wobble hypothesis, J. Mol. Biol. 19 (2) (1966) 548–555.
[44] S.J. Nasvall, P. Chen, G.R. Bjork, The wobble hypothesis revisited: uridine-5-oxyacetic acid is critical for reading of G-ending codons, RNA 13 (12) (2007) 2151–2164.
[45] F.A. Vendeix, A. Dziergowska, E.M. Gustilo, W.D. Graham, B. Sproat, A. Malkiewicz, P.F. Agris, Anticodon domain modifications contribute order to tRNA for ribosome-mediated codon binding, Biochemistry 47 (23) (2008) 6117–6129.
[46] A. Weixlbaumer, F.V.T. Murphy, A. Dziergowska, A. Malkiewicz, F.A. Vendeix, P.F. Agris, V. Ramakrishnan, Mechanism for expanding the decoding capacity of transfer RNAs by modification of uridines, Nat. Struct. Mol. Biol. 14 (6) (2007) 498–502.
[47] S. Anderson, A.T. Bankier, B.G. Barrell, M.H. de Bruijn, A.R. Coulson, J. Drouin, I.C. Eperon, D.P. Nierlich, B.A. Roe, F. Sanger, et al., Sequence and organization of the human mitochondrial genome, Nature 290 (5806) (1981) 457–465.
[48] G. Attardi, Animal mitochondrial DNA: an extreme example of genetic economy, Int. Rev. Cytol. 93 (1985) 93–145.
[49] S. Anderson, M.H. de Bruijn, A.R. Coulson, I.C. Eperon, F. Sanger, I.G. Young, Complete sequence of bovine mitochondrial DNA. Conserved features of the mammalian mitochondrial genome, J. Mol. Biol. 156 (4) (1982) 683–717.
[50] C.N. Jones, K.A. Wilkinson, K.T. Hung, K.M. Weeks, L.L. Spremulli, Lack of secondary structure characterizes the 5′ ends of mammalian mitochondrial mRNAs, RNA 14 (5) (2008) 862–871.
[51] Y. Nakamura, T. Gojobori, T. Ikemura, Codon usage tabulated from international DNA sequence databases: status for the year 2000, Nucleic Acids Res. 28 (1) (2000) 292.
[52] G.R. Bjork, Genetic dissection of synthesis and function of modified nucleosides in bacterial transfer RNA, Prog. Nucleic Acid Res. Mol. Biol. 50 (1995) 263–338.
[53] J. Moriya, T. Yokogawa, K. Wakita, T. Ueda, K. Nishikawa, P.F. Crain, T. Hashizume, S.C. Pomerantz, J.A. McCloskey, G. Kawai, et al., A novel modified nucleoside found at the first position of the anticodon of methionine tRNA from bovine liver mitochondria, Biochemistry 33 (8) (1994) 2234–2239.
[54] A. Sinha, C. Kohrer, M.H. Weber, I. Masuda, V.K. Mootha, Y.M. Hou, U.L. RajBhandary, Biochemical characterization of pathogenic mutations in human mitochondrial methionyl-tRNA formyltransferase, J. Biol. Chem. 289 (47) (2014) 32729–32741.

[55] L. Van Haute, S. Dietmann, L. Kremer, S. Hussain, S.F. Pearce, C.A. Powell, J. Rorbach, R. Lantaff, S. Blanco, S. Sauer, et al., Deficient methylation and formylation of mt-tRNA^{Met} wobble cytosine in a patient carrying mutations in NSUN3, Nat. Commun. 7 (2016) 12039.
[56] N. Takeuchi, M. Kawakami, A. Omori, T. Ueda, L.L. Spremulli, K. Watanabe, Mammalian mitochondrial methionyl-tRNA transformylase from bovine liver. Purification, characterization, and gene structure, J. Biol. Chem. 273 (24) (1998) 15085–15090.
[57] N. Takeuchi, L. Vial, M. Panvert, E. Schmitt, K. Watanabe, Y. Mechulam, S. Blanquet, Recognition of tRNAs by methionyl-tRNA transformylase from mammalian mitochondria, J. Biol. Chem. 276 (23) (2001) 20064–20068.
[58] C.N. Jones, C.I. Jones, W.D. Graham, P.F. Agris, L.L. Spremulli, A disease-causing point mutation in human mitochondrial tRNA^{Met} results in tRNA misfolding leading to defects in translational initiation and elongation, J. Biol. Chem. 283 (49) (2008) 34445–34456.
[59] Y. Bilbille, E.M. Gustilo, K.A. Harris, C.N. Jones, H. Lusic, R.J. Kaiser, M.O. Delaney, L.L. Spremulli, A. Deiters, P.F. Agris, The human mitochondrial tRNA^{Met}: structure/function relationship of a unique modification in the decoding of unconventional codons, J. Mol. Biol. 406 (2) (2011) 257–274.
[60] S.U. Astrom, A.S. Bystrom, Rit1, a tRNA backbone-modifying enzyme that mediates initiator and elongator tRNA discrimination, Cell 79 (3) (1994) 535–546.
[61] T. Suzuki, A. Nagao, T. Suzuki, Human mitochondrial tRNAs: biogenesis, function, structural aspects, and diseases, Annu. Rev. Genet. 45 (2011) 299–329.
[62] C. Takemoto, L.L. Spremulli, L.A. Benkowski, T. Ueda, T. Yokogawa, K. Watanabe, Unconventional decoding of the AUA codon as methionine by mitochondrial tRNA^{Met} with the anticodon f^{5}CAU as revealed with a mitochondrial *in vitro* translation system, Nucleic Acids Res. 37 (5) (2009) 1616–1627.
[63] C. Takemoto, T. Koike, T. Yokogawa, L. Benkowski, L.L. Spremulli, T.A. Ueda, K. Nishikawa, K. Watanabe, The ability of bovine mitochondrial transfer RNA^{Met} to decode AUG and AUA codons, Biochimie 77 (1–2) (1995) 104–108.
[64] C. Takemoto, T. Ueda, K. Miura, K. Watanabe, Nucleotide sequences of animal mitochondrial tRNAs^{Met} possibly recognizing both AUG and AUA codons, Nucleic Acids Symp. Ser. 42 (1999) 77–78.
[65] K. Tomita, T. Ueda, K. Watanabe, 5-Formylcytidine (f^{5}C) found at the wobble position of the anticodon of squid mitochondrial tRNA^{MetCAU}, Nucleic Acids Symp. Ser. 37 (1997) 197–198.
[66] K. Tomita, T. Ueda, S. Ishiwa, P.F. Crain, J.A. McCloskey, K. Watanabe, Codon reading patterns in *Drosophila melanogaster* mitochondria based on their tRNA sequences: a unique wobble rule in animal mitochondria, Nucleic Acids Res. 27 (21) (1999) 4291–4297.
[67] Y. Watanabe, H. Tsurui, T. Ueda, R. Furushima, S. Takamiya, K. Kita, K. Nishikawa, K. Watanabe, Primary and higher order structures of nematode (*Ascaris suum*) mitochondrial tRNAs lacking either the T or D stem, J. Biol. Chem. 269 (36) (1994) 22902–22906.
[68] H. Lusic, E.M. Gustilo, F.A. Vendeix, R. Kaiser, M.O. Delaney, W.D. Graham, V.A. Moye, W.A. Cantara, P.F. Agris, A. Deiters, Synthesis and investigation of the 5-formylcytidine modified, anticodon stem and loop of the human mitochondrial tRNA^{Met}, Nucleic Acids Res. 36 (20) (2008) 6548–6557.
[69] L. Lancaster, H.F. Noller, Involvement of 16S rRNA nucleotides G1338 and A1339 in discrimination of initiator tRNA, Mol. Cell 20 (4) (2005) 623–632.
[70] W.A. Cantara, F.V.T. Murphy, H. Demirci, P.F. Agris, Expanded use of sense codons is regulated by modified cytidines in tRNA, Proc. Natl. Acad. Sci. U.S.A. 110 (27) (2013) 10964–10969.

[71] Z. Oashi, K. Murao, T. Yahagi, D.L. Von Minden, J.A. McCloskey, S. Nishimura, Characterization of C+ located in the first position of the anticodon of *Escherichia coli* $tRNA^{Met}$ as N^4-acetylcytidine, Biochim. Biophys. Acta 262 (2) (1972) 209–213.
[72] T. Muramatsu, S. Yokoyama, N. Horie, A. Matsuda, T. Ueda, Z. Yamaizumi, Y. Kuchino, S. Nishimura, T. Miyazawa, A novel lysine-substituted nucleoside in the first position of the anticodon of minor isoleucine tRNA from *Escherichia coli*, J. Biol. Chem. 263 (19) (1988) 9261–9267.
[73] D. Mandal, C. Kohrer, D. Su, S.P. Russell, K. Krivos, C.M. Castleberry, P. Blum, P.A. Limbach, D. Soll, U.L. RajBhandary, Agmatidine, a modified cytidine in the anticodon of archaeal $tRNA^{Ile}$, base pairs with adenosine but not with guanosine, Proc. Natl. Acad. Sci. U.S.A. 107 (7) (2010) 2872–2877.
[74] R.M. Voorhees, D. Mandal, C. Neubauer, C. Kohrer, U.L. RajBhandary, V. Ramakrishnan, The structural basis for specific decoding of AUA by isoleucine tRNA on the ribosome, Nat. Struct. Mol. Biol. 20 (5) (2013) 641–643.
[75] S. Nakano, T. Suzuki, L. Kawarada, H. Iwata, K. Asano, T. Suzuki, NSUN3 methylase initiates 5-formylcytidine biogenesis in human mitochondrial $tRNA^{Met}$, Nat. Chem. Biol. 12 (7) (2016) 546–551.
[76] B. El Yacoubi, B. Lyons, Y. Cruz, R. Reddy, B. Nordin, F. Agnelli, J.R. Williamson, P. Schimmel, M.A. Swairjo, V. de Crecy-Lagard, The universal YrdC/Sua5 family is required for the formation of threonylcarbamoyladenosine in tRNA, Nucleic Acids Res. 37 (9) (2009) 2894–2909.
[77] F. Juhling, M. Morl, R.K. Hartmann, M. Sprinzl, P.F. Stadler, J. Putz, tRNAdb 2009: compilation of tRNA sequences and tRNA genes, Nucleic Acids Res. 37 (2009) D159–D162 (Database issue).
[78] M.T. Lott, J.N. Leipzig, O. Derbeneva, H.M. Xie, D. Chalkia, M. Sarmady, V. Procaccio, D.C. Wallace, mtDNA variation and analysis using Mitomap and Mitomaster, Curr. Protoc. Bioinformatics 44 (2013) 1.23.21–26.
[79] C.L. Alston, M.C. Rocha, N.Z. Lax, D.M. Turnbull, R.W. Taylor, The genetics and pathology of mitochondrial disease, J. Pathol. (2016).
[80] R.A. de Almeida, M.G. Fraczek, S. Parker, D. Delneri, R.T. O'Keefe, Non-coding RNAs and disease: the classical ncRNAs make a comeback, Biochem. Soc. Trans. 44 (4) (2016) 1073–1078.
[81] J. Vissing, M.B. Salamon, P. Arlien-Soborg, S. Norby, P. Manta, S. DiMauro, H. Schmalbruch, A new mitochondrial $tRNA^{Met}$ gene mutation in a patient with dystrophic muscle and exercise intolerance, Neurology 50 (6) (1998) 1875–1878.
[82] J. Qu, R. Li, X. Zhou, Y. Tong, F. Lu, Y. Qian, Y. Hu, J.Q. Mo, C.E. West, M.X. Guan, The novel A4435G mutation in the mitochondrial $tRNA^{Met}$ may modulate the phenotypic expression of the LHON-associated ND4 G11778A mutation, Invest. Ophthalmol. Vis. Sci. 47 (2) (2006) 475–483.
[83] A. Lombes, D. Bories, E. Girodon, P. Frachon, M.M. Ngo, J. Breton-Gorius, M. Tulliez, M. Goossens, The first pathogenic mitochondrial methionine tRNA point mutation is discovered in splenic lymphoma, Hum. Mutat. 11 (Suppl 1) (1998) S175–S183.
[84] H. Grosjean, E. Westhof, An integrated, structure- and energy-based view of the genetic code, Nucleic Acids Res. 44 (17) (2016) 8020–8040.
[85] T.S. Suzuki, T. Wada, T. Saigo, K. Watanabe, W. Kimitsuna, Taurine as a constituent of mitochondrial tRNAs: new insights into the functions of taurine and human mitochondrial diseases, EMBO J. 21 (23) (2002) 6581–6589.
[86] R. Saxena, B.F. Voight, V. Lyssenko, N.P. Burtt, P.I. de Bakker, H. Chen, J.J. Roix, S. Kathiresan, J.N. Hirschhorn, M.J. Daly, et al., Genome-wide association analysis identifies loci for type 2 diabetes and triglyceride levels, Science 316 (5829) (2007) 1331–1336.

[87] V. Steinthorsdottir, G. Thorleifsson, I. Reynisdottir, R. Benediktsson, T. Jonsdottir, G.B. Walters, U. Styrkarsdottir, S. Gretarsdottir, V. Emilsson, S. Ghosh, et al., A variant in CDKAL1 influences insulin response and risk of type 2 diabetes, Nat. Genet. 39 (6) (2007) 770–775.
[88] F.V. Murphy, V. Ramakrishnan, A. Malkiewicz, P.F. Agris, The role of modifications in codon discrimination by $tRNA^{LysUUU}$, Nat. Struct. Mol. Biol. 11 (12) (2004) 1186–1191.
[89] S.S. Phelps, A. Malkiewicz, P.F. Agris, S. Joseph, Modified nucleotides in $tRNA^{Lys}$ and $tRNA^{Val}$ are important for translocation, J. Mol. Biol. 338 (3) (2004) 439–444.
[90] J.W. Stuart, Z. Gdaniec, R. Guenther, M. Marszalek, E. Sochacka, A. Malkiewicz, P.F. Agris, Functional anticodon architecture of human $tRNA^{Lys3}$ includes disruption of intraloop hydrogen bonding by the naturally occurring amino acid modification, t^6A, Biochemistry 39 (44) (2000) 13396–13404.
[91] B. Mallick, J. Chakrabarti, S. Sahoo, Z. Ghosh, S. Das, Identity elements of archaeal tRNA, DNA Res. 12 (4) (2005) 235–246.
[92] G.R. Björk, Chapter 11: Biosynthesis and function of modified nucleosides, in: D. Söll, U. RajBhandary (Eds.), tRNA: Structure, Biosynthesis, and Function, American Society for Microbiology, Washington, DC, 1995, pp. 165–205.
[93] M. Raba, K. Limburg, M. Burghagen, J.R. Katze, M. Simsek, J.E. Heckman, U.L. Rajbhandary, H.J. Gross, Nucleotide sequence of three isoaccepting lysine tRNAs from rabbit liver and SV40-transformed mouse fibroblasts, Eur. J. Biochem. 97 (1) (1979) 305–318.
[94] P.F. Agris, Wobble position modified nucleosides evolved to select transfer RNA codon recognition: a modified-wobble hypothesis, Biochimie 73 (1991) 1345–1349.
[95] Y. Nakai, N. Umeda, T. Suzuki, M. Nakai, H. Hayashi, K. Watanabe, H. Kagamiyama, Yeast Nfs1p is involved in thio-modification of both mitochondrial and cytoplasmic tRNAs, J. Biol. Chem. 279 (13) (2004) 12363–12368.
[96] T. Suzuki, T. Suzuki, A complete landscape of post-transcriptional modifications in mammalian mitochondrial tRNAs, Nucleic Acids Res. 42 (2014) 7346–7357.
[97] N. Ostrov, M. Landon, M. Guell, G. Kuznetsov, J. Teramoto, N. Cervantes, M. Zhou, K. Singh, M.G. Napolitano, M. Moosburner, et al., Design, synthesis, and testing toward a 57-codon genome, Science 353 (6301) (2016) 819–822.
[98] A.R. Bezerra, A.R. Guimaraes, M.A. Santos, Non-standard genetic codes define new concepts for protein engineering, Life (Basel) 5 (4) (2015) 1610–1628.
[99] T.G. Hagervall, S.C. Pomerantz, J.A. McCloskey, Reduced misreading of asparagine codons by *Escherichia coli* $tRNA^{Lys}$ with hypomodified derivatives of 5-methylaminomethyl-2-thiouridine in the wobble position, J. Mol. Biol. 284 (1) (1998) 33–42.
[100] M. Sugita, K. Shinozaki, M. Sugiura, Tobacco chloroplast $tRNA^{Lys}$(UUU) gene contains a 2.5-kilobase-pair intron: an open reading frame and a conserved boundary sequence in the intron, Proc. Natl. Acad. Sci. U.S.A. 82 (11) (1985) 3557–3561.
[101] W. Fuller, A. Hodgson, Conformation of the anticodon loop in tRNA, Nature 215 (5103) (1967) 817–821.
[102] S. Kurata, A. Weixlbaumer, T. Ohtsuki, T. Shimazaki, T. Wada, Y. Kirino, K. Takai, K. Watanabe, V. Ramakrishnan, T. Suzuki, Modified uridines with C5-methylene substituents at the first position of the tRNA anticodon stabilize U.G wobble pairing during decoding, J. Biol. Chem. 283 (27) (2008) 18801–18811.
[103] K. Watanabe, S. Yokobori, tRNA modification and genetic code variations in animal mitochondria, J. Nucleic Acids 2011 (2011) 623095.
[104] A.S. Kamble, S.B. Sambhare, P.M. Fandilolu, K.D. Sonawane, Structural significance of modified nucleoside 5-taurinomethyl-2-thiouridine, τm^5s^2U, found at 'wobble'

position in anticodon loop of human mitochondrial $tRNA^{Lys}$, Struct. Chem. 27 (3) (2016) 839–854.

[105] P.F. Agris, H. Sierzputowska-Gracz, W. Smith, A. Malkiewicz, E. Sochacka, B. Nawrot, Thiolation of uridine carbon-2 restricts the motional dynamics of the transfer RNA wobble position nucleoside, J. Am. Chem. Soc. 114 (7) (1992) 2652–2656.

[106] G.R. Björk, B. Huang, O.P. Persson, A.S. Bystrom, A conserved modified wobble nucleoside (mcm^5s^2U) in lysyl-tRNA is required for viability in yeast, RNA 13 (8) (2007) 1245–1255.

[107] D.D. Nedialkova, S.A. Leidel, Optimization of codon translation rates via tRNA modifications maintains proteome integrity, Cell 161 (7) (2015) 1606–1618.

[108] N. Manickam, K. Joshi, M.J. Bhatt, P.J. Farabaugh, Effects of tRNA modification on translational accuracy depend on intrinsic codon-anticodon strength, Nucleic Acids Res. 44 (4) (2016) 1871–1881.

[109] A. Rozov, N. Demeshkina, I. Khusainov, E. Westhof, M. Yusupov, G. Yusupova, Novel base-pairing interactions at the tRNA wobble position crucial for accurate reading of the genetic code, Nat. Commun. 7 (2016) 10457.

[110] P.C. Durant, A.C. Bajji, M. Sundaram, R.K. Kumar, D.R. Davis, Structural effects of hypermodified nucleosides in the *Escherichia coli* and human $tRNA^{Lys}$ anticodon loop: the effect of nucleosides s^2U, mcm^5U, mcm^5s^2U, mnm^5s^2U, t^6A, and ms^2t^6A, Biochemistry 44 (22) (2005) 8078–8089.

[111] F.A. Vendeix, F.V.T. Murphy, W.A. Cantara, G. Leszczynska, E.M. Gustilo, B. Sproat, A. Malkiewicz, P.F. Agris, Human $tRNA^{Lys3UUU}$ is pre-structured by natural modifications for cognate and wobble codon binding through keto-enol tautomerism, J. Mol. Biol. 416 (4) (2012) 467–485.

[112] M.J. Johansson, A. Esberg, B. Huang, G.R. Bjork, A.S. Bystrom, Eukaryotic wobble uridine modifications promote a functionally redundant decoding system, Mol. Cell. Biol. 28 (10) (2008) 3301–3312.

[113] M.J. Serra, D.H. Turner, Predicting thermodynamic properties of RNA, Methods Enzymol. 259 (1995) 242–261.

[114] A. Rodriguez-Hernandez, J.L. Spears, K.W. Gaston, P.A. Limbach, H. Gamper, Y.M. Hou, R. Kaiser, P.F. Agris, J.J. Perona, Structural and mechanistic basis for enhanced translational efficiency by 2-thiouridine at the tRNA anticodon wobble position, J. Mol. Biol. 425 (20) (2013) 3888–3906.

[115] T. Numata, Y. Ikeuchi, S. Fukai, T. Suzuki, O. Nureki, Snapshots of tRNA sulphuration via an adenylated intermediate, Nature 442 (7101) (2006) 419–424.

[116] J. Sheng, A. Larsen, B.D. Heuberger, J.C. Blain, J.W. Szostak, Crystal structure studies of RNA duplexes containing s^2U:A and s^2U:U base pairs, J. Am. Chem. Soc. 136 (39) (2014) 13916–13924.

[117] S.S. Ashraf, G. Ansari, R. Guenther, E. Sochacka, A. Malkiewicz, P.F. Agris, The uridine in "U-turn": contributions to tRNA-ribosomal binding, RNA 5 (4) (1999) 503–511.

[118] G.N. Jacobson, P.L. Clark, Quality over quantity: optimizing co-translational protein folding with non-'optimal' synonymous codons, Curr. Opin. Struct. Biol. 38 (2016) 102–110.

[119] T.E. Quax, N.J. Claassens, D. Soll, J. van der Oost, Codon bias as a means to fine-tune gene expression, Mol. Cell 59 (2) (2015) 149–161.

[120] F. Supek, The code of silence: widespread associations between synonymous codon biases and gene function, J. Mol. Evol. 82 (1) (2016) 65–73.

[121] U. Begley, M. Dyavaiah, A. Patil, J.P. Rooney, D. DiRenzo, C.M. Young, D.S. Conklin, R.S. Zitomer, T.J. Begley, Trm9-catalyzed tRNA modifications link translation to the DNA damage response, Mol. Cell 28 (5) (2007) 860–870.

[122] C.T. Chan, W. Deng, F. Li, M.S. DeMott, I.R. Babu, T.J. Begley, P.C. Dedon, Highly predictive reprogramming of tRNA modifications is linked to selective expression of codon-biased genes, Chem. Res. Toxicol. 28 (5) (2015) 978–988.
[123] C.T. Chan, Y.L. Pang, W. Deng, I.R. Babu, M. Dyavaiah, T.J. Begley, P.C. Dedon, Reprogramming of tRNA modifications controls the oxidative stress response by codon-biased translation of proteins, Nat. Commun. 3 (2012) 937.
[124] L. Endres, P.C. Dedon, T.J. Begley, Codon-biased translation can be regulated by wobble-base tRNA modification systems during cellular stress responses, RNA Biol. 12 (6) (2015) 603–614.
[125] M.J. Katz, L. Gandara, A.L. De Lella Ezcurra, P. Wappner, Hydroxylation and translational adaptation to stress: some answers lie beyond the STOP codon, Cell. Mol. Life Sci. 73 (9) (2016) 1881–1893.
[126] J. Fernandez-Vazquez, I. Vargas-Perez, M. Sanso, K. Buhne, M. Carmona, E. Paulo, D. Hermand, M. Rodriguez-Gabriel, J. Ayte, S. Leidel, et al., Modification of $tRNA^{LysUUU}$ by elongator is essential for efficient translation of stress mRNAs, PLoS Genet. 9 (7) (2013) e1003647.
[127] B.P. Anton, S.P. Russell, J. Vertrees, S. Kasif, E.A. Raleigh, P.A. Limbach, R.J. Roberts, Functional characterization of the YmcB and YqeV tRNA methylthiotransferases of *Bacillus subtilis*, Nucleic Acids Res. 38 (18) (2010) 6195–6205.
[128] K.A. Harris, B.G. Bobay, K.L. Sarachan, A.F. Sims, Y. Bilbille, C. Deutsch, D. Iwata-Reuyl, P.F. Agris, NMR-based structural analysis of threonylcarbamoyl-AMP synthase and its substrate interactions, J. Biol. Chem. 290 (33) (2015) 20032–20043.
[129] P.C. Thiaville, B. El Yacoubi, C. Kohrer, J.J. Thiaville, C. Deutsch, D. Iwata-Reuyl, J.M. Bacusmo, J. Armengaud, Y. Bessho, C. Wetzel, et al., Essentiality of threonylcarbamoyladenosine (t^6A), a universal tRNA modification, in bacteria, Mol. Microbiol. 98 (6) (2015) 1199–1221.
[130] P.C. Thiaville, R. Legendre, D. Rojas-Benitez, A. Baudin-Baillieu, I. Hatin, G. Chalancon, A. Glavic, O. Namy, V. de Crecy-Lagard, Global translational impacts of the loss of the tRNA modification t^6A in yeast, Microb. Cell 3 (1) (2016) 29–45.
[131] M. Matuszewski, E. Sochacka, Stability studies on the newly discovered cyclic form of tRNA N^6-threonylcarbamoyladenosine (ct^6A), Bioorg. Med. Chem. Lett. 24 (12) (2014) 2703–2706.
[132] C. Yarian, M. Marszalek, E. Sochacka, A. Malkiewicz, R. Guenther, A. Miskiewicz, P.F. Agris, Modified nucleoside dependent Watson-Crick and wobble codon binding by tRNALysUUU species, Biochemistry 39 (44) (2000) 13390–13395.
[133] K. Miyauchi, S. Kimura, T. Suzuki, A cyclic form of N^6-threonylcarbamoyladenosine as a widely distributed tRNA hypermodification, Nat. Chem. Biol. 9 (2) (2013) 105–111.
[134] F.Y. Wei, T. Suzuki, S. Watanabe, S. Kimura, T. Kaitsuka, A. Fujimura, H. Matsui, M. Atta, H. Michiue, M. Fontecave, et al., Deficit of $tRNA^{Lys}$ modification by Cdkal1 causes the development of type 2 diabetes in mice, J. Clin. Invest. 121 (9) (2011) 3598–3608.
[135] Y. Xu, A.D. MacKerell Jr., L. Nilsson, Structural effects of modified ribonucleotides and magnesium in transfer RNAs, Bioorg. Med. Chem. 24 (20) (2016) 4826–4834.
[136] W.D. Graham, L. Barley-Maloney, C.J. Stark, A. Kaur, C. Stolarchuk, B. Sproat, G. Leszczynska, A. Malkiewicz, N. Safwat, P. Mucha, et al., Functional recognition of the modified human $tRNA^{Lys3UUU}$ anticodon domain by HIV's nucleocapsid protein and a peptide mimic, J. Mol. Biol. 410 (4) (2011) 698–715.
[137] J.L. Spears, X. Xiao, C.K. Hall, P.F. Agris, Amino acid signature enables proteins to recognize modified tRNA, Biochemistry 53 (7) (2014) 1125–1133.
[138] M. Sprinzl, K.S. Vassilenko, Compilation of tRNA sequences and sequences of tRNA genes, Nucleic Acids Res. 33 (Database issue) (2005) D139–D140.

[139] H. Ishikura, Y. Yamada, S. Nishimura, Structure of serine tRNA from *Escherichia coli*. I. Purification of serine tRNA's with different codon responses, Biochim. Biophys. Acta 228 (2) (1971) 471–481.
[140] S.K. Mitra, F. Lustig, B. Akesson, T. Axberg, P. Elias, U. Lagerkvist, Relative efficiency of anticodons in reading the valine codons during protein synthesis *in vitro*, J. Biol. Chem. 254 (14) (1979) 6397–6401.
[141] M.A. Sorensen, J. Elf, E. Bouakaz, T. Tenson, S. Sanyal, G.R. Bjork, M. Ehrenberg, Over expression of a $tRNA^{Leu}$ isoacceptor changes charging pattern of leucine tRNAs and reveals new codon reading, J. Mol. Biol. 354 (1) (2005) 16–24.
[142] S. Yokoyama, T. Watanabe, K. Murao, H. Ishikura, Z. Yamaizumi, S. Nishimura, T. Miyazawa, Molecular mechanism of codon recognition by tRNA species with modified uridine in the first position of the anticodon, Proc. Natl. Acad. Sci. U.S.A. 82 (1985) 4905–4909.
[143] S.J. Nasvall, P. Chen, G.R. Bjork, The modified wobble nucleoside uridine-5-oxyacetic acid in $tRNA^{Procmo5UGG}$ promotes reading of all four proline codons *in vivo*, RNA 10 (10) (2004) 1662–1673.
[144] U. Kothe, M.V. Rodnina, Codon reading by $tRNA^{Ala}$ with modified uridine in the wobble position, Mol. Cell 25 (1) (2007) 167–174.
[145] S. Kimura, T. Suzuki, Iron-sulfur proteins responsible for RNA modifications, Biochim. Biophys. Acta 1853 (6) (2015) 1272–1283.
[146] B. Huang, M.J. Johansson, A.S. Bystrom, An early step in wobble uridine tRNA modification requires the Elongator complex, RNA 11 (4) (2005) 424–436.
[147] A. Esberg, B. Huang, M.J. Johansson, A.S. Bystrom, Elevated levels of two tRNA species bypass the requirement for elongator complex in transcription and exocytosis, Mol. Cell 24 (1) (2006) 139–148.
[148] C. Chen, S. Tuck, A.S. Bystrom, Defects in tRNA modification associated with neurological and developmental dysfunctions in *Caenorhabditis elegans* elongator mutants, PLoS Genet. 5 (7) (2009) e1000561.
[149] J. Cabello-Villegas, M.E. Winkler, E.P. Nikonowicz, Solution conformations of unmodified and $A_{37}N^6$-dimethylallyl modified anticodon stem-loops of *Escherichia coli* $tRNA^{Phe}$, J. Mol. Biol. 319 (5) (2002) 1015–1034.
[150] C. Yarian, H. Townsend, W. Czestkowski, E. Sochacka, A.J. Malkiewicz, R. Guenther, A. Miskiewicz, P.F. Agris, Accurate translation of the genetic code depends on tRNA modified nucleosides, J. Biol. Chem. 277 (19) (2002) 16391–16395.
[151] D. Soll, J.D. Cherayil, R.M. Bock, Studies on polynucleotides. LXXV. Specificity of tRNA for codon recognition as studied by the ribosomal binding technique, J. Mol. Biol. 29 (1) (1967) 97–112.
[152] D. Soll, U.L. RajBhandary, Studies on polynucleotides. LXXVI. Specificity of transfer RNA for codon recognition as studied by amino acid incorporation, J. Mol. Biol. 29 (1) (1967) 113–124.
[153] T. Ikemura, Codon usage and tRNA content in unicellular and multicellular organisms, Mol. Biol. Evol. 2 (1) (1985) 13–34.
[154] K. Wada, Y. Wada, F. Ishibashi, T. Gojobori, T. Ikemura, Codon usage tabulated from the GenBank genetic sequence data, Nucleic Acids Res. 20 (Suppl) (1992) 2111–2118.
[155] S.R. Maloy, V.J. Stewart, R.K. Taylor, Genetic Analysis of Pathogenic Bacteria: A Laboratory Manual, Cold Spring Harbor Laboratory Press, Plainville, NY, 1996.
[156] S.A. Martinis, P. Plateau, J. Cavarelli, C. Florentz, Aminoacyl-tRNA synthetases: a new image for a classical family, Biochimie 81 (7) (1999) 683–700.
[157] P. Auffinger, E. Westhof, Singly and bifurcated hydrogen-bonded base-pairs in tRNA anticodon hairpins and ribozymes, J. Mol. Biol. 292 (3) (1999) 467–483.

[158] P. Auffinger, S. Louise-May, E. Westhof, Molecular dynamics simulations of solvated yeast tRNA^{Asp}, Biophys. J. 76 (1 Pt. 1) (1999) 50–64.
[159] B. Delagoutte, D. Moras, J. Cavarelli, tRNA aminoacylation by arginyl-tRNA synthetase: induced conformations during substrates binding, EMBO J. 19 (21) (2000) 5599–5610.
[160] F.V.T. Murphy, V. Ramakrishnan, Structure of a purine-purine wobble base pair in the decoding center of the ribosome, Nat. Struct. Mol. Biol. 11 (12) (2004) 1251–1252.
[161] K. Chakraburtty, Primary structure of tRNA^{ArgII} of *E. coli* B, Nucleic Acids Res. 2 (10) (1975) 1787–1792.
[162] K. Murao, T. Tanabe, F. Ishii, M. Namiki, S. Nishimura, Primary sequence of arginine transfer RNA from *Escherichia coli*, Biochem. Biophys. Res. Commun. 47 (6) (1972) 1332–1337.
[163] M. Sissler, R. Giege, C. Florentz, Arginine aminoacylation identity is context-dependent and ensured by alternate recognition sets in the anticodon loop of accepting tRNA transcripts, EMBO J. 15 (18) (1996) 5069–5076.
[164] K. Tamura, H. Himeno, H. Asahara, T. Hasegawa, M. Shimizu, *In vitro* study of *E. coli* tRNA^{Arg} and tRNA^{Lys} identity elements, Nucleic Acids Res. 20 (9) (1992) 2335–2339.
[165] G. Jager, R. Leipuviene, M.G. Pollard, Q. Qian, G.R. Bjork, The conserved Cys-X1-X2-Cys motif present in the TtcA protein is required for the thiolation of cytidine in position 32 of tRNA from *Salmonella enterica* serovar Typhimurium, J. Bacteriol. 186 (3) (2004) 750–757.
[166] W.A. Cantara, Y. Bilbille, J. Kim, R. Kaiser, G. Leszczynska, A. Malkiewicz, P.F. Agris, Modifications modulate anticodon loop dynamics and codon recognition of *E. coli* tRNA^{Arg1,2}, J. Mol. Biol. 416 (4) (2012) 579–597.
[167] C. Cheong, G. Varani, I. Tinoco Jr., Solution structure of an unusually stable RNA hairpin, 5'GGAC(UUCG)GUCC, Nature 346 (6285) (1990) 680–682.
[168] P.B. Moore, Structural motifs in RNA, Annu. Rev. Biochem. 68 (1999) 287–300.
[169] S. Sauge-Merle, D. Falconet, M. Fontecave, An active ribonucleotide reductase from *Arabidopsis thaliana*, Eur. J. Biochem. 266 (1) (1999) 62–69.
[170] D. Soll, D.S. Jones, E. Ohtsuka, R.D. Faulkner, R. Lohrmann, H. Hayatsu, H.G. Khorana, Specificity of sRNA for recognition of codons as studied by the ribosomal binding technique, J. Mol. Biol. 19 (2) (1966) 556–573.
[171] S.S. Ashraf, R.H. Guenther, G. Ansari, A. Malkiewicz, E. Sochacka, P.F. Agris, Role of modified nucleosides of yeast tRNA^{Phe} in ribosomal binding, Cell Biochem. Biophys. 33 (3) (2000) 241–252.
[172] J.W. Stuart, K.M. Koshlap, R. Guenther, P.F. Agris, Naturally-occurring modification restricts the anticodon domain conformational space of tRNA^{Phe}, J. Mol. Biol. 334 (5) (2003) 901–918.
[173] T.M. Lowe, S.R. Eddy, tRNAscan-SE: a program for improved detection of transfer RNA genes in genomic sequence, Nucleic Acids Res. 25 (5) (1997) 955–964.
[174] G.M. Garcia, P.K. Mar, D.A. Mullin, J.R. Walker, N.E. Prather, The *E. coli* dnaY gene encodes an arginine transfer RNA, Cell 45 (3) (1986) 453–459.
[175] R.A. Spanjaard, K. Chen, J.R. Walker, J. van Duin, Frameshift suppression at tandem AGA and AGG codons by cloned tRNA genes: assigning a codon to *argU* tRNA and T4 tRNA^{Arg}, Nucleic Acids Res. 18 (17) (1990) 5031–5036.
[176] V. Ramakrishnan, The ribosome emerges from a black box, Cell 159 (5) (2014) 979–984.
[177] T.A. Steitz, A structural understanding of the dynamic ribosome machine, Nat. Rev. Mol. Cell Biol. 9 (3) (2008) 242–253.

[178] K.M. Thompson, S. Gottesman, The MiaA tRNA modification enzyme is necessary for robust RpoS expression in *Escherichia coli*, J. Bacteriol. 196 (4) (2014) 754–761.
[179] J.I. Aubee, M. Olu, K.M. Thompson, The i^6A37 tRNA modification is essential for proper decoding of UUX-Leucine codons during *rpoS and iraP* translation, RNA 22 (5) (2016) 729–742.
[180] D.B. Dunn, The occurrence of 1-methyladenine in ribonucleic acid, Biochim. Biophys. Acta 46 (1961) 198–200.

CHAPTER TWO

The Evolution of Substrate Specificity by tRNA Modification Enzymes

Katherine M. McKenney, Mary Anne T. Rubio, Juan D. Alfonzo[1]
Department of Microbiology, Ohio State Biochemistry Program, The Center for RNA Biology, The Ohio State University, Columbus, OH, United States
[1]Corresponding author: e-mail address: alfonzo.1@osu.edu

Contents

Abstract

All types of nucleic acids in cells undergo naturally occurring chemical modifications, including DNA, rRNA, mRNA, snRNA, and most prominently tRNA. Over 100 different modifications have been described and every position in the purine and pyrimidine bases can be modified; often the sugar is also modified [1]. In tRNA, the function of modifications varies; some modulate global and/or local RNA structure, and others directly impact decoding and may be essential for viability. Whichever the case, the overall importance of modifications is highlighted by both their evolutionary conservation and the fact that organisms use a substantial portion of their genomes to encode modification enzymes, far exceeding what is needed for the de novo synthesis of the canonical nucleotides themselves [2]. Although some modifications occur at exactly the same nucleotide position in tRNAs from the three domains of life, many can be found at

The Enzymes, Volume 41
ISSN 1874-6047
http://dx.doi.org/10.1016/bs.enz.2017.03.002

various positions in a particular tRNA and their location may vary between and within different tRNAs. With this wild array of chemical diversity and substrate specificities, one of the big challenges in the tRNA modification field has been to better understand at a molecular level the modes of substrate recognition by the different modification enzymes; in this realm RNA binding rests at the heart of the problem. This chapter will focus on several examples of modification enzymes where their mode of RNA binding is well understood; from these, we will try to draw general conclusions and highlight growing themes that may be applicable to the RNA modification field at large.

1. INTRODUCTION

The importance of RNA binding percolates all aspects of RNA processing and involves an amazing set of different RNA interaction domains; some are readily recognizable by their degree of sequence conservation, and others are more discreet and escape easy prediction [1]. In general, RNA-binding domains may be divided into two major groups: (1) those belonging to proteins that act alone ("professional RNA-binding proteins"); (2) those that are appended to other proteins, acting as binding modules that work in *cis* or *trans* to anchor other functionalities at the site of action. It is clear that a critical step for the evolution of extant life must have involved the appearance of polypeptides able to interact in some fashion with RNA, facilitating the transition into an RNA–protein world from an all-RNA world. Such developments had to occur after the emergence of templated peptide synthesis, but not too much later. Although early RNA-binding proteins may have been act alone entities, the necessity for modularity must have been imminent [1]. Modularity implies that discreet protein domains, whose sole purpose was to bind RNA, had to fuse with catalytic domains [2,3]; these must have been an indispensable invention for the evolution of RNA processing and certainly for the appearance of RNA modification specificity.

2. THE PROBLEM OF tRNA RECOGNITION

All tRNAs in cells adopt a very similar 3D-folded structure, the characteristic L-shape (Fig. 1), whereby the tRNA can be divided into two functional domains: the codon recognition domain, formed by the coaxial stacking of the D-arm onto the anticodon arm, and the amino acid-accepting domain, formed by the coaxial stacking of the TΨC arm on top of the acceptor arm. These two domains come together as an L-shaped

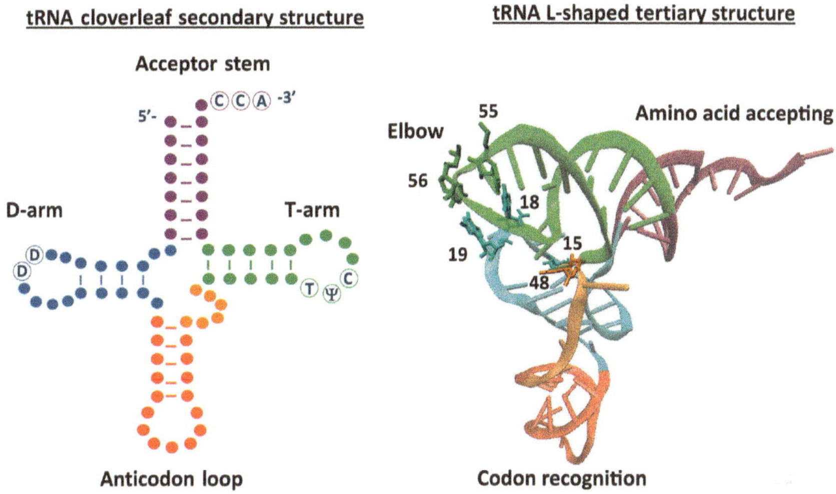

Fig. 1 Characteristic secondary cloverleaf structure of tRNA with relevant stem loops indicated and color-coded (*left*). L-shaped tertiary structure following the same color-coding with the amino acid-accepting arm, codon recognition arm, and the elbow indicated (*right*). The specific base pairs that are needed to form a stable L-shaped structure are numbered and shown as *bonds* in *colors* according to their position in the tRNA (base pairs include nucleotides 15 and 48, 18 and 55, and 19 and 56).

molecule whose structure is stabilized by base-pair interactions between the D-loop, the variable loop, and the aforementioned TΨC loop. Critical to this folding is a stable "elbow structure," which is formed by specific base pairs between several nucleotides: residues 15 of the D-loop and 48 of the variable loop, G_{18} of the D-loop, and Ψ_{55} of the TΨC loop and finally an additional Watson–Crick base pair between G_{19} of the D-loop and the highly conserved C_{56} (of TΨC). It has become apparent that this conserved structure (the elbow) has been used repeatedly for recognition by various tRNA-targeting enzymes such as a number of aminoacyl tRNA synthetases, ribonuclease P, and some tRNA transport factors [4]. In terms of modification, this conserved structure sets a critical baseline for how modification enzymes approach the recognition problem. A sure inference is the fact that if Ψ_{55} is important for proper tRNA folding, the enzyme that catalyzes its formation cannot be dependent on the L-shape structure for recognition. Predictably such enzymes learn to target tRNAs before the invention of the L-shape and inherently must have acted on proto-tRNAs before the elbow structure was adopted. Similarly, methylation of G_9 plays a critical role in tRNA folding [5] and had to be introduced early in the evolution of extant canonical L-shaped tRNAs.

Once the tRNA is folded, then other modifications can be added and the conserved tertiary features of a tRNA can be exploited for recognition. For example, many of the modifications that target the anticodon loop demand such a degree of specificity that mistakes, such as modification of neighboring nucleotides that are not the intended target, could prove detrimental to a cell's well-being. In general, modification enzymes that target the anticodon loop required a fully folded tRNA for activity and it is from the information found in such a structured molecule that exquisite specificity is achieved. This is especially true of anticodon deaminases, where deamination of the wrong nucleotide can potentially reassign the tRNA to a different codon, leading to a miscoding problem. It has been suggested then that tRNA modification enzymes can be divided into two major groups, based on how they recognize the substrate they target: (1) architecture-dependent and (2) architecture-independent enzymes; the former require a fully folded L-shape tRNA for activity, and the latter act earlier and are in general a prerequisite for folding [6]. In turn, many of the architecture-independent enzymes act at sites away from the anticodon arm and play critical roles as structural modulators, while architecture-dependent ones tend to target the anticodon loop, the modifications that they synthesize affecting only local structure and anticodon function. In the following pages, we will highlight how these structural constraints affect the mode of substrate recognition by tRNA-modifying enzymes. We will focus on a few well-characterized examples of how modification enzymes have solved the tRNA-binding problem to discriminate between nearly identical substrates, while only targeting the desired ones. To illustrate this concept, we will focus on three different groups of modifications enzymes: (1) tRNA deaminases, (2) tRNA methyltransferases, and (3) the tRNA-specific pseudouridine synthases.

3. DEAMINASES THAT TARGET tRNA

Historically, the modification field has been divided into two groups: the editors and the modifiers. The term *modification* implies any biological change to a polynucleotide chain that alters its chemical composition beyond that of the canonical nucleotides (G, A, U (T), and C); such changes are programmed rather than stochastic and different from chemical mutagenesis [7]. Thus, RNA editing can be considered a well-defined category within the vast umbrella term of posttranscriptional modifications. Editing *stricto sensu* involves a programmed change of one canonical nucleotide to another (e.g., C to U) or a change of a canonical nucleotide that alters the meaning of the edited RNA beyond what is encoded by the genome (e.g., A-to-I deamination) [7]. Editing may also involve more extensive insertion and deletion of nucleotides as those occurring in *Physarum* and trypanosomatid mitochondria, but this will not be further discussed here.

Among the tRNA deaminases, some edit the first position of the anticodon (position 34), while others target additional positions away from the anticodon, for example, A-to-I deamination at position 37 in eukaryotes, or position 57 or 58 in Archaea [8–10]. The best studied among these enzymes are the adenosine deaminases acting on *t*RNA, or ADAT, where ADAT2/3 deaminates adenosine 34 (A_{34}) to inosine 34 (I_{34}). Inosine expands the decoding capacity of tRNAs allowing a single tRNA to decode multiple codons with the same amino acid. Other ADATs, for example, ADAT1, form inosine at position 37 of the anticodon loop of a single tRNA ($tRNA^{Ala}$) [11]. While I_{37} does not affect decoding specificity, it is important for translational accuracy. In addition, C-to-U changes at other positions have been described in organellar tRNAs from various organisms [12,13], but the enzymes involved have not been identified. Mechanistically these are inferred to occur by hydrolytic deamination. In this section, we will concentrate on the A-to-I enzymes, whose activities have been reconstituted in vitro and mechanisms of tRNA substrate recognition extensively characterized.

Anticodon A-to-I deaminases have been described in Bacteria (ADATa, TadA) and Eukarya (ADAT2/3) [14–18]. The bacterial enzyme is homodimeric in solution and targets a single tRNA, $tRNA^{Arg}{}_{ACG}$, the only A_{34}-containing tRNA in bacteria (Fig. 2). The crystal structure of ADATa

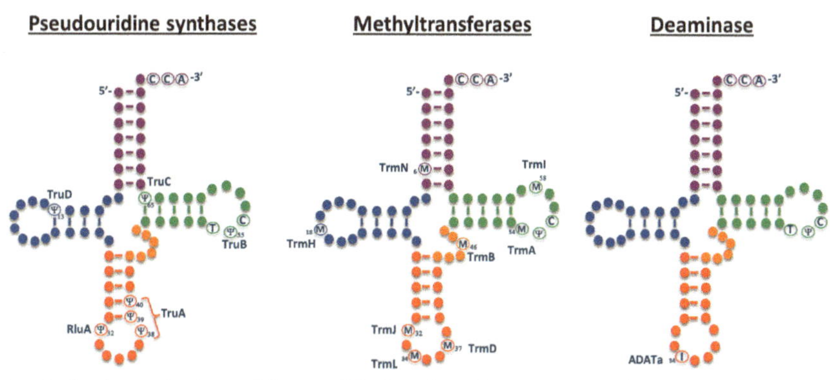

Fig. 2 A schematic view of the modified nucleotide positions within the cloverleaf secondary structure of tRNA and the respective bacterial modification enzymes for pseudourine (Y, *left*), methylation (M, *middle*), and A-to-I deamination (I, *right*).

was reported by several laboratories, which showed the general architecture of the active site and confirmed its similarity with other members of the deaminase superfamily [19–22]. Biochemical analysis demonstrated that this enzyme could efficiently catalyze deamination of minimal substrates representing only the anticodon arm of $tRNA^{Arg}$; a negligible difference was observed in the catalytic efficiency of this enzyme with shorter substrates when compared to a full-length tRNA [19,20]. This observation led to the proposal that ADATa was architecture independent. This may have been the result of its specialization toward a single substrate, which then led to the evolution of a very restrictive RNA-binding domain limited to residues in the proximity of the active site [21]. This proposal was confirmed by the cocrystal structure of ADATa bound to an anticodon arm mimic, where the nucleotide to be deaminated (A_{34}) was replaced by the adenosine analog nebularine [20]. This was necessary to prevent the enzyme from turning over the substrate and enabled trapping of the substrate-bound enzyme. This cocrystal revealed that upon complex formation, the RNA substrate undergoes significant conformational changes. The anticodon nucleotides became splayed out, making them accessible for recognition. The RNA-binding cleft is part of a portion of the dimerization domain and makes sequence-specific contacts with the five anticodon loop nucleotides, including a direct minor groove interaction with the side chain of a conserved Lys106, which simultaneously forms hydrogen bonds with C_{32} and A_{38}. Additionally, a conserved asparagine (Asn138) hydrogen bonds with A_{38}, while a series of positively charged residues (arginines and lysines) make general contacts with backbone phosphates in the anticodon stem. The overall anticodon loop structure poised for catalysis is finally stabilized by specific contacts with G_{36} and U_{33} [20]. The take-home lesson is that indeed, ADATa has built in its active site all that is needed for the specific recognition of a single tRNA, while excluding near cognate substrates.

The homologous enzyme in eukaryotes is ADAT2/3; unfortunately, there is no structural information available for this enzyme, but biochemical data support their heterodimeric structure in solution. The heterodimer is formed by two paralogous subunits encoded by the ADAT2 and ADAT3 genes. Both subunits bear the conserved signature H…CxxxC (where H stand for histidine, C for cysteine, and x for any amino acid) Zn^{2+}-coordinating triad characteristic of members of the cytidine deaminase superfamily, but the conserved proton-shuttling glutamate required for catalysis has been naturally mutated through evolution in the ADAT3 subunit, while it is still present in the TbADAT2 subunit. This led to the suggestion that the ADAT3 subunit solely plays a structural role, a proposal not supported by studies with the *Trypanosoma brucei* enzyme, where ADAT3 coordinates the catalytic Zn^{2+} and in turn partakes in catalysis. Unlike the bacterial ADATa, ADAT2/3 targets seven to eight different tRNAs depending on the organism. Studies with the *T. brucei* enzyme revealed the presence of critical tRNA-binding residues at the C-terminus of TbADAT2, a domain composed of a stretch of positively charged amino acids containing arginines and lysines (the KR domain) [23]. Deletion of the KR domain leads to lack of deamination activity presumably due to the fact that tRNA binding is abrogated. However, the active site still contributes to binding and it is the combination of active-site residues and of the distal KR domain that partly explains substrate specificity. Currently, the contribution of the second subunit (TbADAT3) to binding has not been explored. Regardless, it is clear that unlike bacterial ADATa, eukaryotic ADATs may possess a bipartite-binding domain where active-site residues and more distal domains contribute to substrate recognition.

The described dichotomy between a homodimeric deaminase with a single substrate and a heterodimeric enzyme with multiple substrates sets an interesting scenario that has been suggested for how these enzymes evolved from single-substrate to multisubstrate specificity. Given their sequence conservation, it has been proposed that ADAT3 was the result of gene duplication from ADAT2. This event provided a second copy of an otherwise essential gene, which can now freely mutate by genetic drift. This must have served as the preexisting condition that allowed the further rearrangement of RNA-binding domains away from the active site, which presumably also led to active-site relaxation. Once active-site binding becomes less critical, by virtue that the binding strength is provided by a distal domain, the enzyme was probably abled to accommodate a wider variety of substrates, leading to the appearance of the extant eukaryotic tRNA deaminases. In this realm, gene duplication a purely neutral evolutionary event was a critical step in the advent of multisubstrate deaminases.

4. THE DIVERSITY OF RNA METHYLATIONS, METHYLTRANSFERASES, AND THEIR MODES OF tRNA BINDING

Methylation is one of the most abundant modifications found ubiquitously in a variety of RNAs including tRNA, rRNA, mRNA, tmRNA, snRNA, snoRNA, miRNA, and viral RNA [24–26]. While methylated nucleotides are involved in affecting functional aspects of certain RNAs, they are particularly important in maintaining structural elements as methyl groups can stabilize or modulate the local RNA structure. In either case, introducing methylations precisely in the proper location and substrate is essential. Consequently, RNA methyltransferases acting on multiple substrates have evolved different strategies to ensure that the correct substrate species and individual target nucleotide(s) are selected from a pool of potential choices. This portion of the chapter will focus primarily on tRNA methyltransferases, but will also touch on the methylation of other RNAs.

The majority of methyltransferases use *S*-adenosyl-L-methionine (SAM or AdoMet) as a methyl donor. Methyltransferases that utilize 5,10-methylenetetrahydrofolate or carboxy-*S*-adenosyl methionine as a methyl donor have been identified, but will not be extensively covered here [27,28]. SAM methyltransferases have been categorized into at least five classes (I—V) based on their conserved structural folds [5,29–31]. Most methyltransferases are class I type, harboring a Rossmann fold, but RNA methyltransferases in particular are predominantly confined to class I and class IV types [5,29]. Despite their highly conserved structural fold, the amino acid sequence as well as the mode of substrate binding can differ, often significantly, between enzymes within the same class. Over time, some methyltransferases have gained distinct auxiliary RNA-binding domains, supporting a common theme of modular evolution whereby catalytic domains are combined with different substrate-binding domains to achieve substrate specificity [32–35].

Class I methyltransferases consist of alternating α-helices and parallel β-sheets characteristic of a typical Rossmann fold, apart from the presence of an additional (seventh) antiparallel β-strand [36]. Within this conserved fold, comparative sequence analyses have revealed a set of at least six conserved regions (motifs I–VI) encompassing the SAM-binding domain [29,36]. The first motif harbors a conserved nucleotide-binding site, GxGxG (x is any residue), which forms the first α-helix as it bends underneath SAM [29,30]. The other known conserved SAM-binding residue is an acidic residue at the end of β2, which forms hydrogen bonds with both hydroxyls of the SAM ribose [30]. Recently, in the TlyA ribosomal RNA methyltransferase from *Mycobacterium tuberculosis* an additional novel SAM-binding motif, RxWV, was discovered [37], which affects the structure of a GxGxG motif in a purposeful way [37], and both motifs are required for SAM binding [37]. Still, in general the SAM-binding region normally resides within the N-terminus of many methyltransferases and is assembled primarily from loops following strands 1, 2, and 3, whereas the substrate-binding region is generally located in the C-terminal portion of the β-sheet [29,38]. Beyond that, substrate binding for class I methyltransferases can vary greatly. There are several instances described in which a class I methyltransferases is fused to a known RNA-binding domain as discussed in more detail below [39,40].

Class IV methyltransferases are characterized by the SPOUT domain, which was first discovered by bioinformatics analyses based on the similarity to the SpoU (TrmH) and TrmD tRNA methyltransferases (Fig. 2) [41–46]. The SPOUT domain contains an unusual α/β fold distinguished by a deep trefoil knot formed by the folding of three β-sheets in the C-terminal portion of the SAM-binding motif [5,47,48]. Many SPOUT proteins have acquired additional N- and/or C-terminal extensions, which presumably participate in substrate binding [25,45,49]. Two SPOUT enzymes, TrmL (tRNA Um_{34}/Cm_{34}) and RlmH (rRNA), are considered "minimalist" proteins where substrate binding is limited solely to the SPOUT domain and, in the case of TrmL, no other RNA-binding protein assists in binding [25,49–51]. TrmL acts as a homodimer to catalyze 2′-O-methylation at position 34 on two isoacceptors of $tRNA^{Leu}$ (anticodons: CAA and UAA), and like many RNA-binding enzymes, it uses positively charged residues for substrate recognition [50,51]. Interestingly, only three other SPOUT tRNA methyltransferases are found in bacteria (TrmH, TrmD, and TrmJ); all methylate many more different substrates than TrmL [50]. Presumably, these enzymes acquired nucleic acid-binding domains as N- and C-terminal extensions that support recognition of multiple substrates, whereas the minimal SPOUT enzyme TrmL binds fewer substrates, making these extensions unnecessary. The nucleic acid-binding domains found within SPOUT as well as class I methyltransferases include PUA, THUMP, TRAM (OB-fold), and L30e domains (Fig. 3) [45].

The PUA domain (pseudouridine synthase and archeosine transglycosylase domain) is a highly conserved RNA-binding motif found in all domains of life. The PUA domain is made up of 64–96 amino acids and forms a compact pseudobarrel α/β fold (Fig. 3) [53–55]. In humans, NSUN6 is a class I methylase, which modifies m^5C_{72} of $tRNA^{Cys}$ and $tRNA^{Thr}$ and contains a PUA RNA-binding motif [53,56]. Interestingly, in vitro methylation by NSUN6 requires the 3′-CCA sequence, which is normally a prerequisite for aminoacylation [56]. A previous crystal structure of another PUA-containing enzyme, *Pyrococcus horikoshii* archaeosine tRNA-guanine transglycosylase, revealed that the PUA domain makes a

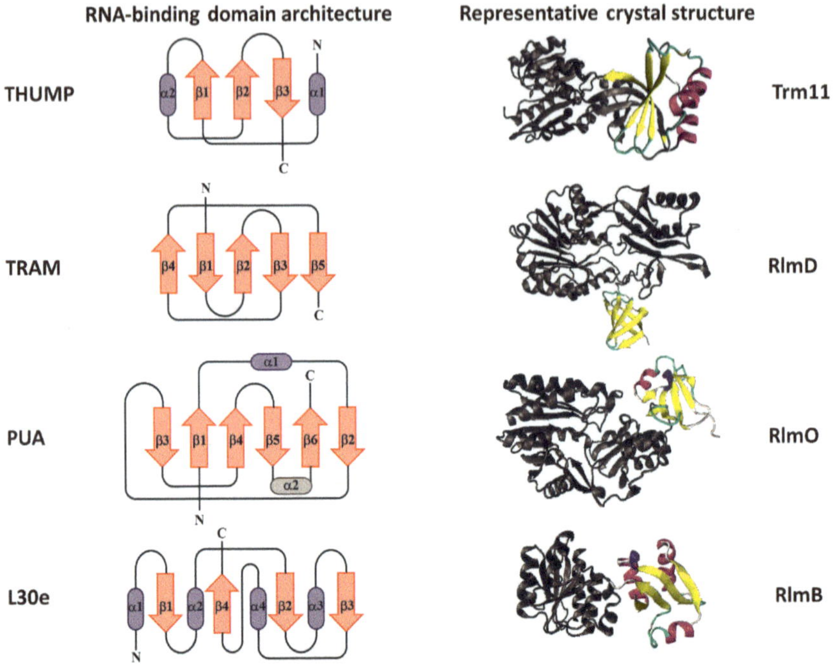

Fig. 3 Topological diagrams of common RNA-binding domains (*left*) and representative crystal structures of RNA methyltransferases containing these domains (*right*). In the topological diagrams the α-helices are shown as *cylinders* in *blue* and β-sheets as *arrows* in *red*. α-Helices that are not well conserved are shown in *gray*. Representative crystal structures are depicted in *cartoon form* with the corresponding domain shown with β-sheets in *yellow* and α-helices in *purple* and *dark blue*. All enzymes are shown here in monomeric form and the protein structure outside of the featured domain is shown in *gray cartoon*. These models were created using the Visual Molecular Dynamics (VMD) program [52].

direct contact with the 3′-CCA of tRNA, indicating a partially conserved mode of binding for enzymes containing this domain [56].

The putative THUMP RNA-binding domain (thiouridine synthase, methyltransferase, and pseudouridine synthase domain) was first discovered by in silico methods and thought to consist of 100–110 amino acids producing an α/β fold [57]. The crystal structure of the *Thermotoga maritima* ThiI 4-thiouridine synthase, in complex with a truncated tRNA, confirmed the role of THUMP domains in RNA binding and revealed a potential molecular ruler mechanism to define substrate specificity [58]. Interestingly, like ThiI, the RNA methyltransferases Trm11, RlmKL, Trm14/TrmN, and RlmN (a radical SAM enzyme) all contain an N-terminal ferredoxin-like (NFLD) domain along with the THUMP domain, implying that THUMP and NFLD together could form a functional binding unit as is seen in Thil [58]. For example, Trm11 from Eukarya and Archaea is comprised of three domains: NFLD, THUMP, and a class I methyltransferase catalytic domain (Fig. 3).

The TRAM domain (Trm2 and MiaB domain) is conserved in Bacteria and Eukarya, but not widely found in Archaea [34]. Some enzymes have recruited the TRAM domain for RNA binding. This domain is predicted to form a simple β-barrel. In methyltransferases, TRAM domains have been identified at the N-terminus of Trm2-like tRNA methyltransferases, the N-terminus of RlmD (formerly RumA, m^5U_{1939} *Escherichia coli* 23S rRNA), and the C-terminus of the 23S RNA-specific uracil O-methyltransferase, FtsJ, from *Halobacterium* (Fig. 3) [32,34]. The *Ec*RlmD enzyme was the first to reveal the structure of the TRAM domain, showing a five-stranded antiparallel β-barrel fold with a Greek key topology [59]. This structure adopts an oligonucleotide binding (OB) fold, which is commonly found in DNA/RNA enzymes [59]. The putative SPOUT methyltransferase MT1 from *Methanobacterium thermoautotrophicum* also contains the OB-fold subdomain unusually located between the center of its β-fold architecture [60].

The L30e domain is an RNA-binding domain commonly found in ribosomal proteins (named for the ribosomal L30 protein) and forms a distinctive three-layer α/β/α sandwich. There are three conserved residues within the L30 protein family, Gly26, Lys28, and Arg52, which are critical for RNA binding. The glycine residue, which lacks a bulky side chain, provides space for the RNA to bind, while the basic residues form hydrogen bonds with nucleotides in the RNA substrate [47]. The SPOUT 2′-O-methyltransferase, RlmB, is comprised of an N-terminal L30e domain extending from the catalytic subunit (Fig. 3) [60]. The RlmB family contains residues corresponding to the three RNA-binding residues in L30 and indeed RlmB has been shown to methylate Gm2251 of 23S rRNA in *E. coli* [60]. A putative SPOUT 2′-O-methyltransferase from *Thermus thermophilus* also appears to have an L30e domain at its N-terminus [61]. While the L30e domain is typically found in ribosomal proteins, methylation of 23S rRNA was not detected in intact ribosomes in *T. thermophilus*, and the putative methyltransferase displays no sequence similarity to L30; consequently, the target RNA is still unknown [61].

5. SEQUENCE vs STRUCTURAL DETERMINANTS FOR SPECIFICITY

There are two general strategies used by methyltransferases for identifying the correct substrate. They either survey structural features or recognize specific sequences within the RNA substrate. The majority of methyltransferases recognize local secondary structure or global tertiary folds, but some act based on a specific consensus sequence where recognition is usually in the context of local structural elements. RNA methyltransferases commonly detect structural features of RNA, as seen with the conserved consensus fold of m^5U methyltransferases [62]. RlmD has three structural domains, an N-terminal TRAM domain, a central α/β domain, and a C-terminal class I methyltransferase domain [62]. RlmD strictly modifies 23S rRNA at a single position (U1939) to m^5U and will not methylate it if the U position is mutated to C [59]. Early studies of another m^5U methyltransferase, TrmA, suggested that it did not recognize a strict consensus sequence on tRNA and it could efficiently modify short versions of the T-stem loop (TSL) [62]. However, a crystal structure of *Ec*TrmA bound to the T-arm revealed a unique colinear base-stacking of G_{53}–A_{58}–G_{57}–C_{56}–U_{55} in a conformation that is not normally seen in a typical unbound tRNA [62]. This presents the U_{54}–U_{55}–C_{56} sequence (where the first U is the methylation target) into the active site [62]. This interaction has been proposed to be a conserved specificity determinant for RlmD and TrmA [62]. Outside of this consensus fold, other RNA-binding elements allow for discrimination between the different substrates (rRNA vs tRNA); for example, unlike RlmD, TrmA does not contain a TRAM domain [32,63].

As previously discussed for deaminases, tRNA methyltransferases are often categorized into two groups based on substrate recognition: those requiring the entire tRNA for activity and those that can act on shorter tRNA substrates. For example, the SPOUT Um_{34}/Cm_{34} *Ec*TrmL methyltransferase specifically recognizes the stem-loop structure of the anticodon stem-loop (ASL) (extended by two nucleotides). *Ec*TrmL also recognizes the specific nucleotides A_{36}–A_{37}–A_{38}, as well as other determinants emphasized in the next section [50]. On the other hand, the *E. coli* TrmJ tRNA Xm_{32} (X is any nucleotide) methyltransferase requires the full-length tRNA for activity [50]. Surprisingly, it detects specific regions within the D-loop rather than the anticodon loop where the nucleotide to be methylated is found [50]. In contrast to *Ec*TrmJ, the archaeal TrmJ from *Sulfolobus acidocaldarius* (*Sa*TrmJ) can act on small tRNA substrates, specifically tRNAs lacking the D- and T-loops [25]. Additionally, *Sa*TrmJ can only methylate C_{32}, whereas bacterial *Ec*TrmJ can modify all four nucleotides at position 32, although A and G are almost never found at that position [25]. Interestingly, the bacterial and archaeal TrmJ are SPOUT methyltransferases, whereas the eukaryotic tRNA Xm_{32} enzyme (Trm7 with auxiliary protein Trm732) is a class I methyltransferase [25]. Likewise, the tRNA m^1G_{37} methyltransferase, TrmD (SPOUT), does not require the L-shaped tertiary structure of tRNA, but instead recognizes the anticodon arm structure and also uses G_{36}–G_{37} as a sequence determinant [50]. Unlike its bacterial counterpart, the eukaryotic m^1G_{37} methyltransferase Trm5, an unrelated class I enzyme, surveys the entirety of the tRNA tertiary structure to identify the correct substrate. It is noteworthy that both TrmJ and Trm7 as well as TrmD and Trm5 are just a few examples of RNA modification enzymes that have undergone convergent evolution, whereas two unrelated enzymes (in this case SPOUT vs Rossmann-fold methyltransferases) have evolved independently yet perform the same catalytic reaction. This further highlights the functional significance of tRNA modifications in all domains of life.

6. PREEXISTING MODIFICATIONS AND THEIR REQUIREMENT FOR SPECIFICITY

Another aspect of specificity in RNA modification enzymes is the order of processing events and the role played by other modifications. Having a systematic sequence of processing events or addition of modifications can ensure the proper structure and function of RNA and allows for quality control. Of course, these events can be dictated by compartmentalization, perhaps best demonstrated by retrograde transport systems where some modifications require reimport into the nucleus.

Several methylations, in addition to other modifications, require the RNA to be processed before it can be introduced. This can be due to impediment via steric hindrance in the preprocessed substrate where processing can be necessary to form the proper sequence or structural recognition elements. An interesting case is the splicing of eukaryotic tRNA introns. The wybutosine (yW) biosynthesis pathway in eukaryotes requires the addition of m^1G_{37} by Trm5 as the first of several steps [64,65]. In yeast, the tRNA must be spliced before the methylation reaction can take place; however, the splicing machinery is located on the cytoplasmic side of the outer mitochondrial membrane and it is in this context that intron removal takes place [48,65–69]. The Trm5 enzyme, however, is a nuclear enzyme [65]. Therefore, $tRNA^{Phe}$ must be imported back into the nucleus after splicing in order to be methylated [65,66]. Afterward the tRNA is reexported to the cytoplasm to become further modified to yW. In contrast, some methylations are actually intron dependent. In yeast and humans, for example, the multisite m^5C tRNA methyltransferase, Trm4, requires an intron for methylation at positions 34 and 40 [70,71]. Trm4 does not survey the overall L-shaped tRNA structure, but rather has been suggested to be dependent on specific nucleotide sequence and base pairs within the intron. However, Trm4-mediated m^5C modification at positions 48 and 49 is unaffected by the presence or absence of an intron in yeast.

The majority of RNA methyltransferases modify canonical nucleotides, but they can also methylate more complex substrates like nucleotides with preexisting modifications. RlmH (SPOUT) from *E. coli* specifically methylates position 1915 of 23S rRNA exclusively if that position has a preexisting pseudouridine (ψ) [31]. It is the only methylated pseudouridine found in bacterial RNA to date, although several have been found in Eukarya ($m^1\Psi$, Ψm, and $m^1acp^3\Psi$) [31]. Modifications at positions outside of the target methylation site also influence methylation events. For example, *Ec*TrmL methylates pyrimidines but not purines at the wobble position and requires prior formation of i^6A at position 37 [50]. The sequence A_{36}–A_{37}–A_{38} is also important for *Ec*TrmL methylation, but it is unclear whether this sequence is necessary for recognition by the enzyme itself or because of the i^6A_{37} modification [50]. In any case, the addition of i^6A_{37} into an otherwise unmodified in vitro transcribed $tRNA^{Leu}$ is sufficient to recruit TrmL to the tRNA [50]. A similar situation occurs in yeast where i^6A_{37} is a prerequisite for m^3C_{32} formation in $tRNA^{ser}$ [72]. These results provide evidence for an anticodon loop circuitry where modifications within this loop are highly interrelated.

Methylation can also act as a requirement for subsequent modifications at the same position or elsewhere in the RNA. The methylation of m^1A is found universally at position 58 of tRNA and is additionally present at several other positions within the T-loop depending on the organism [73]. In Archaea, certain tRNAs contain m^1A at position 58 or 57 as a mandatory step in the biosynthesis of m^1I_{57}, produced by adenosine-to-inosine (A-to-I) deamination. The methylation is catalyzed by the tetrameric archaeal TrmI enzyme and is strictly required for deamination to occur [10,73]. The orthologous m^1A_{58} site-specific methyltransferase in yeast also forms a tetramer, but is comprised of two different evolutionary-related subunits (further discussed in the next section). A similar, yet unconventional, two-step pathway was recently discovered in *T. brucei* involving the TbADAT2/3 deaminase discussed previously. In addition to its role in the conserved A-to-I deamination at position 34, TbADAT2/3 also converts C-to-U at position 32 in trypanosomes [16,74], but it requires a methylated substrate, m^3C_{32}, and the corresponding methyltransferase, TbTrm140, for this activity [75,76]. Remarkably, the methyltransferase

TbTrm140 also requires the presence of TbADAT2/3 to form the methylated substrate prior to deamination [75,76]. Therefore, these enzymes are mutually dependent on each other for function [75,76].

In an analogous scenario and adding further complexity to the yW biosynthesis in $tRNA^{phe}$ highlighted earlier, it has been suggested that one of the enzymes in the pathway, Tyw1, depends on prior formation of Gm_{34} and Cm_{32} by the Trm7 enzyme (along with its associated partner proteins) [72,77]. Therefore, the entire proposed pathway consists of export of the intron-containing $tRNA^{Phe}$ precursor from the nucleus to the cytoplasm for splicing and 2′-O-methylation by Trm7, followed by import back into the nucleus for m^1G_{37} and lastly reexport into the cytoplasm for the final steps to produce yW_{37} (Fig. 4) [77].

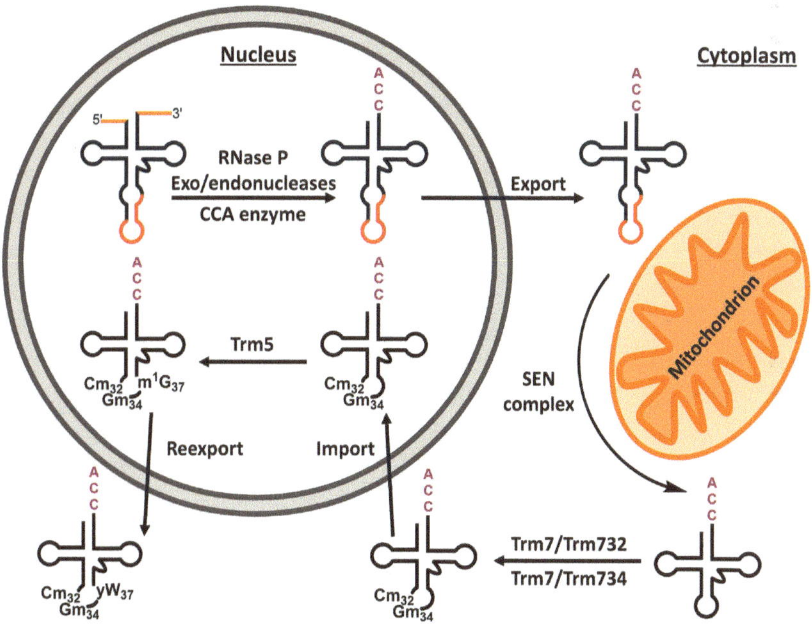

Fig. 4 Processing pathway of $tRNA^{Phe}$ in *Saccharomyces cerevisiae.* $tRNA^{phe}$ is transcribed in the nucleus with 5′- and 3′-extensions (*orange*) and an intron (*red*). The extensions are removed and a 3′-CCA sequence is added in the nucleus. The tRNA is exported to the cytoplasm and undergoes splicing by the splicing endonuclease (SEN) protein complex located to the outer mitochondrial membrane facing the cytoplasm. The spliced tRNA is 2′-O-methylated by Trm7/Trm732 for Cm_{32} and Trm7/Trm734 for Gm_{34}. The tRNA is then imported back into the nucleus where it is methylated to m^1G_{37} by Trm5, initiating the first step of the yW pathway. The tRNA is reexported where it undergoes further biosynthesis steps to form yW when the presence of Cm_{32} and Gm_{34} serve as key determinants.

Besides acting as an obligatory intermediate for additional modifications, methylations can positively or negatively influence other modifications and vice versa. The modification m^7G_{37} positively impacts Gm_{18} and m^1G_{37}, while Ψ_{55} negatively affects the levels of Gm_{18}, $m^5S^2U_{54}$, and m^1A_{58} in *T. thermophilus* [72]. In trypanosomes, it has been proposed that upon mitochondrial import, editing of $tRNA^{Trp}$ relies on the interplay between several modifications; it is 2′-O-methylated at position 32, thiolated at position 33 to form S^2U_{33}, and edited from C_{34} to U_{34}, while the unedited C_{34} is methylated to Cm_{34}. Interestingly, downregulation of the thiolation activity leads to almost complete editing at position 34. Therefore, methylation may serve as a positive determinant for C_{34} to U_{34} editing, while thiolation negatively affects the editing reaction helping maintain two functional forms of $tRNA^{Trp}$, edited and unedited [78].

7. MULTISUBUNIT METHYLTRANSFERASES

One noteworthy aspect of RNA methyltransferases is their tendency to recruit additional partner proteins, which often play a part in substrate binding, catalytic efficiency, stability, or localization. Furthermore, oligomerization of the same subunit can also vary between methylation enzymes from different domains of life or even within the same domain affecting substrate specificity or catalysis. It has been proposed that the formation of homodimers could act as a possible step in the evolution of heterodimeric enzymes whereby (as discussed previously) a gene duplication event may arise and lead to functional specialization of each subunit over time [36,79]. In general, SPOUT methyltransferases form homodimers where each monomer partakes in substrate binding and the active site is present at the dimer interface. However, the SPOUT enzyme, Trm10, acts as a monomeric enzyme, suggesting that catalysis and substrate recognition for this enzyme may differ from canonical SPOUT methyltransferases [5,80,81]. Unlike the SPOUT methyltransferases, the class I methyltransferases are normally found as monomers, although there are few exceptions such as *Bacillus subtilis* TrmB, which will be discussed later in this section.

Some enzymes recruit additional proteins to mediate substrate binding. The yeast 2′-O-methyltransferase Trm7 interacts with an associated protein Trm732 to methylate C_{32} and with another protein, Trm734 for methylation of X_{34} on $tRNA^{Phe}$, $tRNA^{Leu}$, and $tRNA^{Trp}$ [77]. In this case, Trm7 is the catalytic subunit, which uses either Trm732 or Trm734 to direct substrate specificity. The adaptor protein Trm112 provides another example in which a methyltransferase requires an unrelated protein factor for activity. Trm112 interacts with multiple tRNA methyltransferases including Trm9 and Trm11 to form mcm^5U_{34} and m^2G_{10}, respectively, in eukaryotic tRNAs [82]. Initial studies in yeast showed that deletion of Trm11 displayed a minimal growth defect, whereas Trm112 deletion resulted in a severe growth phenotype, indicating its importance in other cellular functions [82]. In fact, Trm112 associates with several different factors involved in other cellular activities; for instance, it provides structural stability for another methyltransferase, Bud32, which methylates G1575 of 18S rRNA in eukaryotes [83–85]. Therefore, Trm112 performs multiple functional roles in tRNA modification as well as other activities in the cell.

In yeast, the m^1A_{58} is a common but nonuniversal modification as it appears in 21 of 32 sequenced tRNAs [86]. This methylation is catalyzed by a heterotetrameric enzyme composed of the subunits Trm6 (Gcd10p) and Trm61 (Gcd14p); mutation of the putative SAM-binding residues of Trm61 abolished activity, while in vitro neither Trm61 nor Trm6 could by themselves bind tRNA [86]. This suggested that Trm61 was responsible for catalysis, whereas Trm6 aided in tight binding to the tRNA substrate in the context of the other subunit. Mutational studies later demonstrated that residues in both Trm61 and Trm6 subunits participate in tRNA binding [87]. As mentioned earlier the orthologous bacterial/archaeal methyltransferase TrmI forms a tetramer, which may be conserved for its role in recognition, binding, or activity. However, the binding interface appears to diverge between prokaryotic TrmI and eukaryotic Trm6/Trm61, indicating that these enzymes evolved different means of substrate recognition and binding [87].

In eukaryotes, Trm8 and Trm82 form a complex which modifies position 46 of seven different tRNAs with m^7G [88]. While Trm8 contains the putative catalytic site and has tRNA-binding ability, Trm82 does not engage in either catalysis or tRNA binding [79]. Instead, the formation of the Trm8–Trm82 complex causes Trm8 to undergo a considerable conformational change forming the correct architecture for catalysis [5]. Although both Trm8/Trm82 enzymes are conserved among Eukarya, Bacteria utilize a single subunit enzyme, TrmB, a homolog of Trm8. Interestingly, there are differences even among bacterial TrmB, where *E. coli* TrmB is a monomeric enzyme, whereas *B. subtilis* forms homodimers [36]. The dimerization of *Bs*TrmB is also unusual for a class I methyltransferase as its structural elements deviate from the expected Rossmann fold [36]. This finding supports the evolution of obligatory dimers observed for Trm8/Trm82 as well as a variety of other dimeric methyltransferases like those listed earlier.

Rather than enlisting other proteins, some methyltransferases, lacking intrinsic substrate specificity, utilize guide RNAs to direct catalysis. These enzymes form complexes with small C/D box guide RNAs that direct methylation to the target nucleotide. This mechanism is seen primarily in eukaryotic rRNA and small nuclear RNA as well as archaeal 2′-O-methylation of rRNA and in some cases tRNA [89]. The need for RNA-dependent methylation of tRNA is not conserved as 2′-O-methyltransferase of positions 32 and 34 of yeast $tRNA^{Trp}$ and of position 18 from Bacteria (TrmH) and Eukarya (Trm3) does not involve guide RNAs [73]. Methylation via guide RNAs is yet to be seen in Bacteria, suggesting that this system likely evolved after the Bacteria split from Archaea and Eukarya.

8. PSEUDOURIDINE SYNTHASES

Pseudouridine (Ψ) is the most common modified nucleotide found in RNAs from all three kingdoms of life [90–93]. In 1951, as the first modified nucleosides were discovered in RNA hydrolysates, pseudouridine was identified and became known as the fifth nucleoside, 5-ribosyl uracil [94–97]. Pseudouridine is a C5-glycoside isomer of uridine and the only C–C bond joining a nucleobase to a sugar known in nucleic acids [95,97]. Formation of pseudouridine is catalyzed by pseudouridine synthases (Pus or Ψ synthases), a class of RNA-modifying enzymes responsible for the posttranscriptional site-specific isomerization of uridine to pseudouridine in a wide array of RNA substrates including rRNA, mRNA, snoRNA, snRNA, 5S RNA, and tRNA [98–103]. Notably, these enzymes are unable to isomerize free nucleotides. Upon isomerization, there is no detectable release of free uridine and the reaction occurs without the need for ATP or any other cofactor [104,105]. Although a general isomerization mechanism is established, the specific chemical mechanism by which these enzymes catalyze the reaction has been a challenging problem [105–110]. A clear outcome of pseudouridylation is the formation of an extra hydrogen bond donor at the newly formed N1 position, imparting structural stability to the target RNA, which has in some cases proven important for translational efficiency and accuracy [111,112].

9. CLASSIFICATION OF RNA PSEUDOURIDINE SYNTHASES AND THE CONSERVED CATALYTIC CORE DOMAIN

The first tRNA Ψ synthases were identified from *Salmonella typhimurium* and *E. coli* and are the product of the *truA* gene [98,99,113].

Thereafter, the Ψ synthases have been classified into families based on limited sequence similarities and named using the bacterial nomenclature. The founding members of each family: TruA, TruB, RluA, RsuA, TruD are also present in Archaea and Eukarya (Fig. 1) [100,114,115]. The sixth and newest family, not found in Bacteria but found in Archaea and Eukarya, includes the human Pus10 (pseudouridine synthase) [116–118]. Thus, the original nomenclature was based on the target substrate. For instance, TruA and TruB modify tRNA, while RluA and RsuA act on large and small subunit ribosomal RNA, respectively [104,119]. However, there are unique exceptions where a pseudouridine synthase has multisubstrate RNA specificity; the RluA family of Ψ synthases, for example, catalyzes Ψ_{746} in the large ribosomal 23S rRNA and Ψ_{32} in $tRNA^{Phe}$ [120]. Pseudouridine formation is catalyzed either by these independent proteins or by H/ACA snoRNA-dependent ribonucleoprotein particles [101,103,121–124]. The eukaryotic synthases (Pus), originally discovered in *Saccharomyces cerevisiae*, have been named numerically based on the order of their discovery, where Pus4p and Pus7p are the orthologs of *E. coli* TruB and TruD enzymes [125–129], respectively.

Crystal structures have been determined for proteins belonging to each of the six families and their structural diversity underscores the significant role that protein dynamics must play in RNA substrate recognition [104,119,130]. The underlying theme is that despite low level of primary structural similarity among Ψ synthases, all adopt exquisitely superimposable folds with three-dimensional structural features that share a common catalytic core and a cleft comprised of a conserved set of active-site residues [104,110,114,119,130] (Fig. 5). In addition to a conserved catalytic aspartate, studies of the first four families (TruA, TruB, RsuA, and RluA) of Ψ synthases have revealed a second important residue: Tyr, in *E. coli* and archaeal Pus10; Asp in hPus10; and Glu in TruD [115,117,131–135]. This semiconserved residue is proposed to aid in the formation of a buried salt bridge along with the universal catalytic aspartate [104,130]. Moreover, the tyrosine residue may also help pack the substrate via the uracil base. In the fifth family of Ψ synthases, which includes TruD and *S. cerevisiae* Pus7, the tyrosine is not conserved, but mutational and enzymatic analyses indicate that a strictly conserved glutamate unique to this family likely participates as a general base [132].

The domain structure of Ψ synthases has been classified into three major conserved motifs I, II, and III [35,114]. Conserved among all Ψ synthases, the conserved catalytic core is found in Motif II and features an absolutely conserved aspartate residue. Having minimal sequence similarity to other

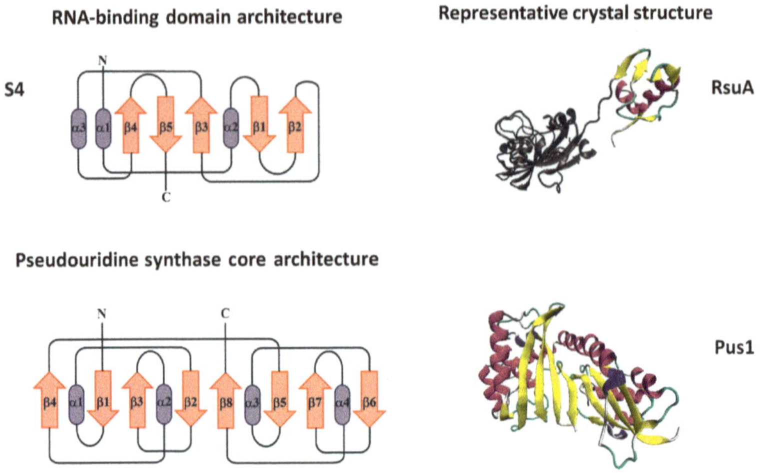

Fig. 5 Topological diagram of the S4 RNA-binding domain (*top left*) and representative crystal structure of RsuA containing an S4 domain (*top right*). The protein structure outside of the featured domain is shown in *gray cartoon*. Topological diagram of the core domain of pseudouridine synthases based off of the *E. coli* enzyme TruA (*bottom left*). The core of pseudouridine synthases is comprised of two domains each with a βαββαβ arrangement forming an alternating eight-stranded β-sheet. The arrangement, but not necessarily the connectivity shown here, is conserved among pseudouridine synthases. Representative crystal structure of human Pus1 (*bottom right*). These models were created using the Visual Molecular Dynamics (VMD) program [52].

synthases, TruA and its homologs appear to be the most distantly related [104]. The RluA and RsuA family are most closely related to each other as they share three conserved sequence motifs (motifs I, II, and III). Only motifs I and II were found in the TruB family. The structure of TruB revealed that motif II is part of the active site and that motif I plays an architectural role or reinforcing motif II, in agreement with site-directed mutagenesis studies of motif I residues [136–138]. Moreover, although TruB lacks motif III, the sequences of the TruB family of enzymes can be aligned with those from the RluA and RsuA families [139]. The TruD family, specific for pseudouridylation of position 13 in tRNA, has motifs I, II, and III and is remarkable because the crystal structure determination of TruD revealed a two-domain structure. TruD has a catalytic domain that differs in sequence, but is structurally very similar to the catalytic domain of other pseudouridine synthases and a second large domain with a novel α/β fold not found in any other protein [139–141]. Interestingly, all TruD homologs are present in all phyla and are all circularly permuted, meaning that they have spatially conserved structural elements with other synthases, but located at the opposite ends of the enzyme [142]. In other words, structural features found at the C-terminus of other synthases are at the N-terminus in TruD [142]. Therefore, in a parsimonious evolutionary scenario, it is proposed that TruD was the first to diverge from all other synthases [104,140].

The mechanism by which the active-site residues manage to disconnect, rotate, and reconnect the uracil base to the C5 carbon is still unclear [105–107,109,110]. One possibility is a nucleophilic attack by the aspartate on the sugar or the substrate uracil [105,107,110]. The chemical mechanism for this reaction would minimally involve three steps: (1) cleavage of the N-glycosidic linkage of the target uridine, (2) rotation of the uracil ring to position C5 of the pyrimidine near C1′ of the ribosyl group of the RNA, and (3) formation of a new C1′–C5 carbon–carbon bond [105,107,110]. Traditional analysis of pseudouridine formation uses a tritium release assay; however, the quench-flow technique has been adapted to further study pseudouridine formation [143]. The results from studying three *E. coli* enzymes showed that single-turnover rate constants for pseudouridylation by TruB, RluA, and TruA were determined to be 0.5, 0.7, and $0.35\,s^{-1}$, respectively [143]. Thus, the reaction chemistry may be the rate-limiting step, since these rate constants were reported to be similar to the k_{cat} tested under multiple-turnover conditions [143]. Although both the single- and multiple-turnover rate constants have been measured, the detailed mechanistic steps involved in catalysis still remain unsolved [119].

10. RNA BINDING: CONTRIBUTIONS FROM PERIPHERAL DOMAINS

The core Ψ synthase fold is embellished by divergent peripheral structural elements in some of the Ψ synthase enzymes [104,105]. These elements often provide additional independently folded domains that may confer unique substrate specificity. For example, RsuA, which catalyzes pseudouridine formation at position 516 in 16S rRNA important for ribosome assembly, possesses an appended domain at its N-terminus with similarity to the RNA-binding domain of ribosomal protein S4 (Fig. 5) [115,144,145]. Despite the conservation with S4, the N-terminal domain of RsuA and S4 binds rRNA at different locations [115,144]. A total of six Arg and two Lys residues on the surface of the N-terminal domain contribute the set of positively charged residues that bind rRNA in both proteins. Thus, it is the flexible linker that connects the central catalytic domain to this N-terminus that may provide the required flexibility to adjust the catalytic domain relative to a specific substrate [144]. The *Haemophilus influenza* RsuA also contains an N-terminal S4-like α3/β4 domain followed by a catalytic domain [146]. Whereas the individual domains of *E. coli* and *H. influenza* RsuA are structurally similar, the relative spatial placement among the two structures differs greatly [146]. The *E. coli* structure displays an extended open conformation with no direct contacts between the domains, while the *H. influenza* RsuA is in a closed conformation with a large interface between the two domains [146]. Domain closure involves several basic and polar residues forming a putative RNA-binding cleft. It is proposed that this relative repositioning of the S4 and catalytic domains is used to modulate the shape and size of the rRNA-binding site in pseudouridine synthases possessing S4 domains (Fig. 5) [146].

Unique to the Pus10 family of synthases is the presence of an accessory N-terminal THUMP RNA-binding domain (Fig. 3), also found in tRNA methyltransferases, as described earlier, and in other RNA modification enzymes [57,117]. The N-terminal THUMP domain of Pus10 contributes to tRNA binding by a proposed induced-fit mechanism for pseudouridine formation. In archaeal Pus10a, in addition to the aforementioned catalytic aspartate, a forefinger loop and an Arg and a Tyr residue are essential for Ψ_{54}, but not for Ψ_{55} formation [118]. Since the forefinger loop is needed for both rRNA and tRNA Ψ synthase activities of RluA, but only for tRNA Ψ_{54} activity of Pus10, archaeal Pus10 proteins are proposed to use two distinct mechanisms for substrate uridine recognition, binding, and catalysis [134].

hPus10 is a crescent-shaped molecule with two domains, including the universally conserved Ψ synthase catalytic domain [117]. Superposition of the catalytic domains of Pus10 and other Ψ synthases shows that the full set of conserved Ψ synthase active-site residues are present in Pus10, indicating that it likely employs a catalytic mechanism similar to other Ψ synthases [116]. The Pus10 active site is located in a deep pocket of a basic cleft adjacent to flexible thumb and forefinger loops, which could provide further RNA substrate stabilization during binding [117]. Modeling studies support the role of the N-terminal domain as an accessory RNA-binding domain, with a cleft between the catalytic and accessory domains large enough and electrostatically compatible to accommodate an RNA [117]. Homology models of archaeal Pus10 proteins based on the crystal structure of human Pus10 have revealed that subtle structural differences between all Pus10 proteins [118]. While most archaeal Pus10 produce both tRNA Ψ_{54}

and Ψ_{55}, some produce only Ψ_{55}. The archaeal Pus10 synthase enzymes from *Methanocaldococcus jannaschii* and *Pyrococcus furiosus* can isomerize tRNA at two positions Ψ_{54} and Ψ_{55} with the two modifications occurring independently [134]. Furthermore, unlike bacterial TruB and yeast Pus4, archaeal Pus10 does not require a U54:A58 reverse-Hoogsteen base pair and a pyrimidine at position 56 [118]. Within the Pus10 family, another interesting observation is that the Pus10 gene is essential in *Haloferax volcanii*, whereas the orthologous tRNA Ψ_{55} synthase genes are nonessential in yeast (Pus4) and *E. coli* (TruB) [118,133,143].

Within the TruB family of Ψ synthases, *E. coli* TruB and yeast Pus4 are functional as monomers and specifically targets U_{55} in the TSL of almost all tRNAs. For recognition and binding to its RNA substrate, the *E. coli* TruB PUA RNA-binding domain is present at the C-terminus, which differs from the S4 RNA-binding domain of RsuA found at its N-terminus. Although the RNA-binding domains of both proteins are present at antipodal termini, upon the superimposition of the central catalytic domain of RsuA and TruB structures, both PUA and S4 RNA-binding domains were located on the same structural side of the respective proteins [104,137,144].

As discussed earlier, Ψ synthases share common three-dimensional structural features and generally exist as monomeric proteins [104,105,147,148]. The one exception is the TruA family whose members are dimeric in solution with two positively charged RNA-binding clefts along its surface; each cleft contains the highly conserved aspartic acid at its center [122,148,149]. TruA utilizes its dimerization interface to form the key domain critical for binding via the conserved tRNA elbow. Furthermore, TruA is distinct from other Ψ synthases in its substrate specificity and modifies multiple tRNAs (for example, 17 tRNAs in *E. coli*) that contain uridines at positions 38, 39, and 40 [122,148,149]. The TruA enzymes are also more compact, in that they lack the extra peripheral domains at both the N-terminal and C-terminal regions. Since both the *T. thermophilus* and *E. coli* TruA enzymes consist of only the catalytic core domain in the monomeric structure, it is proposed that the other monomer in the active dimer of TruA probably plays a role similar to that of the PUA domain of TruB or the S4 domain of RsuA [122,148,149]. On the other hand, PusI (in the TruA family) modifies positions 38, 39, and 40 in the anticodon stem of tRNA but acts as a monomer [107,150–152]. The PusI monomer harbors a 10-stranded β-sheet formed by the interaction of the N-terminus and C-terminus domains. In the crystal structure of PusI, an unexpected topological relationship was revealed between both the N-terminal and C-terminal domains

[59,107]. In the β1 and β2 strands of the βαββαβ core topology of PusI, RNA binding is mediated by a classical RNA recognition motif, a well-characterized domain found in several RNA-binding proteins [39,153,154].

11. MULTISUBSTRATE SPECIFICITY OF PSEUDOURIDINE SYNTHASES

TruB is a single site-specific enzyme targeting position 55 in all *E. coli* tRNAs with the exception of the initiator $tRNA^{iMet}$ [112,155,156]. Therefore, this enzyme faces a different challenge in discriminating against a single nonsubstrate $tRNA^{iMet}$ in a sea of true substrates. This enzyme recognizes the reverse-Hoogsteen base pair between T_{54} and A_{58} found in a majority of tRNAs; surprisingly, this base pair is absent in initiator $tRNA^{Met}$ [136,137]. In the cocrystal structure of TruB with tRNA, the enzyme specifically stacks a conserved histidine residue on the loop side of this pair, while disrupting stacking of the T-loop nucleotides in the folded tRNA [136,137]. The crystal structures of *E. coli* TruB bound to a 22-nucleotide TSL mini-helix portion of tRNA and an intact tRNA with 5-fluorouridine replacing U55 also revealed that this position is flipped out and bound to the enzymes' catalytic cleft [136,137]. In addition, three T-loop nucleotides were also flipped out of the helical stack, thereby disrupting the tertiary base-pairing interaction between the D- and T-loops and affecting the global fold of the tRNA. TruB uses its "thumb" to bind the major groove of the tRNA substrate [136,137]. Upon RNA binding, the thumb was observed to undergo structural reorganization [136,137,157–159].

The global tRNA structure is not a recognition element for the TruB Ψ synthase, as mutations in the tRNA that disrupt the T- and D-loop interaction have no effect on pseudouridylation [159]. However, sequence determinants for pseudouridylation at position 55 reside around the T-loop, whereby TSL mini-helices are efficient pseudouridylation substrates. Two substrate-free structures of TruB unveiled unusual protein flexibility involving hinge-bending movement in the central β-sheet of the catalytic module. Hinge bending is proposed to act as a clamp for positioning the substrate [159]. Moreover, the β-sheet-bending movement affects the shape the catalytic cleft and creates disordering of the active site. Based on the significant conformational change observed between the structures of RNA-bound and -free forms, a proposed model for tRNA recognition includes a combination of rigid docking followed by induced-fit binding [104,159].

Human Pus1 (hPus1), a member of the TruA family, pseudouridylates specific uridine residues in several noncoding RNAs, including the U2 spliceosomal RNA and steroid receptor activator RNA in addition to tRNAs [107,160,161]. Since it was not possible to obtain crystals of the full-length hPusI protein, a minimal system using a truncated protein composed of the complete catalytic core and of a minimal tRNA substrate showed comparable pseudouridylation activity to that seen with full-length tRNA [107,160,161]. This result supports the fact that the disordered N- and C-terminal extensions, which were removed in the truncated protein, are not essential for enzymatic activity, as was predicted a priori by the RONN server in silico [107,160,161]. The hPus1 enzyme has a fold similar to that found in bacterial Ψ synthases, with a central domain that contains the catalytic Asp146. It adopts a mixed α/β fold with an extended central antiparallel four-stranded N-terminal β-sheet that packs against a six-stranded C-terminal β-sheet decorated by helices and loops as found in other Ψ synthases [107,160,161].

Remarkably, it was the analyses of three separate crystal structures of hPus1 that facilitated the identification of conformational differences observed in flexible regions of the protein, which have been proposed to be important for substrate recognition by induced fit. A flexible hinge at the base of the sheet allows the enzyme to open and close around an electropositive active-site cleft [161]. Although hPusI has no N- or C-terminal peripheral domains, it has a positively charged electrostatic surface extending from the N-terminus of the catalytic domain to the active site of the enzyme [161]. The N-terminal domain has been suggested as the RNA-binding site. Unique to hPus1 are two long α-helices inserted at the C-terminus of the core domain [161]. These two C-terminal α-helices are proposed to specifically permit only a specific binding orientation of tRNA, which differs from that of TruA [104,161]. The structures of the catalytic core domain of hPusI also revealed that the C-terminus forms the walls of the RNA-binding surface by extending along the back and top of the central β-sheet of the catalytic core domain [104,161]. The walls of the cleft are formed by two structures: (1) the forefinger loop formed by N-terminal domain loops and (2) the thumb loop at the C-terminal domain loops. Docking of tRNA to hPus1 in a productive orientation requires only minor conformational changes in the enzyme and tRNA [161]. The docked tRNA is bound by the electropositive surface of the protein—employing a completely different binding mode than that seen for the *E. coli* TruA homolog in complex with tRNA [161].

Among the six identified RNA Ψ synthase families, the RluA family of RNA Ψ synthases appears to be the most complex based on their difference in RNA substrate specificity [120,162]. The notable feature of the RluA enzyme is in its dual specificity for Ψ_{32} in several tRNAs as well as Ψ746 in the 23S rRNA [120,162]. RluA can pseudouridylate shorter RNA substrates, including a tRNA ASL and also a stem-loop RNA that contains sequences flanking the target uridine at position 746 from 23S rRNA [163,164]. A number of nonrelated RNA-binding proteins use a "direct readout" to recognize RNA in a sequence-specific manner [165], reminiscent of what was discussed with bacterial ADATa in the deaminase section of this chapter. Given that the sequence ΨUXXAAA (X is any nucleotide) is found in all RluA substrates, it was expected that RluA Ψ synthase would also use a similar strategy for substrate recognition [120,155,162]. However, structural analyses revealed that RluA induces the reorganization of a 21-nucleotide ASL of $tRNA^{Phe}$ [162]. The drastic structural change forces the anticodon loop to form a reverse-Hoogsteen base pair with three flipped-out nucleotides, instead of the canonical U-turn conformation [162]. Presumably, RluA indirectly reads out the RNA structure by probing RNA loops for their ability to adopt the reorganized fold [162]. This was further supported by results from structure-guided mutagenesis, sequence conservation, and cocrystal structure [162]. Upon association with RNA, the solvent-accessible surface of the free RluA enzyme becomes buried. Sharing the same structural context as the TruB thumb, the RluA thumb differs only in having a more compact thumb loop structure to interact with its substrate's major groove [162]. It is proposed that the ASL RNA is positioned with its axis parallel to the β-sheet core of RluA [162].

The TruA active site only has a small region of positively charged residues, while PusI also has a larger positive binding surface from two extended α-helices dedicated to binding the substrate RNA. This would concur with the role of *E. coli* TruA in modifying ASLs from many tRNAs, while the hPusI has more substrate promiscuity in having a wider variety of RNAs as substrates. From the differences in their approach toward achieving substrate specificity, TruA and TruB are proposed to have evolved independently from their common ancestor [104,162]. TruB, having single-site specificity, uses accessory domains along with induced-fit conformational changes to actively sample the substrate sequence near the target site, while TruA utilizes dimerization to provide a scaffold for recognition of the conserved elbow structure of tRNA, while maintaining multisite selection of its substrate by retaining flexibility in the ASL

[148,158]. This has suggested a molecular ruler mechanism as a strategy for how TruA recognizes its target [148,149]. TruA binds the ASL through the initial docking of the tRNA elbow and the D stem, which differs from the TruB enzyme, which directly docks the TSL into the active site [104,149,158,162]. Consequently, TruB induces the thumb domain to clasp onto nucleotides near the target site and inserts His43 into the emptied space left behind by flipping out the base, whereas the double-stranded stem is always kept away from the active site [104,162]. In contrast to the extensive interactions between TruB and the neighboring region of the target site, TruA has more limited contact with the ASL near the target site and relies more on secondary structure recognition of the substrate to achieve U_{55} site specificity [104,158,162].

Interestingly, upon juxtaposition of the RNA-bound structure RluA with that of TruB, the cores of the two enzymes superimpose well, yet the position of the RNA substrate in their active site differs by as much as a 60-degree angle [104,162]. Presumably, the difference may be accounted for by the presence of the peripheral PUA domain in RluA, and lack of a forefinger loop in TruB, in contrast to RluA [104,162]. Thus, the forefinger loop in combination with an appended binding domain can drastically change how the enzyme binds the substrate [104,162]. Furthermore, structural comparison of RluA with the distantly related TruA shows a strikingly similar conformation including both forefinger and thumb loops despite little sequence similarity [104,162]. Juxtaposition of the RNA-bound RluA with the structure-free RsuA shows that RsuA possesses a forefinger structure, but not a thumb, as opposed to TruB which lacks a forefinger loop, but does have a thumb loop [104,162]. The use of such structural protrusions (forefinger and thumb loop) for recognition had already been proposed in the study of RluC and RluD, which together with RluB and RluE are part of the bacterial RluA family of synthases [104,105,112,166]. However, members of the bacterial RluA family have different RNA specificities in comparison to their yeast counterparts. Four members of the RluA family are present in both *E. coli* (RluA, RluC, RluD, and TruC) and *S. cerevisiae* (Pus5p, Pus6p, Rib2/Pus8p, and Pus9p) [104,105,112,166]. Three of the four *E. coli* enzymes modify 23S ribosomal RNA: RluC, RluD, and RluA (which also modifies position 32 in tRNAs) [105,112]. In notable contrast, only one of the four *S. cerevisiae* enzymes, Pus5p, modifies mitochondrial 21S rRNA [129]. In conclusion, despite a degree of domain conservation between this group of enzymes, their variability in domain arrangement makes it difficult to determine their intended targets a priori.

12. CONCLUSION

RNA modification enzymes act on a great diversity of substrates, and despite much research in recent years, still little is known about their mechanism for substrate recognition and specifically as to what their RNA-binding domains look like. In the particular case of tRNA, the matured substrate is highly structured and a general theme is that some modifications act early contributing to tRNA folding; others act after the near-native structure has formed and are thus limited to recognizing salient features of the 3D L-shape tRNA. Therefore, although all fully folded tRNAs are very similar in the 3D structure, modification enzymes as a family survey a much larger structural landscape depending on how early or late they act in the folding pathway. In this chapter, we have attempted to illustrate well-known examples of how tRNA-modifying enzymes solve the binding problem. We have also highlighted the exquisite specificity that is needed to only target a subset of substrates in a vast pool of very similar nonsubstrates. A general principle that comes from this is that through evolution RNA-binding domains have been appended to the ends of modification enzymes and in doing so have provided a degree of active-site relaxation. By contrast, another family of tRNA-interacting enzymes, the aminoacyl tRNA synthetases, use a more limited set of features for tRNA recognition. For instance, studies using in vitro aminoacylation revealed the acceptor stem as a key identity element. Studies with portions of $tRNA^{Ala}$ and yeast AlaRS showed that the first three base pairs of the acceptor stem were important for recognition [167–169]. In addition, the anticodon loop often bears identity elements for the selection of the correct tRNA by the cognate synthetase, and as such, it directly serves as a link between the protein synthesis and the genetic code [167,169,170]. Thus, in general, synthetases primarily recognize features of the anticodon loops and acceptor stems of tRNAs [171–179]. In the case of modification enzymes, to modify tRNAs at different stages of folding, active-site flexibility was necessary to accommodate a diversity of substrates. For instance, with deaminases, as discussed earlier, it may have also allowed a given enzyme to perform more than one reaction, as seen with the *T. brucei* ADAT2/3 enzyme, which can catalyze both A-to-I and C-to-U deamination. Likewise, it may have been that the modularity imparted by appended binding domains also provided the driving force for the evolution of multisubstrate specificity. Given the bewildering number of modified substrates, the modification field is indeed a fertile ground to further explore the limits of RNA binding and the basis for substrate specificity by this large family of enzymes.

ACKNOWLEDGMENTS

We thank NIH Grant GM084065 to J.D.A. and a Graduate Fellowship from The Ohio State University Center for RNA Biology to K.M. for support.

REFERENCES

[1] B.M. Lunde, C. Moore, G. Varani, RNA-binding proteins: modular design for efficient function, Nat. Rev. Mol. Cell Biol. 8 (6) (2007) 479–490.
[2] V. Blanc, N. Navaratnam, J.O. Henderson, S. Anant, S. Kennedy, A. Jarmuz, J. Scott, N.O. Davidson, Identification of GRY-RBP as an apolipoprotein B RNA-binding protein that interacts with both apobec-1 and apobec-1 complementation factor to modulate C to U editing, J. Biol. Chem. 276 (13) (2001) 10272–10283.
[3] A. Mehta, M.T. Kinter, N.E. Sherman, D.M. Driscoll, Molecular cloning of apobec-1 complementation factor, a novel RNA-binding protein involved in the editing of apolipoprotein B mRNA, Mol. Cell. Biol. 20 (5) (2000) 1846–1854.
[4] J. Zhang, A.R. Ferré-D'Amaré, The tRNA elbow in structure, recognition and evolution, Life (Basel, Switzerland) 6 (1) (2016) 3.
[5] W.E. Swinehart, J.E. Jackman, Diversity in mechanism and function of tRNA methyltransferases, RNA Biol. 12 (4) (2015) 398–411.
[6] H. Grosjean, J. Edqvist, K.B. Stråby, R. Giegé, Enzymatic formation of modified nucleosides in tRNA: dependence on tRNA architecture, J. Mol. Biol. 255 (1) (1996) 67–85.
[7] M.W. Gray, Diversity and evolution of mitochondrial RNA editing systems, IUBMB Life 55 (4–5) (2003) 227–233.
[8] Z. Paris, I.M.C. Fleming, J.D. Alfonzo, Determinants of tRNA editing and modification: avoiding conundrums, affecting function, Semin. Cell Dev. Biol. 23 (3) (2012) 269–274.
[9] L. Randau, B.J. Stanley, A. Kohlway, S. Mechta, Y. Xiong, D. Söll, A cytidine deaminase edits C to U in transfer RNAs in Archaea, Science 324 (5927) (2009) 657–659.
[10] H. Grosjean, F. Constantinesco, D. Foiret, N. Benachenhou, A novel enzymatic pathway leading to 1-methylinosine modification in Haloferax volcanii tRNA, Nucleic Acids Res. 23 (21) (1995) 4312–4319.
[11] A. Gerber, H. Grosjean, T. Melcher, W. Keller, Tad1p, a yeast tRNA-specific adenosine deaminase, is related to the mammalian pre-mRNA editing enzymes ADAR1 and ADAR2, EMBO J. 17 (16) (1998) 4780–4789.
[12] A. Janke, S. Pääbo, Editing of a tRNA anticodon in marsupial mitochondria changes its codon recognition, Nucleic Acids Res. 21 (7) (1993) 1523–1525.
[13] J.D. Alfonzo, V. Blanc, A.M. Estévez, M.A.T. Rubio, L. Simpson, C to U editing of the anticodon of imported mitochondrial tRNA(Trp) allows decoding of the UGA stop codon in Leishmania tarentolae, EMBO J. 18 (24) (1999) 7056–7062.
[14] J. Wolf, A.P. Gerber, W. Keller, TadA, an essential tRNA-specific adenosine deaminase from Escherichia coli, EMBO J. 21 (14) (2002) 3841–3851.
[15] A.P. Gerber, W. Keller, An adenosine deaminase that generates inosine at the wobble position of tRNAs, Science 286 (5442) (1999) 1146–1149.

[16] M.A.T. Rubio, I. Pastar, K.W. Gaston, F.L. Ragone, C.J. Janzen, G.A.M. Cross, F.N. Papavasiliou, J.D. Alfonzo, An adenosine-to-inosine tRNA-editing enzyme that can perform C-to-U deamination of DNA, Proc. Natl. Acad. Sci. U.S.A. 104 (19) (2007) 7821–7826.
[17] A.G. Torres, D. Piñeyro, M. Rodríguez-Escribà, N. Camacho, O. Reina, A. Saint-Léger, L. Filonava, E. Batlle, L. Ribas De Pouplana, Inosine modifications in human tRNAs are incorporated at the precursor tRNA level, Nucleic Acids Res. 43 (10) (2015) 5145–5157.
[18] W. Zhou, D. Karcher, R. Bock, Identification of enzymes for adenosine-to-inosine editing and discovery of cytidine-to-uridine editing in nucleus-encoded transfer RNAs of Arabidopsis, Plant Physiol. 166 (4) (2014) 1985–1997.
[19] J. Kim, V. Malashkevich, S. Roday, M. Lisbin, V.L. Schramm, S.C. Almo, Structural and kinetic characterization of Escherichia coli TadA, the wobble-specific tRNA deaminase, Biochemistry 45 (20) (2006) 6407–6416.
[20] H.C. Losey, A.J. Ruthenburg, G.L. Verdine, Crystal structure of Staphylococcus aureus tRNA adenosine deaminase TadA in complex with RNA, Nat. Struct. Mol. Biol. 13 (2) (2006) 153–159.
[21] Y. Elias, R.H. Huang, Biochemical and structural studies of A-to-I editing by tRNA: A34 deaminases at the wobble position of transfer RNA, Biochemistry 44 (36) (2005) 12057–12065.
[22] M. Kuratani, R. Ishii, Y. Bessho, R. Fukunaga, T. Sengoku, M. Shirouzu, S. Sekine, S. Yokoyama, Crystal structure of tRNA adenosine deaminase (TadA) from Aquifex aeolicus, J. Biol. Chem. 280 (16) (2005) 16002–16008.
[23] F.L. Ragone, J.L. Spears, J.M. Wohlgamuth-Benedum, N. Kreel, F.N. Papavasiliou, J.D. Alfonzo, The C-terminal end of the Trypanosoma brucei editing deaminase plays a critical role in tRNA binding, RNA 17 (7) (2011) 1296–1306.
[24] Y. Motorin, M. Helm, RNA nucleotide methylation, Wiley Interdiscip. Rev. RNA 2 (5) (2011) 611–631.
[25] R.J. Liu, T. Long, M. Zhou, X.L. Zhou, E.D. Wang, tRNA recognition by a bacterial tRNA Xm32 modification enzyme from the SPOUT methyltransferase superfamily, Nucleic Acids Res. 43 (15) (2015) 7489–7503.
[26] H. Hori, Methylated nucleosides in tRNA and tRNA methyltransferases, Front. Genet. 5 (May) (2014) 144.
[27] R. Yamagami, K. Yamashita, H. Nishimasu, C. Tomikawa, A. Ochi, C. Iwashita, A. Hirata, R. Ishitani, O. Nureki, H. Hori, The tRNA recognition mechanism of folate/FAD-dependent tRNA methyltransferase (TrmFO), J. Biol. Chem. 287 (51) (2012) 42480–42494.
[28] J. Kim, H. Xiao, J.B. Bonanno, C. Kalyanaraman, S. Brown, X. Tang, N.F. Al-Obaidi, Y. Patskovsky, P.C. Babbitt, M.P. Jacobson, et al., Structure-guided discovery of the metabolite carboxy-SAM that modulates tRNA function, Nature 498 (7452) (2013) 123–126.
[29] P.Z. Kozbial, A.R. Mushegian, Natural history of S-adenosylmethionine-binding proteins, BMC Struct. Biol. 5 (2005) 19.
[30] H.L. Schubert, R.M. Blumenthal, X. Cheng, Many paths to methyltransfer: a chronicle of convergence, Trends Biochem. Sci. 28 (6) (2003) 329–335.
[31] R. Ero, L. Peil, A. Liiv, J. Remme, Identification of pseudouridine methyltransferase in Escherichia coli, RNA 14 (10) (2008) 2223–2233.
[32] J. Urbonavicius, G. Jäger, G.R. Björk, Amino acid residues of the Escherichia coli tRNA(m5U54)methyltransferase (TrmA) critical for stability, covalent binding of tRNA and enzymatic activity, Nucleic Acids Res. 35 (10) (2007) 3297–3305.
[33] K.M. McKenney, J.D. Alfonzo, From prebiotics to probiotics: the evolution and functions of tRNA modifications, Life (Basel, Switzerland) 6 (1) (2016) 13.

[34] V. Anantharaman, E.V. Koonin, L. Aravind, TRAM, a predicted RNA-binding domain, common to tRNA uracil methylation and adenine thiolation enzymes, FEMS Microbiol. Lett. 197 (2) (2001) 215–221.
[35] L. Aravind, E.V. Koonin, Novel predicted RNA-binding domains associated with the translation machinery, J. Mol. Evol. 48 (3) (1999) 291–302.
[36] I. Zegers, D. Gigot, F. van Vliet, C. Tricot, S. Aymerich, J.M. Bujnicki, J. Kosinski, L. Droogmans, Crystal structure of Bacillus subtilis TrmB, the tRNA (m7G46) methyltransferase, Nucleic Acids Res. 34 (6) (2006) 1925–1934.
[37] M.A. Witek, E.G. Kuiper, E. Minten, E.K. Crispell, G.L. Conn, A novel motif for S-adenosyl-L-methionine binding by the ribosomal RNA methyltransferase TlyA from mycobacterium tuberculosis, J. Biol. Chem. 292 (5) (2017) 1977–1987.
[38] J.L. Martin, F.M. McMillan, SAM (dependent) I AM: the S-adenosylmethionine-dependent methyltransferase fold, Curr. Opin. Struct. Biol. 12 (6) (2002) 783–793.
[39] C. Pastore, I. Topalidou, F. Forouhar, A.C. Yan, M. Levy, J.F. Hunt, Crystal structure and RNA binding properties of the RNA recognition motif (RRM) and AlkB domains in human AlkB homolog 8 (ABH8), an enzyme catalyzing tRNA hypermodification, J. Biol. Chem. 287 (3) (2012) 2130–2143.
[40] W.L. Towns, T.J. Begley, Transfer RNA methytransferases and their corresponding modifications in budding yeast and humans: activities, predications, and potential roles in human health, DNA Cell Biol. 31 (4) (2012) 434–454.
[41] V. Anantharaman, E.V. Koonin, L. Aravind, SPOUT: a class of methyltransferases that includes spoU and trmD RNA methylase superfamilies, and novel superfamilies of predicted prokaryotic RNA methylases, J. Mol. Microbiol. Biotechnol. 4 (1) (2002) 71–75.
[42] H.J. Ahn, H.-W. Kim, H.-J. Yoon, B. Lee Il, S.W. Suh, J.K. Yang, Crystal structure of tRNA(m1G37)methyltransferase: insights into tRNA recognition, EMBO J. 22 (11) (2003) 2593–2603.
[43] K. Lim, H. Zhang, A. Tempczyk, W. Krajewski, N. Bonander, J. Toedt, A. Howard, E. Eisenstein, O. Herzberg, Structure of the YibK methyltransferase from Haemophilus influenzae (HI0766): a cofactor bound at a site formed by a knot, Proteins 51 (1) (2003) 56–67.
[44] P.A. Elkins, J.M. Watts, M. Zalacain, A. Van Thiel, P.R. Vitazka, M. Redlak, C. Andraos-Selim, F. Rastinejad, W.M. Holmes, Insights into catalysis by a knotted TrmD tRNA methyltransferase, J. Mol. Biol. 333 (5) (2003) 931–949.
[45] K. Tkaczuk, S. Dunin-Horkawicz, E. Purta, J.M. Bujnicki, Structural and evolutionary bioinformatics of the SPOUT superfamily of methyltransferases, BMC Bioinformatics 8 (2007) 73.
[46] O. Nureki, K. Watanabe, S. Fukai, R. Ishii, Y. Endo, H. Hori, S. Yokoyama, Deep knot structure for construction of active site and cofactor binding site of tRNA modification enzyme, Structure 12 (4) (2004) 593–602.
[47] G. Michel, V. Sauvé, R. Larocque, Y. Li, A. Matte, M. Cygler, The structure of the RlmB 23S rRNA methyltransferase reveals a new methyltransferase fold with a unique knot, Structure 10 (10) (2002) 1303–1315.
[48] T. Christian, Y.-M. Hou, Distinct determinants of tRNA recognition by the TrmD and Trm5 methyl transferases, J. Mol. Biol. 373 (3) (2007) 623–632.
[49] B. Van Laer, M. Roovers, L. Wauters, J.M. Kasprzak, M. Dyzma, E. Deyaert, R.K. Singh, A. Feller, J.M. Bujnicki, L. Droogmans, et al., Structural and functional insights into tRNA binding and adenosine N1-methylation by an archaeal Trm10 homologue, Nucleic Acids Res. 44 (2) (2016) 940–953.
[50] M. Zhou, T. Long, Z.-P. Fang, X.-L. Zhou, R.-J. Liu, E.-D. Wang, Identification of determinants for tRNA substrate recognition by *Escherichia coli* C/U34 2′-O-methyltransferase, RNA Biol. 12 (8) (2015) 900–911.

[51] R.J. Liu, M. Zhou, Z.P. Fang, M. Wang, X.L. Zhou, E.D. Wang, The tRNA recognition mechanism of the minimalist SPOUT methyltransferase TrmL, Nucleic Acids Res. 41 (16) (2013) 7828–7842.
[52] W. Humphrey, A. Dalke, K. Schulten, VMD: visual molecular dynamics, J. Mol. Graph. 14 (1) (1996) 33–38.
[53] C.S. Cerrudo, P.D. Ghiringhelli, D.E. Gomez, Protein universe containing a PUA RNA-binding domain, FEBS J. 281 (1) (2014) 74–87.
[54] I. Pérez-Arellano, J. Gallego, J. Cervera, The PUA domain—a structural and functional overview, FEBS J. 274 (19) (2007) 4972–4984.
[55] L.H.G. Larsen, A. Rasmussen, A.M.B. Giessing, G. Jogl, F. Kirpekar, Identification and characterization of the Thermus thermophilus 5-methylcytidine (m5C) methyltransferase modifying 23 S ribosomal RNA (rRNA) base C1942, J. Biol. Chem. 287 (33) (2012) 27593–27600.
[56] S. Haag, A.S. Warda, J. Kretschmer, M.A. Günnigmann, C. Höbartner, M.T. Bohnsack, NSUN6 is a human RNA methyltransferase that catalyzes formation of m5C72 in specific tRNAs, RNA 21 (9) (2015) 1532–1543.
[57] L. Aravind, E.V. Koonin, THUMP—a predicted RNA-binding domain shared by 4-thiouridine, pseudouridine synthases and RNA methylases, Trends Biochem. Sci. 26 (4) (2001) 215–217.
[58] P. Neumann, K. Lakomek, P.T. Naumann, W.M. Erwin, C.T. Lauhon, R. Ficner, Crystal structure of a 4-thiouridine synthetase-RNA complex reveals specificity of tRNA U8 modification, Nucleic Acids Res. 42 (10) (2014) 6673–6685.
[59] T.T. Lee, S. Agarwalla, R.M. Stroud, Crystal structure of RumA, an iron-sulfur cluster containing E. coli ribosomal RNA 5-methyluridine methyltransferase, Structure 12 (3) (2004) 397–407.
[60] N. Leulliot, M.T. Bohnsack, M. Graille, D. Tollervey, H. Van Tilbeurgh, The yeast ribosome synthesis factor Emg1 is a novel member of the superfamily of alpha/beta knot fold methyltransferases, Nucleic Acids Res. 36 (2) (2008) 629–639.
[61] O. Nureki, M. Shirouzu, K. Hashimoto, R. Ishitani, T. Terada, M. Tamakoshi, T. Oshima, M. Chijimatsu, K. Takio, D.G. Vassylyev, et al., An enzyme with a deep trefoil knot for the active-site architecture, Acta Crystallogr. D Biol. Crystallogr. 58 (7) (2002) 1129–1137.
[62] A. Alian, T.T. Lee, S.L. Griner, R.M. Stroud, J. Finer-Moore, Structure of a TrmA-RNA complex: a consensus RNA fold contributes to substrate selectivity and catalysis in m5U methyltransferases, Proc. Natl. Acad. Sci. U.S.A. 105 (19) (2008) 6876–6881.
[63] H. Walbott, N. Leulliot, H. Grosjean, B. Golinelli-Pimpaneau, The crystal structure of Pyrococcus abyssi tRNA (uracil-54, C5)-methyltransferase provides insights into its tRNA specificity, Nucleic Acids Res. 36 (15) (2008) 4929–4940.
[64] P.J. Sample, L. Kořený, Z. Paris, K.W. Gaston, M.A.T. Rubio, I.M.C. Fleming, S. Hinger, E. Horáková, P.A. Limbach, J. Lukeš, et al., A common tRNA modification at an unusual location: the discovery of wyosine biosynthesis in mitochondria, Nucleic Acids Res. 43 (8) (2015) 4262–4273.
[65] T. Ohira, T. Suzuki, Retrograde nuclear import of tRNA precursors is required for modified base biogenesis in yeast, Proc. Natl. Acad. Sci. U.S.A. 108 (26) (2011) 10502–10507.
[66] A.K. Hopper, Transfer RNA post-transcriptional processing, turnover, and subcellular dynamics in the yeast *Saccharomyces cerevisiae*, Genetics 194 (1) (2013) 43–67.
[67] W.-K. Huh, J.V. Falvo, L.C. Gerke, A.S. Carroll, R.W. Howson, J.S. Weissman, E.K. O'Shea, Global analysis of protein localization in budding yeast, Nature 425 (6959) (2003) 686–691.

[68] T. Yoshihisa, K. Yunoki-Esaki, C. Ohshima, N. Tanaka, T. Endo, Possibility of cytoplasmic pre-tRNA splicing: the yeast tRNA splicing endonuclease mainly localizes on the mitochondria, Mol. Biol. Cell 14 (8) (2003) 3266–3279.
[69] S. Goto-Ito, T. Ito, M. Kuratani, Y. Bessho, S. Yokoyama, Tertiary structure checkpoint at anticodon loop modification in tRNA functional maturation, Nat. Struct. Mol. Biol. 16 (10) (2009) 1109–1115.
[70] Y. Motorin, H. Grosjean, Multisite-specific tRNA: m5C-methyltransferase (Trm4) in yeast Saccharomyces cerevisiae: identification of the gene and substrate specificity of the enzyme, RNA 5 (8) (1999) 1105–1118.
[71] B. Brzezicha, M. Schmidt, I. Makałowska, A. Jarmołowski, J. Pieńkowska, Z. Szweykowska-Kulińska, Identification of human tRNA: m5C methyltransferase catalysing intron-dependent m5C formation in the first position of the anticodon of the pre-tRNA(CAA)Leu, Nucleic Acids Res. 34 (20) (2006) 6034–6043.
[72] A.G. Arimbasseri, J. Iben, F. Wei, K. Rijal, K. Tomizawa, M. Hafner, R.J. Maraia, Evolving specificity of tRNA 3-methyl-cytidine-32 (m3C32) modification: a subset of tRNAsSer requires N6-isopentenylation of A37, RNA 22 (9) (2016) 1400–1410.
[73] M. Roovers, A primordial RNA modification enzyme: the case of tRNA (m1A) methyltransferase, Nucleic Acids Res. 32 (2) (2004) 465–476.
[74] M.A.T. Rubio, F.L. Ragone, K.W. Gaston, M. Ibba, J.D. Alfonzo, C to U editing stimulates A to I editing in the anticodon loop of a cytoplasmic threonyl tRNA in Trypanosoma brucei, J. Biol. Chem. 281 (1) (2006) 115–120.
[75] M.A.T. Rubio, K.W. Gaston, K.M. McKenney, I.M.C. Fleming, Z. Paris, P.A. Limbach, J.D. Alfonzo, Editing and methylation at a single site by functionally interdependent activities, Nature 542 (7642) (2017) 494–497.
[76] W.T. Yewdell, J. Chaudhuri, Molecular biology: RNA editing packs a one-two punch, Nature 542 (7642) (2017) 420–421.
[77] M.P. Guy, B.M. Podyma, M.A. Preston, H.H. Shaheen, K.L. Krivos, P.A. Limbach, A.K. Hopper, E.M. Phizicky, Yeast Trm7 interacts with distinct proteins for critical modifications of the tRNAPhe anticodon loop, RNA 18 (10) (2012) 1921–1933.
[78] M. Rubio, J.D. Alfonzo, Editing and modification in trypanosomatids: the reshaping of non-coding RNAs, Top. Curr. Genet. 12 (2005) 71–86.
[79] M.P. Guy, E.M. Phizicky, Two-subunit enzymes involved in eukaryotic post-transcriptional tRNA modification, RNA Biol. 11 (12) (2014) 1608–1618.
[80] Z. Shao, W. Yan, J. Peng, X. Zuo, Y. Zou, F. Li, D. Gong, R. Ma, J. Wu, Y. Shi, et al., Crystal structure of tRNA m1G9 methyltransferase Trm10: insight into the catalytic mechanism and recognition of tRNA substrate, Nucleic Acids Res. 42 (1) (2014) 509–525.
[81] M.A. Kurowski, J.M. Sasin, M. Feder, J. Debski, J.M. Bujnicki, Characterization of the cofactor-binding site in the SPOUT-fold methyltransferases by computational docking of S-adenosylmethionine to three crystal structures, BMC Bioinformatics 4 (2) (2003) 9.
[82] S.K. Purushothaman, J.M. Bujnicki, H. Grosjean, B. Lapeyre, Trm11p and Trm112p are both required for the formation of 2-methylguanosine at position 10 in yeast tRNA, Mol. Cell. Biol. 25 (11) (2005) 4359–4370.
[83] R. Sardana, A.W. Johnson, The methyltransferase adaptor protein Trm112 is involved in biogenesis of both ribosomal subunits, Mol. Biol. Cell 23 (21) (2012) 4313–4322.
[84] S. Figaro, L. Wacheul, S. Schillewaert, M. Graille, E. Huvelle, R. Mongeard, C. Zorbas, D.L.J. Lafontaine, V. Heurgue-Hamard, Trm112 is required for Bud23-mediated methylation of the 18S rRNA at position G1575, Mol. Cell. Biol. 32 (12) (2012) 2254–2267.

[85] J. Létoquart, E. Huvelle, L. Wacheul, G. Bourgeois, C. Zorbas, M. Graille, V. Heurgué-Hamard, D.L.J. Lafontaine, Structural and functional studies of Bud23-Trm112 reveal 18S rRNA N7-G1575 methylation occurs on late 40S precursor ribosomes, Proc. Natl. Acad. Sci. U.S.A. 111 (51) (2014) e5518–e5526.

[86] J. Anderson, L. Phan, A.G. Hinnebusch, The Gcd10p/Gcd14p complex is the essential two-subunit tRNA(1-methyladenosine) methyltransferase of *Saccharomyces cerevisiae*, Proc. Natl. Acad. Sci. U.S.A. 97 (10) (2000) 5173–5178.

[87] S.G. Ozanick, J.M. Bujnicki, D.S. Sem, J.T. Anderson, Conserved amino acids in each subunit of the heteroligomeric tRNA m1A58 Mtase from Saccharomyces cerevisiae contribute to tRNA binding, Nucleic Acids Res. 35 (20) (2007) 6808–6819.

[88] A. Alexandrov, M.R. Martzen, E.M. Phizicky, Two proteins that form a complex are required for 7-methylguanosine modification of yeast tRNA, RNA 8 (10) (2002) 1253–1266.

[89] J. Somme, B. Van Laer, M. Roovers, J. Steyaert, W. Versées, L. Droogmans, Characterization of two homologous 2′-O-methyltransferases showing different specificities for their tRNA substrates, RNA 20 (8) (2014) 1257–1271.

[90] S. Dunin-Horkawicz, A. Czerwoniec, M.J. Gajda, M. Feder, H. Grosjean, J.M. Bujnicki, MODOMICS: a database of RNA modification pathways, Nucleic Acids Res. 34 (Database issue) (2006) d145–d149.

[91] W.A. Cantara, P.F. Crain, J. Rozenski, J.A. McCloskey, K.A. Harris, X. Zhang, F.A.P. Vendeix, D. Fabris, P.F. Agris, The RNA modification database, RNAMDB: 2011 update, Nucleic Acids Res. 39 (Database issue) (2011) d195–d201.

[92] M.A. Machnicka, K. Milanowska, O.O. Osman, E. Purta, M. Kurkowska, A. Olchowik, W. Januszewski, S. Kalinowski, S. Dunin-Horkawicz, K.M. Rother, et al., MODOMICS: a database of RNA modification pathways—2013 update, Nucleic Acids Res. 41 (Database issue) (2013) d262–d267.

[93] H. Grosjean, R. Benne, Modification and Editing of RNA, American Society of Microbiology, Washington, DC, 1998.

[94] W.E. Cohn, E. Volkin, Nucleoside-5′-phosphates from ribonucleic acid, Nature 167 (4247) (1951) 483–484.

[95] F.F. Davis, F.W. Allen, Ribonucleic acids from yeast which contain a fifth nucleotide, J. Biol. Chem. 227 (2) (1957) 907–915.

[96] C.T. Yu, F.W. Allen, Studies on an isomer of uridine isolated from ribonucleic acids, Biochim. Biophys. Acta 32 (1959) 393–406.

[97] W.E. Cohn, 5-Ribosyl uracil, a carbon-carbon ribofuranosyl nucleoside in ribonucleic acids, Biochim. Biophys. Acta 32 (1959) 569–571.

[98] R. Cortese, H.O. Kammen, S.J. Spengler, B.N. Ames, Biosynthesis of pseudouridine in transfer ribonucleic acid, J. Biol. Chem. 249 (4) (1974) 1103–1108.

[99] M.S. Ciampi, F. Arena, R. Cortese, Biosynthesis of pseudouridine in the in vitro transcribed tRNATyr precursor, FEBS Lett. 77 (1) (1977) 75–82.

[100] J. Ofengand, Ribosomal RNA pseudouridines and pseudouridine synthases, FEBS Lett. 514 (1) (2002) 17–25.

[101] J. Ge, Y.-T. Yu, RNA pseudouridylation: new insights into an old modification, Trends Biochem. Sci. 38 (4) (2013) 210–218.

[102] S.C. Pomerantz, J.A. McCloskey, Detection of the common RNA nucleoside pseudouridine in mixtures of oligonucleotides by mass spectrometry, Anal. Chem. 77 (15) (2005) 4687–4697.

[103] Y.-T. Yu, U.T. Meier, RNA-guided isomerization of uridine to pseudouridine—pseudouridylation, RNA Biol. 11 (12) (2014) 1483–1494.

[104] T. Hamma, A.R. Ferré-D'Amaré, Pseudouridine synthases, Chem. Biol. 13 (11) (2006) 1125–1135.

[105] F. Spenkuch, Y. Motorin, M. Helm, Pseudouridine: still mysterious, but never a fake (uridine)! RNA Biol. 11 (12) (2015) 1540–1554.
[106] M.K. McDonald, E.J. Miracco, J. Chen, Y. Xie, E.G. Mueller, The handling of the mechanistic probe 5-fluorouridine by the pseudouridine synthase TruA and its consistency with the handling of the same probe by the pseudouridine synthases TruB and RluA, Biochemistry 50 (3) (2011) 426–436.
[107] N. Czudnochowski, G.W. Ashley, D.V. Santi, A. Alian, J. Finer-Moore, R.M. Stroud, The mechanism of pseudouridine synthases from a covalent complex with RNA, and alternate specificity for U2605 versus U2604 between close homologs, Nucleic Acids Res. 42 (3) (2014) 2037–2048.
[108] R.G. Silva, V.L. Schramm, Uridine phosphorylase from Trypanosoma cruzi: kinetic and chemical mechanisms, Biochemistry 50 (42) (2011) 9158–9166.
[109] G.R. Veerareddygari, S.K. Singh, E.G. Mueller, The pseudouridine synthases proceed through a glycal intermediate, J. Am. Chem. Soc. 138 (25) (2016) 7852–7855.
[110] E.G. Mueller, Chips off the old block, Nat. Struct. Biol. 9 (5) (2002) 320–322.
[111] J. Ofengand, A. Malhotra, J. Remme, N.S. Gutgsell, M. Del Campo, S. Jean-Charles, L. Peil, Y. Kaya, Pseudouridines and pseudouridine synthases of the ribosome, Cold Spring Harb. Symp. Quant. Biol. 66 (2001) 147–159.
[112] N.S. Gutgsell, M.P. Deutscher, J. Ofengand, The pseudouridine synthase RluD is required for normal ribosome assembly and function in *Escherichia coli*, RNA 11 (7) (2005) 1141–1152.
[113] H.O. Kammen, C.C. Marvel, L. Hardy, E.E. Penhoet, Purification, structure, and properties of Escherichia coli tRNA pseudouridine synthase I, J. Biol. Chem. 263 (5) (1988) 2255–2263.
[114] E.V. Koonin, Pseudouridine synthases: four families of enzymes containing a putative uridine-binding motif also conserved in dUTPases and dCTP deaminases, Nucleic Acids Res. 24 (12) (1996) 2411–2415.
[115] J. Conrad, L. Niu, K. Rudd, B.G. Lane, J. Ofengand, 16S ribosomal RNA pseudouridine synthase RsuA of Escherichia coli: deletion, mutation of the conserved Asp102 residue, and sequence comparison among all other pseudouridine synthases, RNA 5 (6) (1999) 751–763.
[116] M. Roovers, C. Hale, C. Tricot, M.P. Terns, R.M. Terns, H. Grosjean, L. Droogmans, Formation of the conserved pseudouridine at position 55 in archaeal tRNA, Nucleic Acids Res. 34 (15) (2006) 4293–4301.
[117] C.J. McCleverty, M. Hornsby, G. Spraggon, A. Kreusch, Crystal structure of human Pus10, a novel pseudouridine synthase, J. Mol. Biol. 373 (5) (2007) 1243–1254.
[118] P. Gurha, R. Gupta, Archaeal Pus10 proteins can produce both pseudouridine 54 and 55 in tRNA, RNA 14 (12) (2008) 2521–2527.
[119] F. Spenkuch, Y. Motorin, M. Helm, Pseudouridine: still mysterious, but never a fake (uridine)! RNA Biol. 11 (12) (2014) 1540–1554.
[120] S. Raychaudhuri, L. Niu, J. Conrad, B.G. Lane, J. Ofengand, Functional effect of deletion and mutation of the Escherichia coli ribosomal RNA and tRNA pseudouridine synthase RluA, J. Biol. Chem. 274 (27) (1999) 18880–18886.
[121] T. Hamma, A.R. Ferré-D'Amaré, The box H/ACA ribonucleoprotein complex: interplay of RNA and protein structures in post-transcriptional RNA modification, J. Biol. Chem. 285 (2) (2010) 805–809.
[122] S. Hur, R.M. Stroud, J. Finer-Moore, Substrate recognition by RNA 5-methyluridine methyltransferases and pseudouridine synthases: a structural perspective, J. Biol. Chem. 281 (51) (2006) 38969–38973.
[123] T. Kiss, E. Fayet-Lebaron, B.E. Jády, Box H/ACA small ribonucleoproteins, Mol. Cell 37 (5) (2010) 597–606.

[124] W.A. Decatur, X. Liang, D. Piekna-Przybylska, M.J. Fournier, Identifying effects of snoRNA-guided modifications on the synthesis and function of the yeast ribosome, Methods Enzymol. 425 (2007) 283–316.
[125] T. Samuelsson, M. Olsson, Transfer RNA pseudouridine synthases in Saccharomyces cerevisiae, J. Biol. Chem. 265 (15) (1990) 8782–8787.
[126] X. Ma, X. Zhao, Y.-T. Yu, Pseudouridylation (Psi) of U2 snRNA in S. cerevisiae is catalyzed by an RNA-independent mechanism, EMBO J. 22 (8) (2003) 1889–1897.
[127] H.F. Becker, Y. Motorin, R.J. Planta, H. Grosjean, The yeast gene YNL292w encodes a pseudouridine synthase (Pus4) catalyzing the formation of psi55 in both mitochondrial and cytoplasmic tRNAs, Nucleic Acids Res. 25 (22) (1997) 4493–4499.
[128] K. Hellmuth, H. Grosjean, Y. Motorin, K. Deinert, E. Hurt, G. Simos, Cloning and characterization of the Schizosaccharomyces pombe tRNA:pseudouridine synthase Pus1p, Nucleic Acids Res. 28 (23) (2000) 4604–4610.
[129] A. Urban, I. Behm-Ansmant, C. Branlant, Y. Motorin, RNA sequence and two-dimensional structure features required for efficient substrate modification by the Saccharomyces cerevisiae RNA:{Psi} -synthase Pus7p, J. Biol. Chem. 284 (9) (2009) 5845–5858.
[130] A.R. Ferré-D'Amaré, RNA-modifying enzymes, Curr. Opin. Struct. Biol. 13 (1) (2003) 49–55.
[131] K. Phannachet, Y. Elias, R.H. Huang, Dissecting the roles of a strictly conserved tyrosine in substrate recognition and catalysis by pseudouridine 55 synthase, Biochemistry 44 (47) (2005) 15488–15494.
[132] C.M. Chan, R.H. Huang, Enzymatic characterization and mutational studies of TruD—the fifth family of pseudouridine synthases, Arch. Biochem. Biophys. 489 (1–2) (2009) 15–19.
[133] K. Chatterjee, I.K. Blaby, P.C. Thiaville, M. Majumder, H. Grosjean, Y.A. Yuan, R. Gupta, V. de Crécy-Lagard, The archaeal COG1901/DUF358 SPOUT-methyltransferase members, together with pseudouridine synthase Pus10, catalyze the formation of 1-methylpseudouridine at position 54 of tRNA, RNA 18 (3) (2012) 421–433.
[134] A. Joardar, S. Jana, E. Fitzek, P. Gurha, M. Majumder, K. Chatterjee, M. Geisler, R. Gupta, Role of forefinger and thumb loops in production of Ψ54 and Ψ55 in tRNAs by archaeal Pus10, RNA 19 (9) (2013) 1279–1294.
[135] V. Ramamurthy, S.L. Swann, J.L. Paulson, C.J. Spedaliere, E.G. Mueller, Critical aspartic acid residues in pseudouridine synthases, J. Biol. Chem. 274 (32) (1999) 22225–22230.
[136] C. Hoang, A.R. Ferré-D'Amaré, Cocrystal structure of a tRNA Ψ55 pseudouridine synthase: nucleotide flipping by an RNA-modifying enzyme, Cell 107 (7) (2001) 929–939.
[137] C. Hoang, C.S. Hamilton, E.G. Mueller, A.R. Ferré-D'Amaré, Precursor complex structure of pseudouridine synthase TruB suggests coupling of active site perturbations to an RNA-sequestering peripheral protein domain, Protein Sci. 14 (8) (2005) 2201–2206.
[138] C.J. Spedaliere, C.S. Hamilton, E.G. Mueller, Functional importance of motif I of pseudouridine synthases: mutagenesis of aligned lysine and proline residues, Biochemistry 39 (31) (2000) 9459–9465.
[139] Y. Kaya, M. Del Campo, J. Ofengand, A. Malhotra, Crystal structure of TruD, a novel pseudouridine synthase with a new protein fold, J. Biol. Chem. 279 (18) (2004) 18107–18110.
[140] Y. Kaya, J. Ofengand, A novel unanticipated type of pseudouridine synthase with homologs in bacteria, archaea, and eukarya, RNA 9 (6) (2003) 711–721.

[141] U.B. Ericsson, P. Nordlund, B.M. Hallberg, X-ray structure of tRNA pseudouridine synthase TruD reveals an inserted domain with a novel fold, FEBS Lett. 565 (1–3) (2004) 59–64.
[142] C. Hoang, A.R. Ferre-D'Amare, Crystal structure of the highly divergent pseudouridine synthase TruD reveals a circular permutation of a conserved fold, RNA 10 (7) (2004) 1026–1033.
[143] J.R. Wright, L.C. Keffer-Wilkes, S.R. Dobing, U. Kothe, Pre-steady-state kinetic analysis of the three Escherichia coli pseudouridine synthases TruB, TruA, and RluA reveals uniformly slow catalysis, RNA 17 (12) (2011) 2074–2084.
[144] J. Sivaraman, V. Sauvé, R. Larocque, E.A. Stura, J.D. Schrag, M. Cygler, A. Matte, Structure of the 16S rRNA pseudouridine synthase RsuA bound to uracil and UMP, Nat. Struct. Biol. 9 (5) (2002) 353–358.
[145] B.L. Staker, P. Korber, J.C. Bardwell, M.A. Saper, Structure of Hsp15 reveals a novel RNA-binding motif, EMBO J. 19 (4) (2000) 749–757.
[146] A. Matte, G.V. Louie, J. Sivaraman, M. Cygler, S.K. Burley, Structure of the pseudouridine synthase RsuA from Haemophilus influenzae, Acta Crystallogr. F Struct. Biol. Cryst. Commun. 61 (Pt 4) (2005) 350–354.
[147] F. Lecointe, G. Simos, A. Sauer, E.C. Hurt, Y. Motorin, H. Grosjean, Characterization of yeast protein Deg1 as pseudouridine synthase (Pus3) catalyzing the formation of psi 38 and psi 39 in tRNA anticodon loop, J. Biol. Chem. 273 (3) (1998) 1316–1323.
[148] S. Hur, R.M. Stroud, How U38, 39, and 40 of many tRNAs become the targets for pseudouridylation by TruA, Mol. Cell 26 (2) (2007) 189–203.
[149] X. Dong, Y. Bessho, R. Shibata, M. Nishimoto, M. Shirouzu, S. Kuramitsu, S. Yokoyama, Crystal structure of tRNA pseudouridine synthase TruA from Thermus thermophilus HB8, RNA Biol. 3 (3) (2006) 115–122.
[150] J. Chen, J.R. Patton, Cloning and characterization of a mammalian pseudouridine synthase, RNA 5 (3) (1999) 409–419.
[151] V. Arluison, M. Buckle, H. Grosjean, Pseudouridine synthetase Pus1 of Saccharomyces cerevisiae: kinetic characterisation, tRNA structural requirement and real-time analysis of its complex with tRNA, J. Mol. Biol. 289 (3) (1999) 491–502.
[152] H. Grosshans, F. Lecointe, H. Grosjean, E. Hurt, G. Simos, Pus1p-dependent tRNA pseudouridinylation becomes essential when tRNA biogenesis is compromised in yeast, J. Biol. Chem. 276 (49) (2001) 46333–46339.
[153] G.M. Daubner, A. Cléry, F.H.-T. Allain, RRM-RNA recognition: NMR or crystallography...and new findings, Curr. Opin. Struct. Biol. 23 (1) (2013) 100–108.
[154] F.H. Allain, P.W. Howe, D. Neuhaus, G. Varani, Structural basis of the RNA-binding specificity of human U1A protein, EMBO J. 16 (18) (1997) 5764–5772.
[155] K. Nurse, J. Wrzesinski, A. Bakin, B.G. Lane, J. Ofengand, Purification, cloning, and properties of the tRNA psi 55 synthase from Escherichia coli, RNA 1 (1) (1995) 102–112.
[156] S.M. Kinghorn, C.P. O'Byrne, I.R. Booth, I. Stansfield, Physiological analysis of the role of truB in Escherichia coli: a role for tRNA modification in extreme temperature resistance, Microbiology 148 (11) (2002) 3511–3520.
[157] K. Phannachet, R.H. Huang, Conformational change of pseudouridine 55 synthase upon its association with RNA substrate, Nucleic Acids Res. 32 (4) (2004) 1422–1429.
[158] H. Pan, S. Agarwalla, D.T. Moustakas, J. Finer-Moore, R.M. Stroud, Structure of tRNA pseudouridine synthase TruB and its RNA complex: RNA recognition through a combination of rigid docking and induced fit, Proc. Natl. Acad. Sci. U.S.A. 100 (22) (2003) 12648–12653.

[159] B.N. Chaudhuri, S. Chan, L.J. Perry, T.O. Yeates, Crystal structure of the apo forms of psi 55 tRNA pseudouridine synthase from Mycobacterium tuberculosis: a hinge at the base of the catalytic cleft, J. Biol. Chem. 279 (23) (2004) 24585–24591.
[160] L. Huang, M. Pookanjanatavip, X. Gu, D.V. Santi, A conserved aspartate of tRNA pseudouridine synthase is essential for activity and a probable nucleophilic catalyst, Biochemistry 37 (1) (1998) 344–351.
[161] N. Czudnochowski, A.L. Wang, J. Finer-Moore, R.M. Stroud, In human pseudouridine synthase 1 (hPus1), a C-terminal helical insert blocks tRNA from binding in the same orientation as in the Pus1 bacterial homologue TruA, consistent with their different target selectivities, J. Mol. Biol. 425 (20) (2013) 3875–3887.
[162] C. Hoang, J. Chen, C.A. Vizthum, J.M. Kandel, C.S. Hamilton, E.G. Mueller, A.R. Ferré-D'Amaré, Crystal structure of pseudouridine synthase RluA: indirect sequence readout through protein-induced RNA structure, Mol. Cell 24 (4) (2006) 535–545.
[163] J. Cabello-Villegas, E.P. Nikonowicz, Solution structure of psi32-modified anticodon stem-loop of Escherichia coli tRNAPhe, Nucleic Acids Res. 33 (22) (2005) 6961–6971.
[164] C.S. Hamilton, T.M. Greco, C.A. Vizthum, J.M. Ginter, M.V. Johnston, E.G. Mueller, Mechanistic investigations of the pseudouridine synthase RluA using RNA containing 5-fluorouridine, Biochemistry 45 (39) (2006) 12029–12038.
[165] J.R. Williamson, Induced fit in RNA-protein recognition, Nat. Struct. Biol. 7 (10) (2000) 834–837.
[166] M. Del Campo, Y. Kaya, J. Ofengand, Identification and site of action of the remaining four putative pseudouridine synthases in Escherichia coli, RNA 7 (11) (2001) 1603–1615.
[167] R. Giegé, M. Sissler, C. Florentz, Universal rules and idiosyncratic features in tRNA identity, Nucleic Acids Res. 26 (22) (1998) 5017–5035.
[168] R.W. Chambers, On the recognition of tRNA by its aminoacyl-tRNA ligase, Prog. Nucleic Acid Res. Mol. Biol. 11 (1971) 489–525.
[169] R. Giegé, The early history of tRNA recognition by aminoacyl-tRNA synthetases, J. Biosci. 31 (4) (2006) 477–488.
[170] L.L. Kisselev, The role of the anticodon in recognition of tRNA by aminoacyl-tRNA synthetases, Prog. Nucleic Acid Res. Mol. Biol. 32 (1985) 237–266.
[171] M.E. Saks, J.R. Sampson, J.N. Abelson, The transfer RNA identity problem: a search for rules, Science 263 (5144) (1994) 191–197.
[172] J.J. Perona, A. Hadd, Structural diversity and protein engineering of the aminoacyl-tRNA synthetases, Biochemistry 51 (2012) 8705–8729.
[173] A. Yaremchuk, I. Kriklivyi, M. Tukalo, S. Cusack, Class I tyrosyl-tRNA synthetase has a class II mode of cognate tRNA recognition, EMBO J. 21 (14) (2002) 3829–3840.
[174] J. Cavarelli, D. Moras, Recognition of tRNAs by aminoacyl-tRNA synthetases, FASEB J. 7 (1) (1993) 79–86.
[175] C.W. Carter Jr., Cognition, mechanism, and evolutionary relationships in aminoacyl-tRNA synthetases, Annu. Rev. Biochem. 62 (1) (1993) 715–748.
[176] M. Ibba, D. Soll, Aminoacyl-tRNA synthesis, Annu. Rev. Biochem. 69 (2000) 617–650.
[177] C.W. Carter, R. Wolfenden, tRNA acceptor-stem and anticodon bases embed separate features of amino acid chemistry, RNA Biol. 13 (2) (2016) 145–151.
[178] T.J. Bullwinkle, M. Ibba, Emergence and evolution, Top. Curr. Chem. 344 (2014) 43–87.
[179] P.J. Beuning, K. Musier-Forsyth, Transfer RNA recognition by aminoacyl-tRNA synthetases, Biopolymers 52 (1) (1999) 1–28.

CHAPTER THREE

TrmD: A Methyl Transferase for tRNA Methylation With m^1G37

Ya-Ming Hou*,[1], Ryuma Matsubara*, Ryuichi Takase*, Isao Masuda*, Joanna I. Sulkowska†

*Thomas Jefferson University, Philadelphia, PA, United States
†Center of New Technologies, University of Warsaw, Warsaw, Poland
[1]Corresponding author: e-mail address: ya-ming.hou@jefferson.edu

Contents

Abstract

TrmD is an *S*-adenosyl methionine (AdoMet)-dependent methyl transferase that synthesizes the methylated m^1G37 in tRNA. TrmD is specific to and essential for bacterial growth, and it is fundamentally distinct from its eukaryotic and archaeal counterpart Trm5. TrmD is unusual by using a topological protein knot to bind AdoMet. Despite its restricted mobility, the TrmD knot has complex dynamics necessary to transmit the signal of AdoMet binding to promote tRNA binding and methyl transfer. Mutations in the TrmD knot block this intramolecular signaling and decrease the synthesis of m^1G37-tRNA, prompting ribosomes to +1-frameshifts and premature termination of protein synthesis. TrmD is unique among AdoMet-dependent methyl transferases in that it requires Mg^{2+} in the catalytic mechanism. This Mg^{2+} dependence is important for regulating Mg^{2+} transport to *Salmonella* for survival of the pathogen in the host cell. The strict conservation of TrmD among bacterial species suggests that a better characterization of its enzymology and biology will have a broad impact on our understanding of bacterial pathogenesis.

The Enzymes, Volume 41
ISSN 1874-6047
http://dx.doi.org/10.1016/bs.enz.2017.03.003

1. INTRODUCTION

All natural transfer RNA (tRNA) molecules contain posttranscriptional modifications to achieve their biological functions. More than 100 such posttranscriptional modifications have been identified to date. Those localized to the elbow region of the L-shaped tRNA tertiary structure generally stabilize the folding of the nucleic acid, while those localized at or near the anticodon ensure translational accuracy during protein synthesis on the ribosome. Each of these posttranscriptional modifications is synthesized in a pathway involving one or multiple enzymes. While the greatest majority of these enzymes are nonessential for life, acting for example as a chaperone to modulate tRNA activity [1], a very small number of these enzymes are absolutely required for cell growth and survival. TrmD is an example of one of these essential enzymes [2–8], responsible for methyl transfer from AdoMet to the N^1 position of the G37 base to synthesize m^1G37 on tRNA [9] (Fig. 1). The methylated m^1G37 is on the 3′-side of the anticodon, and it

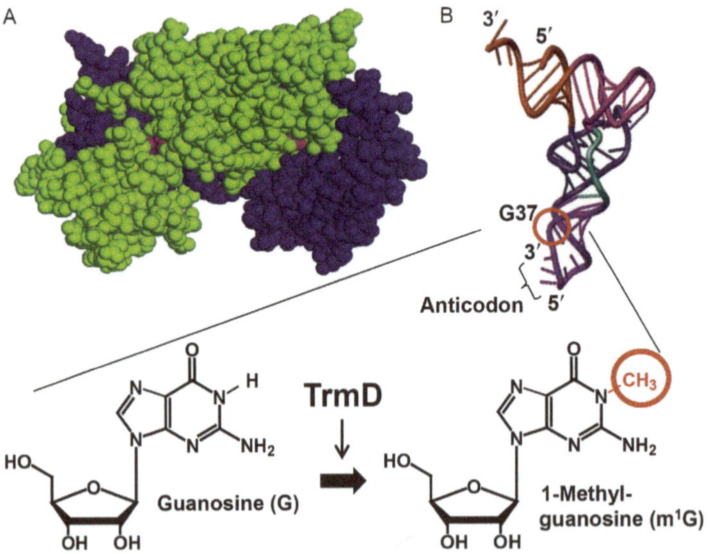

Fig. 1 TrmD structure and activity. (A) The obligated dimer structure of TrmD (PDB: IUAK), showing chain A in *green* and chain B in *blue* and the two AdoMet molecules in *pink* at each dimer interface. (B) TrmD catalyzes methylation to the N^1 position of G37 in tRNA, using AdoMet as the methyl donor and producing m^1G37 on the 3′-side of the tRNA anticodon. The position of G37 and synthesis of m^1G37 on the L-shaped tRNA tertiary structure is *circled*.

is necessary for suppressing tRNA frameshifting during protein synthesis on the ribosome [10–12]. Unlike missense errors, frameshifting errors are almost always lethal, because they change the translational reading frame and introduce premature termination codons. TrmD is broadly conserved in sequence and structure among bacterial species, in both Gram (+) and Gram (−), but it is absent from the eukaryotic and archaeal domains [13,14]. Instead, eukaryotes such as humans and archaeal organisms use Trm5 to synthesize m^1G37-tRNA. Intriguingly, while catalyzing the same chemical reaction, TrmD and Trm5 are fundamentally distinct from each other in nearly all aspects of the methyl transfer reaction, including the global architecture of the enzyme [15–19], the binding of AdoMet [20], the interaction with tRNA [21], the discrimination of the G37 base [22], the control of the rate-limiting step [23], and the catalytic mechanism [24–27]. TrmD and Trm5 are therefore considered as an analogous pair of enzymes that share no structural homology [28], but use the same chemical substrates and produce the same chemical products. As such, TrmD is ranked as a high-priority antimicrobial target [29], providing the attractive potential to isolate TrmD-specific drugs that would have no interaction with Trm5, thus limiting potential side effects. To fully explore this potential, a better understanding of TrmD structure and function and its biological roles in bacterial pathogenesis is necessary. Here, we highlight the most salient features of TrmD and provide an example illustrating how it regulates bacterial pathogenesis.

2. GLOBAL STRUCTURE OF TrmD

TrmD is an obligated homodimer that places each active site at the dimer interface (Fig. 1A) [15,16]. This global structure is different from the structure of Trm5, which is active as a monomer [14]. TrmD is strongly conserved in sequence among evolutionarily diverse bacterial species. High-resolution crystal structures of TrmD in a binary complex with AdoMet [15], or with the unreactive analog sinefungin [15], or with the reaction product AdoHcy (*S*-adenosyl homocysteine) [16] were known for more than 10 years before a ternary complex with sinefungin and a bound tRNA [17] became available. Most of these structures are based on *Haemophilus influenzae* TrmD (*Hi*TrmD), which shares high sequence identify (>83%) and similarity (>93%) with *Escherichia coli* TrmD (*Ec*TrmD), the best-characterized member of the TrmD family [20,23,30,31]. The close sequence conservation between these two enzymes permits exploration of the enzyme's structure–function relationship in depth.

In all of the available structures of the TrmD dimer (e.g., Ref. [17]), each monomeric chain is made up of three distinct domains: an N-terminal domain (residues 1–160 in *Hi*TrmD and *Ec*TrmD) for binding AdoMet, a C-terminal domain for binding tRNA (residues 169–246), and a flexible linker in between (residues 161–168) (Fig. 2A). However, while both monomeric chains A and B are capable of binding AdoMet at the same time, only chain B is capable of binding tRNA by positioning the G37 base within its flexible linker [17]. This positioning organizes the otherwise disordered flexible linker into a helical structure. The resulting stoichiometry of one TrmD dimer binding with two AdoMet molecules but only one tRNA molecule is consistent between biochemical and structural studies (Fig. 2B) [23]. In this stoichiometry, the N-terminal domain of chain A is paired with the C-terminal domain of chain B to assemble the active site (Fig. 2C). Implicit in this stoichiometry is an asymmetry, in which only the AdoMet in chain A is active for methyl transfer, whereas the AdoMet in chain B is inactive. This half-of-the-sites asymmetry, which can alternate the active AdoMet between the two chains, provides one extreme example of negative cooperativity [32]. In contrast to positive cooperativity, where ligand binding to chain A activates ligand binding to chain B, negative cooperativity entails that the ligand bound to chain A prevents another ligand from binding to chain B. Model studies show that negative cooperativity can regulate ligand binding or enzyme activity between the two chains, depending on the substrate concentration [33]. When substrates are saturating, negative cooperativity modulates the two chains slowly, so that the inactivation of chain B by chain A is slow and gradual. Conversely, when substrates are limiting, the modulation becomes ultrasensitive, so that the inactivation of chain B by chain A is rapid and highly responsive to small fluctuations of substrate concentrations [33]. How TrmD responds to changes in substrate levels remains an important open question.

3. THE TREFOIL KNOT OF TrmD AND MUTATIONS IN THE KNOT

In both monomeric chains of TrmD, AdoMet is bound in the N-terminal domain to the deep cleft of a trefoil knot fold [17], which is a topological knot that involves three crossings of the protein backbone through a loop (Fig. 3A). Proteins with a knotted fold are rare (~1.5% in

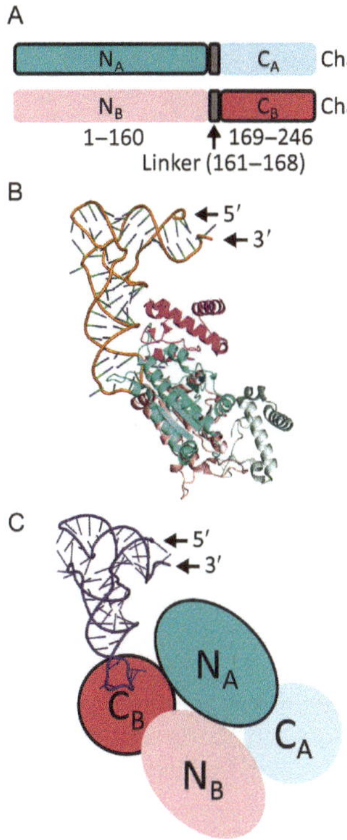

Fig. 2 Domain structure of TrmD. (A) Each chain of the TrmD dimer has three distinct domains: an N-terminal domain (residues 1–160) for binding AdoMet, a flexible linker (residues 161–168) for binding G37 of tRNA, and a C-terminal domain (residues 169–246). In the dimer of TrmD, one active site is formed by the N-terminal domain of chain A (N_A, *dark blue* with a *black outline*) and the C-terminal of chain B (C_B, *dark pink* with a *black* outline). The other active site, which is not operational, is formed by the N-terminal domain of chain B (N_B, *light pink* without an outline) with the C-terminal domain of chain A (C_A, *light blue* without an outline). The sequence and length of each domain is highly conserved among TrmD enzymes across diverse organisms of the bacterial domain. (B) Each TrmD dimer (chain A in *blue* and chain B in *pink*) binds one tRNA molecule in the crystal structure of the ternary complex (PDB: 4YVI). The G37 base of the tRNA is inserted to the active site formed by the N_A domain and the C_B domain. (C) A cartoon diagram of the TrmD dimer, showing that one active site is assembled between the N_A (in *dark blue*) and the C_B (in *dark pink*), while the anticodon region of tRNA (in *purple*) is bound to the flexible linker of chain B. The *darker color* with a *black outline* indicates the active unit of the dimer structure.

the Protein Data Bank), but they occur across a wide range of families, containing either a trefoil (3_1), figure-of-eight (4_1), Gordian (5_2), or stevedore (6_1) knot in their structure [34]. While protein knots are generally proposed to enhance protein stability [35], the trefoil knot in TrmD was shown to be required for methyl transfer [31].

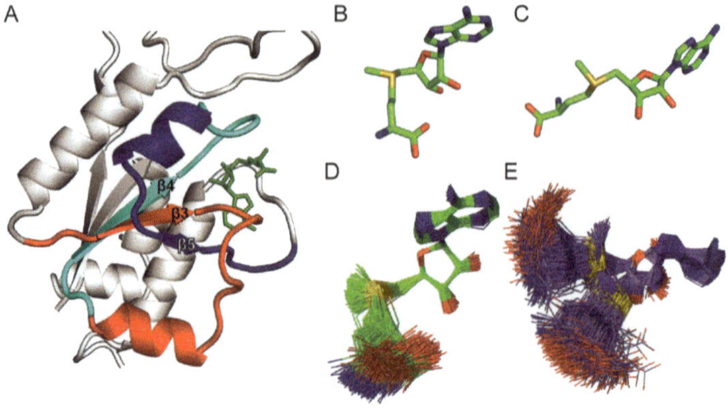

Fig. 3 The trefoil knot structure of TrmD. (A) The trefoil knot is formed starting with the central β3 strand, which turns into β4 through a loop. The β4 makes a turn into β5, which makes a circular insertion through the loop to come out with another loop that binds the adenine base of AdoMet (PDB: 4YVI). The three β stands β3, β4, and β5, together with β1 and β2 (not labeled), form the central β sheet in the N-terminal domain of each monomeric chain in TrmD. (B) The bent conformation of AdoMet in the catalytically active monomer. (C) The open and extended conformation of AdoMet in the catalytically inactive monomer, which is similar to the structure observed in Trm5. (D) The bent conformation of AdoMet has rigid constraints and maintains the bent shape. (E) The open conformation of AdoMet has high flexibility and can extend from the bent shape to the open shape.

The TrmD trefoil knot consists of three β-strands at the central β-sheet (Fig. 3A). This knot starts with β3, which is followed by a loop that turns at the back of β3 and emerges into β4. The end of β4 is followed by another loop that turns into β5, which makes a circular insertion into the knot by crossing over β3 and coming out of the knot with a loop that binds the adenine ring of the methyl donor. In each TrmD trefoil knot, AdoMet is bound in an unusual bent shape [17], which constrains the adenosine and methionine moiety of the methyl donor to face each other (Fig. 3B). In contrast, the majority of methyl transferases (e.g., Trm5) bind AdoMet in the open space of a dinucleotide fold [18,19], which allows the two moieties to extend apart from each other (Fig. 3C). However, although both trefoil knots in TrmD adopt AdoMet in the bent shape in crystal structures, molecular simulation analysis indicates that the two knots have different dynamics and mobility [31]. The active knot is more constrained and binds AdoMet rigidly in the bent shape (Fig. 3D), whereas the inactive knot is more dynamic and binds AdoMet in a range of shapes from the bent to the extended open conformation (Fig. 3E). The different dynamics of the two knots reflects the asymmetry of their relationship.

Additional molecular simulation analysis shows that the trefoil knot is required for AdoMet to adopt the bent shape and that the bent shape is required for TrmD to catalyze methyl transfer [31]. Without the knot, as found in the crystal structure of *Aquifex aeolicus* TrmD [36], AdoMet cannot bend and can only exist in the open shape. Without being in the bent shape, AdoMet would be positioned in a spatial geometry incompatible with the position of the G37 base and unfavorable for methyl transfer. The importance of the trefoil knot is further manifested in its ability to transmit the signal of AdoMet binding in the bent shape, via intramolecular motions, to the tRNA site to stabilize the nucleic acid substrate binding. tRNA stabilization in the binding site then activates a signal transmitted to the active site to promote methyl transfer [31]. These intramolecular motions are not random, but are dedicated to harnessing the AdoMet binding energies to promote tRNA binding and to activate methyl transfer. Indeed, mutations that disrupt AdoMet binding also destabilize tRNA binding and decrease methyl transfer, while mutations that destabilize tRNA binding have little effect on AdoMet binding but directly affect methyl transfer [31]. Thus, the interaction of the trefoil knot with AdoMet in the bent shape promotes the initial step of the methyl transfer reaction. It is this interaction that determines the positioning of G37-tRNA on the enzyme and the assembly of the active site to catalyze methyl transfer.

Additional simulation analysis shows that the TrmD trefoil knot is important for asymmetric catalysis, because it mediates AdoMet signaling across the dimer interface [31]. This intermolecular signaling confers the asymmetry between the two active sites, such that one is catalytically active while the other is inactive. A mutation that disrupts the stability of the trefoil knot eliminates this asymmetry and equalizes the two active sites, making both AdoMet molecules mobile and both monomeric chains capable of binding one tRNA [31]. Intriguingly, despite enabling the two chains to methylate tRNA simultaneously, the mutant is compromised in both the affinity of tRNA binding and the rate of methyl transfer relative to the native and asymmetric enzyme (by 40- and 30-fold, respectively), resulting in a severe loss of catalytic efficiency (by 1200-fold) [31]. Thus, the asymmetry mediated by the trefoil knot at the dimer interface is important for producing the maximum efficiency of methyl transfer by TrmD.

The structural and molecular simulation analyses to date emphasize the notion that the trefoil knot of TrmD is required for the catalytic mechanism in three ways. It is the structure that enables AdoMet to achieve the bent shape necessary for methyl transfer. It is an organized protein fold within the enzyme that captures the free energy of AdoMet binding to stabilize and orient tRNA binding and to facilitate methyl transfer at the active site. It is also the mechanism for cross-chain communication between the two monomers of TrmD to coordinate with each other for the highest efficiency of methyl transfer. In addition, a mutation within the trefoil knot, which was isolated from a genetic analysis [37], renders the enzyme temperature sensitive and catalytically compromised [38]. This mutation, S88L, occurs at the beginning of the central β3 strand, which leads the way into the trefoil knot. The S88L mutation affects a position highly conserved among Gram (−), but not Gram (+), TrmD enzymes [38]. The temperature sensitivity induced by the mutation supports the general notion that a knot structure confers thermal stability to protein enzymes [35].

4. SYNTHESIS OF m^1G37-tRNA BY TrmD

The synthesis of m^1G37-tRNA by TrmD is a posttranscriptional event. The efficient methylation by the enzyme on transcripts of tRNA [23], lacking any posttranscriptional modifications, supports the notion that the enzyme catalyzes a primary reaction on tRNA without the requirement for any prior modifications. This is unlike secondary modification reactions on tRNA, where a prior modification is necessary. The synthesis of a complex modified base, such as cmo^5U (5-carboxymethoxy uridine) or $mcmo^5$ (5-methoxy-carbonyl methoxy uridine) at the tRNA wobble position, is initiated with the primary reaction, followed by a series of secondary reactions [39,40]. To understand the structural basis of how TrmD recognizes its tRNA substrate, a comprehensive analysis of individual domains of the nucleic acid was performed [21]. This analysis revealed that TrmD requires only a stem–loop structure, in contrast to the requirement for the complete tRNA structure by Trm5 [21]. Specifically, TrmD recognizes the anticodon stem–loop (ASL) of tRNA, which consists of 5 bp of the stem and 7 nucleotides in the anticodon loop. However, this simple ASL is not sufficient, but an extension of the stem to 9 bp is necessary [21]. The extended structure recapitulates the composite structure of the D stem (typically 4 bp) coaxially stacking on the ASL in the native tRNA L-shape structure, which is the core structure of the vertical arm of the L-shape. This notion is intriguing, suggesting that TrmD recognizes the D-ASL vertical arm of the L in a model consistent with the ternary crystal structure of TrmD in complex with tRNA [17]. Indeed, TrmD makes contact only with the vertical arm of the L in the crystal structure, and the contact is mediated by interactions with the phosphodiester backbone of the nucleic acid in a mechanism known as "indirect readout" [41]. It is in contrast to a "direct readout" mechanism, where enzyme makes contact with nucleobases of the nucleic acid substrate.

Given that TrmD only requires the D-ASL structure for recognition of the tRNA substrate, the possibility exists that the m^1G37 methylation may occur before the complete transcription of a tRNA. No data are available to date to address this possibility. The only available information is that the m^1G37 methylation by TrmD does not need any other prior modification, aminoacylation, or even CCA addition to tRNA [21]. We do not yet know if the methylation can occur cotranscriptionally, or before tRNA processing at the 5′- or 3′-end. More research is necessary to understand the TrmD reaction relative to the biogenesis of tRNA processing and maturation.

TrmD synthesizes the methylated m^1G37 on bacterial tRNAs that contain both G37 and a preceding G36, the 3′-nucleotide of the anticodon [22]. Analysis of the TrmD–tRNA complex reveals that, after G37 is flipped out from the anticodon loop and is recognized by the flexible linker of chain B, the open space left between positions 35 and 38 is occupied by G36 in a stacked position stabilized by a pocket near the trefoil knot in chain A [17] (Fig. 4A). Thus, while G37 is the substrate for methylation, G36 provides additional interactions with the enzyme to induce further conformational changes that strengthen the positioning of G37. Importantly, the resulting m^1G37 has the ability to remodel the structure of the tRNA anticodon loop relative to the unmethylated G37. In X-ray crystal structures of the ASL domain of $tRNA^{Pro/CGG}$ (CGG: anticodon) in complex with a *Thermus thermophilus* ribosome [42], the unmethylated G37 prevents the ASL from establishing the typical hydrogen-bond interaction between the O^2 atom of U32 and the N^6 atom of A38, thus rendering the nucleotides from 30 to 32 on the 5′-side of the ASL disordered and invisible (Fig. 4B). In contrast, the methylated m^1G37 is able to remodel the ASL structure and restore the U32–A38 interaction to the conformation consistent with a cognate codon–anticodon pair (Fig. 4B). The lack of structural ordering of the ASL as induced by the unmethylated G37 is striking and is also observed in a mutant structure of ASL containing an extra G37.5 nucleotide (the mutant denoted as [ASL + G37.5] = ASL with an insertion of G between G37 and A38) (Fig. 4B). This mutant [ASL + G37.5] structure is prone to +1-frameshifting [42], most likely because the G37.5 nucleotide forms an aberrant base pair with U32 by displacing A38 away from the

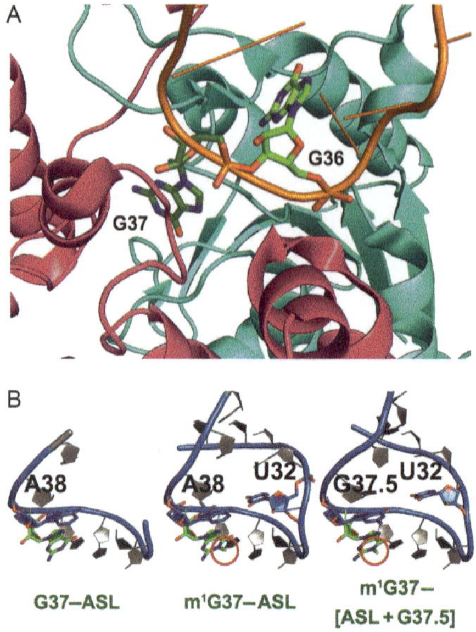

Fig. 4 G37 in TrmD and in the tRNA anticodon stem–loop structure. (A) G37 binding to TrmD organizes the flexible linker of chain B (*magenta*) and projects G36 to the trefoil knot of chain A (*cyan*) to stabilize the entire tRNA molecule bound to the enzyme (PDB: 4YVI). (B) The unmethylated G37 in the anticodon stem–loop (ASL) structure of $tRNA^{Pro/CGG}$ (G37–ASL) prevents the interaction between U32 and A38 and renders the 5′-side of the ASL disordered and invisible (*left*, PDB: 4P70). The methylated m^1G37 remodels the ASL structure (m^1G37–ASL) to allow U32–A38 base pairing as in a canonical structure (*middle*, PDB: 4LT8). The insertion of G37.5 to the ASL, resulting in the structure of m^1G37-[ASL + G37.5], disrupts U32–A38 pairing to form the aberrant U32–G37.5 pairing in a disordered structure (*right*, PDB: 4L47) similar to that in the G37–ASL structure.

normal position (Fig. 4B). Comparison of all three examples indicates that a signature for tRNA to shift to the +1-frame is the loss of the U32–A38 base pair in the ALS. By restoring this base pair, m^1G37 provides a mechanism to suppress +1-frameshift.

5. RIBOSOME FRAMESHIFTING IN THE ABSENCE OF TrmD

The TrmD product m^1G37-tRNA is an integral component of translational reading frame accuracy in bacteria [10–12]. In normal conditions, each bacterial ribosome decodes three mRNA nucleotides into a single amino acid at a rapid rate of ~20 residues per second, moving the associated tRNA from the A-site (the aminoacyl-tRNA site), to the P-site (the peptidyl-tRNA site), and through the E-site (the exit site). However, despite these complex and dynamic movements, an *E. coli* ribosome makes infrequent frameshifts in either the 5′ or 3′ direction. The shift by one nucleotide in the 5′ direction results in the ribosome moving backward with a −1 frameshift, while the shift by one nucleotide in the 3′ direction results in the ribosome moving forward with a +1 frameshift. Notably, bacteria *do* use frameshifts as a "programmed" mechanism to regulate gene expression. Examples include the programmed +1 frameshift for expression of the *prfB* gene (for expression of release factor RF2) [43,44] and the −1 frameshift for expression of the *dnaX* gene (for expression of DNA polymerase III) [45,46]. However, nonprogrammed frameshifts are considered translational errors, usually arising from shifting of a tRNA–ribosome complex on slippery mRNA sequences. Unlike missense errors, which replace one amino acid with another but still permit continued synthesis to the full-length protein, a frameshift error is deleterious, changing the reading frame and introducing premature termination codons. The frequency of nonprogrammed frameshifts is typically low, estimated to be less than one per 30,000 amino acids (or less than 0.003%) over all sequence contexts [47]. The methylated m^1G37-tRNA is a key suppressor of +1 frameshift errors.

One of the most frameshift-prone mRNA sequences is CC[C/U]-[C/U], which codes for proline (Pro). In *E. coli* and throughout bacterial organisms, all three isoacceptors of $tRNA^{Pro}$ contain m^1G37 in the natural form (Fig. 5). The others that naturally contain m^1G37 are the GAG and CAG isoacceptors of $tRNA^{Leu}$ and the CCG isoacceptor of $tRNA^{Arg}$. The consistent association of m^1G37 with $tRNA^{Pro}$ emphasizes the importance of the methylation for translation of Pro codons. Of the three isoacceptors of $tRNA^{Pro}$, the UGG isoacceptor reads all four Pro codons (5′-CCN) through the use of cmo^5U34 at the wobble position. This isoacceptor is essential for growth [48], and without m^1G37, it is the most shift-prone among the three [11]. For example, in the absence of m^1G37, the UGG isoacceptor can read the mRNA sequence CC[C/U]-N either in the 0-frame or the +1-frame with a similar free energy of stabilization, indicating a minimum energetic penalty for the tRNA to shift to the +1-frame. The next most shift-prone tRNA is the GGG isoacceptor [10], which can read the mRNA sequence CC[C/U]-[C/U] with identical stability in the 0- and +1-frame. Among the total sense codons in protein-coding genes, CC[C/U]-N occurs 17,000 times in the K12 genome of *E. coli* and 20,000 times in the LT2 genome of *Salmonella enterica serovar*

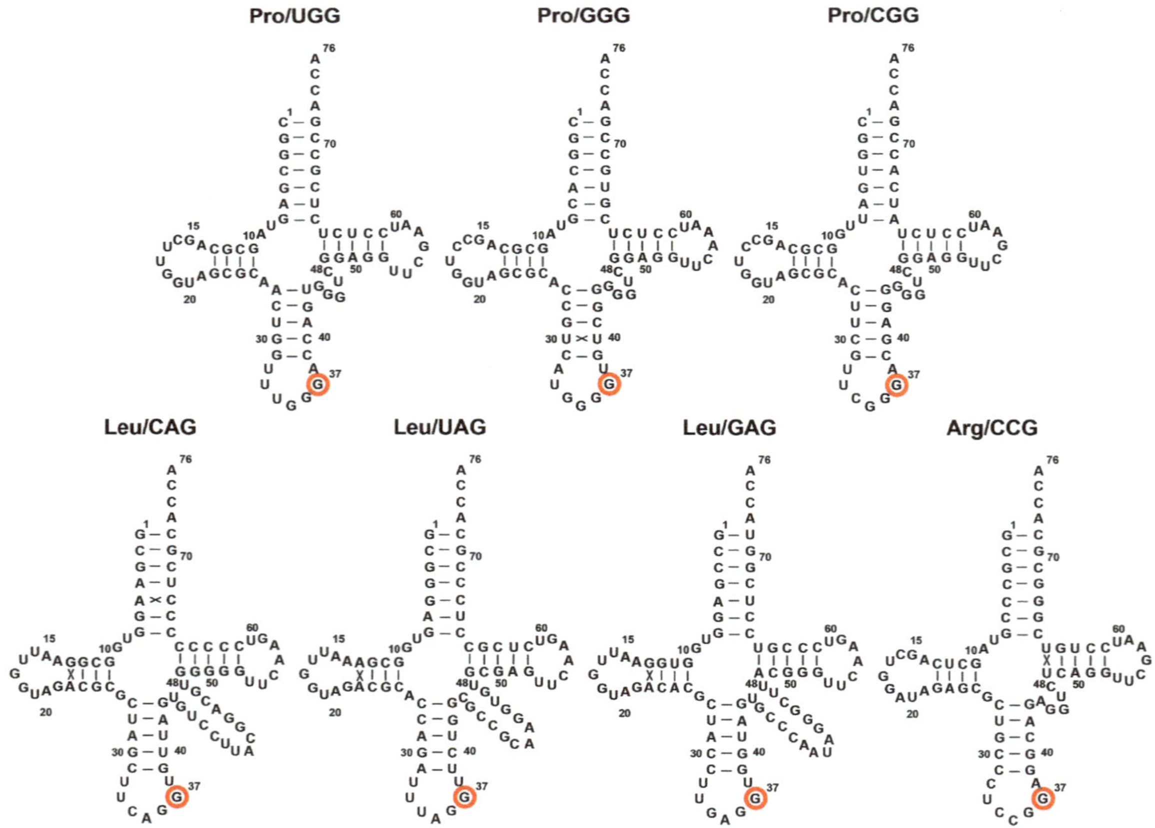

Fig. 5 Sequence and cloverleaf structure of *E. coli* tRNA species that are substrates for TrmD, including the Leu/UAG isoacceptor that is most likely a substrate as well. The G37 base to be methylated to m^1G37 is marked with a *red circle*. The numbering is based on the structure of yeast $tRNA^{Phe}$ [88].

Typhimurium (hereafter referred to as *Salmonella*). These high frequencies of occurrence pose a major challenge for translating ribosomes to suppress +1-frameshifting. Even the less frequent codon sequence CC[C/U]-[C/U] occurs 2300 times in the *E. coli* genome. In each natural occurrence, the CC[C/U]-N or CC[C/U]-[C/U] sequence can be directly next to the start codon AUG at the second position, or within a short distance from the start codon, or further downstream from the start codon.

Using the CC[C/U]-[C/U] sequence as an example, a cell-based reporter assay has examined its propensity of inducing ribosomal frameshifting at various positions throughout a reporter gene [10]. The results show that the sequence has the highest propensity of inducing frameshifts at the second codon position, when cells are lacking m^1G37-$tRNA^{Pro}$ [10]. At this position, the frequency of +1-frameshifting is ~1% when cells synthesize m^1G37-tRNA, but the frequency is raised by almost 10-fold when cells lose the methylation. At any of the downstream positions, the effect of m^1G37-tRNA is smaller (three- to four-fold). Thus, while m^1G37-tRNA is important for suppressing +1-frameshifting throughout all sequence contexts, it has the strongest effect at the second codon. In a broader perspective, the CC[C/U]-N sequence is also slippery, and it is read by the UGG isoacceptor in the absence of m^1G37. A genome-wide analysis of *E. coli* K12 bacteria identifies 48 protein-coding genes with the CC[C/U]-N sequence at the second codon, representing a frequency of 1.1% (out of 4289 genes). Some of these genes are essential themselves (e.g., *lolB*, a conserved outer membrane protein in Gram-negative bacteria). Maintenance of reading frame accuracy of these genes during translation on the ribosome is expected to depend on the presence of m^1G37-$tRNA^{Pro}$ (Fig. 6).

The second codon in bacterial mRNAs is unique. Its translation requires stable positioning of the initiator tRNA at the ribosomal P-site. This stabilization is enforced by EF-P, the translation factor that stimulates the first peptide bond formation [49]. Indeed, the cell-based reporter assay shows that EF-P also suppresses +1-frameshifting and that it is most effective at the second codon [10]. Notably, the suppression of +1-frameshifts by EF-P is an action on one Pro codon within the context of a slippery sequence and it is distinct from the other function of the factor, which is to release ribosome stalling when encountering poly-Pro codons [50,51].

The importance of m^1G37 and EF-P for suppressing +1-frameshifts at the second codon position is further demonstrated in kinetic assays, which used purified ribosomes and protein factors to reconstitute an *E. coli*

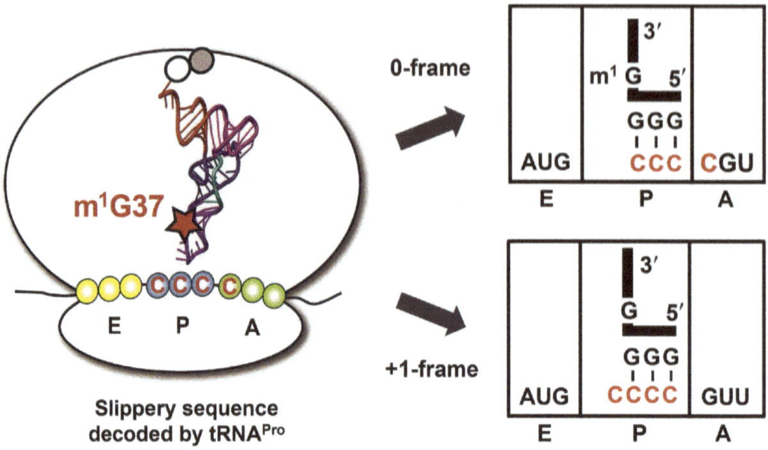

Fig. 6 Suppression of ribosomal +1-frameshifting by m^{1}G37-tRNA. On a slippery mRNA sequence AUG-CCC-C, the methylated m^{1}G37-tRNA^{Pro} maintains the correct reading frame (0-frame), whereas the unmethylated G37-tRNA^{Pro} has a high propensity to shift to the +1-frame, which most frequently occurs during the tRNA sitting at the P-site next to an empty A-site. The tRNA is shown in the L-shape with the anticodon and G37 or m^{1}G37 highlighted.

translational apparatus programmed by an mRNA [10,11]. These assays show that both m^{1}G37 and EF-P execute suppression of +1-frameshifts most effectively at the second codon position relative to other positions. Among all natural posttranscriptional modifications in tRNA^{Pro}, m^{1}G37 is the single determinant that suppresses +1-frameshifts [10]. However, while m^{1}G37 alone is sufficient to completely arrest +1-frameshifts of the UGG isoacceptor, it is insufficient for the GGG isoacceptor and requires the assistance of EF-P [10]. The kinetic assays also reveal that +1-frameshifts can occur in one of two mechanisms. The slow mechanism is during tRNA^{Pro} sitting at the P-site next to an empty A-site (Fig. 6). This slow shift occurs on a timescale 10^3-fold slower relative to peptide bond formation [10], indicating that it is relevant only when cells are starved of nutrients and cannot synthesize sufficient amounts of Pro-tRNA^{Pro} to load up the A-site. In contrast, the fast mechanism is during tRNA^{Pro} translocation from the A-site to the P-site and it occurs on a timescale comparable to that of peptide bond formation [10], indicating that it has the potential to shift the reading frame even in normal cellular condition.

Both cell-based and kinetic assays have highlighted the propensity of the second codon position to promote +1-frameshifts. This codon position is associated with the first translocation after the first peptide bond is made, and the translating ribosome is ready to move through the second codon to position it at the P-site and open up the A-site for the third codon. The first translocation has a higher probability of shifting on the registry of the mRNA relative to later translocation events for several reasons. The first translocation lacks an E-site tRNA, which helps to maintain the reading frame [52,53], it retains residual interaction with the Shine–Dalgarno sequence [54], and it involves the structurally and dynamically unfavorable movements of transitioning the initiator tRNA from the P-site to the E-site [55]. The first translocation is also the defining moment for the translating ribosome to enter the elongation phase, and it is important for positioning the ribosome on the correct reading frame for downstream synthesis of full-length protein. Both TrmD and EF-P are strictly conserved in bacteria: while TrmD is essential for cell growth [10], EF-P is required for robust cell viability for most bacteria [56]. The dedicated action of TrmD and EF-P at the first translocation indicates an evolutionarily conserved process to take action as early as possible to safeguard the reading frame accuracy of protein synthesis.

6. Mg^{2+} DEPENDENCE FOR METHYL TRANSFER

Methyl transfer by TrmD requires Mg^{2+} in the catalytic mechanism [26]. This is highly unusual, because the methyl group of AdoMet is already positively charged and can be easily transferred without the need for metal ions. Indeed, the majority of AdoMet-dependent methyl transferases (e.g., Trm5) require no metal ion [26]. Some exceptions *do* exist, for instance, for methyl transferases that use Mg^{2+} to catalyze *O*-methyl transfer to catechols and flavonoids [57–59], and N^6-methyl transfer to adenine in DNA [60,61]. In these exceptions, metal ions are used to modulate the selection and binding of the target substrates but not catalysis. The Mg^{2+} dependence of TrmD is also distinct from many nucleic acid metabolic enzymes (e.g., DNA and RNA polymerase, restriction enzymes, and catalytic ribozymes) [62,63]. While the latter enzymes use Mg^{2+} in their reaction, they use the metal ion to both stabilize the nucleic acid or nucleotide substrate and promote catalysis. Instead, TrmD uses Mg^{2+} exclusively to promote catalysis in the transition state of methyl transfer [26]. The chemical novelty of this transition state is that Mg^{2+} is a direct ligand to the nucleobase G37 in the tRNA, whereas DNA and RNA polymerases and related enzymes coordinate the metal ion with the phosphodiester backbone of the nucleic acid substrate.

In addition to Mg^{2+}, TrmD can also use Ca^{2+} and Mn^{2+} as an active ion, but not Ni^{2+} or Co^{2+} [26]. Using *Ec*TrmD as an example, analysis of the enzyme activity as a function of metal ion concentration shows that the Mg^{2+}-dependent activity is maximal at the intracellular concentration of the metal ion in *E. coli* (2–3 mM), but that the Ca^{2+}- and Mn^{2+}-dependent activity is below 1% (for Ca^{2+}) or around 20% (for Mn^{2+}) of the maximum at the intracellular concentration of the respective metal ion (0.1 and 80 μM) [64,65]. Thus, Mg^{2+} is the only physiologically relevant metal ion, given that TrmD is essential for cell growth and a reduction of its activity to below 20% is sufficient to cause cell death [38]. Interestingly, the binding stoichiometry of Mg^{2+} to TrmD, as measured by the enzyme activity as a function of the metal ion, shows that it is one metal ion required for the activity of one dimer [26]. This is consistent with the stoichiometry of one methyl transfer per dimer [23].

Direct fluorescence-based binding assays with AdoMet or tRNA show that the single Mg^{2+} required for methyl transfer is not necessary for substrate binding [26]. Even inactive metal ions such as Co^{2+} or Ni^{2+}, or even the metal-chelating EDTA, promote the same binding affinity. Instead, the single Mg^{2+} required for methyl transfer is involved in the abstraction of the N^1 proton from G37-tRNA, which is likely the rate-limiting step of the TrmD-catalyzed methyl transfer [26]. The Mg^{2+} reaction has a solvent deuterium kinetic isotope effect (KIE) of 5–6, the Mn^{2+} reaction has a KIE of 1.8, and the Ca^{2+} reaction has a KIE of 3–4 [26]. The different magnitudes of KIE with the three metal ions reflect differences in their ability to stabilize a common species in the deprotonation of G37. The largest KIE for the Mg^{2+} reaction indicates that the reaction is most sensitive to perturbation of the proton abstraction, whereas the smaller KIEs for the Mn^{2+} and Ca^{2+} reactions suggest that other steps (e.g., rearrangement of catalytic residues) may have become partially rate limiting.

Metal rescue experiments show that Mg^{2+} stabilizes the developing negative charge on the O^6 of G37 during proton abstraction [26]. The principle of metal rescue is that hard metal ions (e.g., Mg^{2+}) prefer coordination with oxygen, whereas soft metal ions (e.g., Co^{2+}) prefer coordination with sulfur [63]. Indeed, substitution of O^6 of G37-tRNA with S^6 reduces the rate of the Mg^{2+} reaction. In contrast, while Co^{2+} is inactive with O^6 of G37-tRNA, it is fully active with S^6 of G37-tRNA [63]. These results demonstrate a direct role of the metal ion in the coordination with the atom at the 6th position of G37. Further, in the fully active pair of Co^{2+}-TrmD with the s^6G37-tRNA substrate, the formation of a metal–thiolate bond is indicated by a unique UV–visible absorption spectral change for the charge transfer from the sulfur to the transition metal ion with d-shell electrons [63]. Further analysis shows that the R154A and D169A substitutions each abolish the charge transfer, indicating that the native residue of each is involved in catalysis [63]. This is supported by the tRNA-bound crystal structure of the enzyme [17], where the strictly conserved R154 provides a positive charge to stabilize the O^6 of G37, whereas the negative charged D169 (or E169 in some bacterial species) provides the general base that abstracts the N^1 proton from G37.

The catalytic role of Mg^{2+} to TrmD is transient, with a K_d (Mg^{2+}) of 0.68 ± 0.03 mM for the *E. coli* enzyme [26] and a K_d (Mg^{2+}) of 0.42 ± 0.02 mM for the *Salmonella* enzyme. None of the available crystal structures of TrmD contain a metal ion even though some complexes were soaked with 0.2 M Ca^{2+} [20]. The transient nature of Mg^{2+} is consistent with the proposed catalytic mechanism involving G37-tRNA [26]. In this mechanism (Fig. 7), D169 is the general base to abstract the N^1 proton from

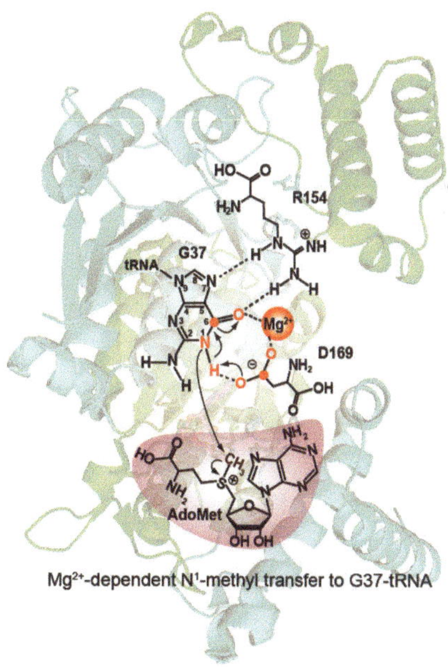

Fig. 7 The proposed Mg^{2+} in the catalytic mechanism of TrmD, involving an eight-membered ring that consists of the N^1, C6, and O^6 of the guanine ring, the hydrogen bond to the imino group of N^1, the two coordination bonds of Mg^{2+}, and the three atoms of the carboxylate of the general base D169. Atomic numbering is indicated for guanine.

G37, while the deprotonation is accompanied by developing electron density on the O^6 of G37. The developing negative charge on O^6 of G37 is stabilized through coordination with Mg^{2+} and by hydrogen-bond interaction with the side chain of R154. The charge stabilization of O^6 in turn facilitates Mg^{2+} to coordinate with the general base D169 and to help it to align more properly for proton abstraction. The activated N^1 nucleophile is then poised for nucleophilic attack on the sulfonium center of AdoMet, resulting in synthesis of m^1G37-tRNA and release of AdoHcy. The rate-limiting step is assigned to the action of D169, rather than to the protonation of the leaving group, due to the importance of D169 and the increase of activity as the proton concentration is lowered. The geometry of this catalytic step involves an eight-membered ring that may be nonplanar, such that Mg^{2+} can be the last member to join the transition state, providing a control for methyl transfer only when both G37-tRNA and AdoMet are appropriately bound.

7. REGULATION OF Mg^{2+} TRANSPORT IN *SALMONELLA*

The biological function of the Mg^{2+} dependence of TrmD has been implicated in regulating metal ion transport in *Salmonella* [66]. This pathogen is the etiologic agent of human gastroenteritis, causing 1.2 million illnesses, 23,000 cases of hospitalization, and 450 deaths in the United States each year. Mg^{2+} transport is important for *Salmonella* survival, infection, and virulence in host cells. The host compartment for *Salmonella* is low in pH, low in Mg^{2+}, and contains host antimicrobial peptides. This environment activates the PhoPQ two-component system to activate transcription of virulence genes, as well as transcription of the Mg^{2+} transporter *mgtA* gene and the *mgtCBR* virulence operon [67–71]. Transcription of *mgtA* and *mgtCBR* is then further regulated by translation of the 5′-leader of each mRNA in a mechanism that determines transcription attenuation of the downstream gene [72–77] (Fig. 8A). In this mechanism, the 5′-leader of *mgtA* is *mgtL*, coding for a small peptide (Fig. 8A). Rapid translation of the 5′-leader *mgtL* would expose a Rho-termination (rut: *Rho*-utilization) site, which terminates transcription ahead of the structural gene *mgtA*, whereas slow or stalled translation of *mgtL* would induce a conformational change of the 5′-leader mRNA and preclude the rut site, allowing transcription to continue through *mgtA* [76,78]. This translation-coupled transcription attenuation mechanism is reminiscent of the regulatory function of the short open reading frame in the *trp*, *his*, and other amino acid biosynthetic operons [79], in which the efficiency of translation of the leader peptide

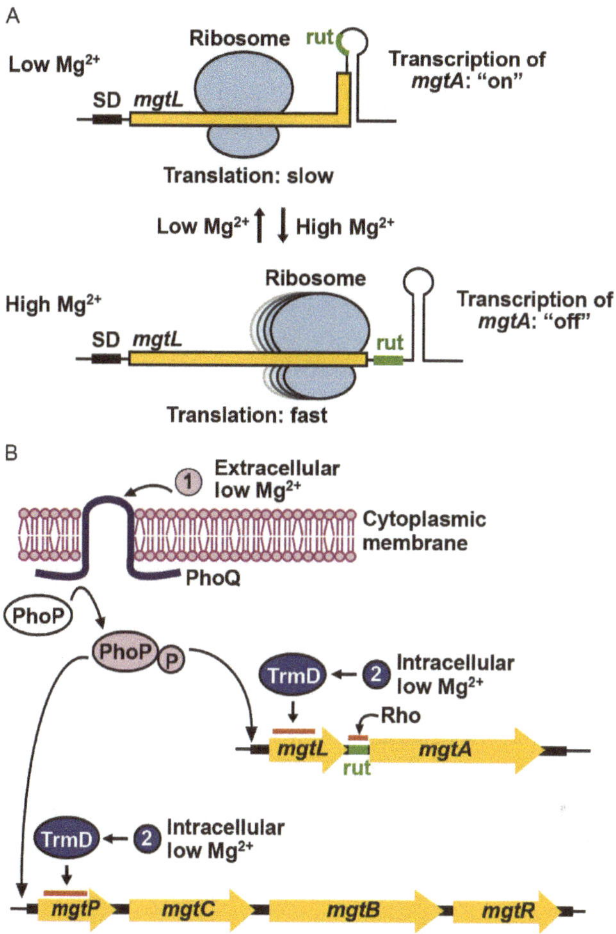

Fig. 8 A proposed mechanism of Mg^{2+} sensing of *Salmonella*. (A) Regulation of *mgtA* transcription in response to Mg^{2+} level is mediated by the translation speed of the 5′-leader mRNA *mgtL* sequence, which codes for a short peptide. At low Mg^{2+}, ribosomal translation of *mgtL* is slow, sequestering the rut sequence in a stem–loop structure and excluding the Rho-dependent transcription termination to permit transcription of *mgtA* in the "on" state. At high Mg^{2+}, ribosomal translation of *mgtL* is fast, exposing the rut sequence and resulting in the Rho-dependent transcription termination of *mgtA*. (B) The two-component system PhoPQ provides a sensor for extracellular Mg^{2+}. At low extracellular Mg^{2+}, the activated PhoQ phosphorylates PhoP to turn on transcription of the Mg^{2+} transporter gene *mgtA* as well as the virulence operon *mgtCBR*, where *mgtB* is another Mg^{2+} transporter gene. Transcription of *mgtA* is then regulated by translation of the 5′-leader sequence *mgtL*, which encodes a small peptide. Rapid translation of *mgtL* exposes the Rho-utilization (rut) sequence and terminates transcription ahead of *mgtA*. In contrast, slow and incomplete translation of *mgtL* sequesters the rut sequence and allows transcription into the structural gene. The Mg^{2+}-dependent methyl transferase activity of TrmD can act as an intracellular sensor that regulates the speed of translation of *mgtL*.

determines whether transcription is terminated at the rut site ahead of the structural gene or allowed to read-through the rut site and into the structural gene. Understanding how both transcription and translation regulate the expression of *mgtA*, the major Mg^{2+} transporter gene [80], is necessary to improve our management of *Salmonella* infection.

For *mgtA* expression, the 5′-leader *mgtL* contains multiple codons that require the TrmD product m^1G37-tRNA for efficient and accurate translation (Fig. 8B). These include codons for Pro, Leu, and Arg [72]. The dominance of Pro codons in the short MgtL peptide (4 out of 17 residues) is striking. Indeed, the reduction of the level of Pro stalls translation of *mgtL* and activates transcription of *mgtA* [72]. However, the physiological connection between the abundance of Pro and the availability of Mg^{2+} is not obvious. Instead, recent data show that it is the efficiency of translation of the Pro codons in *mgtL* that is subject to Mg^{2+} regulation [66]. The efficiency of translation involves not just the level of Pro, but also the quality of translation when Pro-$tRNA^{Pro}$ is reading a codon on the ribosome. In fact, the data suggest that the quality of translation of Pro-$tRNA^{Pro}$ is worthy of more consideration. For example, different types of mutations in *mgtL* have the ability to influence the expression of *mgtA* [66]. A mutation in *mgtL* resulting in the sequence of three Pro codons in a row, which is a strong signal to stall the translating ribosome [50,51], activates transcription of *mgtA*. In contrast, substitution of three of the four natural Pro codons in *mgtL* reduces transcription of *mgtA*. Deletion of EF-P, which would promote +1-frameshifts and ribosome stalling, induces the expression of *mgtA*. This result of EF-P, which is related to the quality of translation of Pro codons, argues against the level of Pro in determining the response to Mg^{2+}. Additionally, lesions in *rpmA* and *rpmE,* encoding ribosomal proteins L27 and L37, respectively, activate transcription of *mgtA*, possibly by reducing ribosomal protein synthesis. Most importantly, the *trmD-S88L* mutation, which reduces the synthesis of m^1G37-tRNA [38], activates the transcription of *mgtA* to a constitutively high level.

Of all of the mutations that were examined recently [66], only the mutation in TrmD is associated with an Mg^{2+}-dependent activity, suggesting a role for TrmD in Mg^{2+} sensing for regulation of the metal ion transport in *Salmonella.* This role is consistent with the observation that, while the substitution of Pro codons in *mgtL* abolishes the response to Pro, it does not abolish the response to Mg^{2+}, but simply reduce the response [72]. The reduction of the response can be explained by the presence of the remaining Leu and Arg codons in *mgtL* that are still subject to the regulation by the

Mg^{2+}-dependent activity of TrmD. The data suggest that TrmD can be an intracellular sensor for Mg^{2+} (Fig. 8B). In high Mg^{2+}, TrmD would be active and translation of the 5′-leader *mgtL* would be fast, leading to transcription termination ahead of the structural gene. In limiting Mg^{2+}, however, TrmD would be inactivated and the deficiency in m^1G37-tRNA would halt the translation of the 5′-leader *mgtL,* favoring transcription read-through into the *mgtA* gene.

The key advantage for the proposed mechanism is the ability of *Salmonella* to directly sense changes in the intracellular demand for Mg^{2+}. Notably, when cell growth is rapid, intracellular Mg^{2+} is depleted by cell divisions, even though the metal ion is abundant outside. The coordination of TrmD sensing of intracellular Mg^{2+} with the PhoPQ sensing of the extracellular Mg^{2+} constitutes a dual-sensing model. The regulation of bacteria K^+ homeostasis is an example of a dual-sensing model, where one single enzyme (KdpD) with two separate domains for sensing extra- and intracellular K^+ is used to regulate the transcription of the main K^+ transporter gene [81]. The potential synergy between TrmD sensing and PhoPQ sensing should confer the bacteria with robust homeostasis of Mg^{2+} for fitness in host cells. If this model is supported, it will have wide-reaching impact on bacterial pathogenesis, because the PhoPQ system [82,83], the 5′-leader of *mgtA*, and the TrmD enzyme are all broadly conserved among pathogenic species of *Enterobacteriaceae.*

8. CONCLUSION

TrmD is an important and critical enzyme for bacteria. It is ranked as a high-priority antimicrobial drug target [29], due to its essentiality for bacterial growth, strict conservation across a wide range of bacterial species, different structure and mechanism from its human counterpart Trm5, and the possession of an AdoMet binding site, where drug-like small molecules can bind to. However, while the pharmaceutical industry has attempted to target TrmD, progress has stalled [84], due to the lack of a TrmD-specific strategy. Insights into such a strategy are now beginning to emerge based on data from the extensive enzymology and cell-based studies as summarized here. The new focus should be to understand how TrmD differs from Trm5 and to determine the genes whose translation is most sensitive to targeting of TrmD. The unique trefoil knot of TrmD is a major difference from the active site of Trm5, and it is also distinct from the trefoil knot of other methyl transferases [26]. The trefoil knot of TrmD binds AdoMet in the unusual bent shape, and this binding is the basis to initiate substrate signaling to catalyze methyl transfer. The bent shape offers structural novelty and diversity from the conventional AdoMet analogs, and it should be explored for drug design. The Mg^{2+}-dependent catalysis of TrmD is unique among known AdoMet-dependent methyl transferases. This Mg^{2+} dependence appears to have a role in regulating the transport and homeostasis of the metal ion when *Salmonella* infects its host. Given that successful drug targeting of HIV integrase has been achieved by disrupting the divalent ion interaction at the active site [85–87], drug targeting of the metal ion-binding site in TrmD would provide an attractive antimicrobial strategy. More efforts should be made to understand the Mg^{2+} dependence of TrmD and its biology in bacteria. The TrmD product m^1G37-tRNA is the major suppressor of +1-frameshifts at Pro codons, but not all Pro codons share the same propensity of inducing frameshifting. Identifying Pro codons in bacterial genes that are most dependent on TrmD for translation would be a major step forward to develop a TrmD-specific strategy that will improve human health in the global population.

ACKNOWLEDGMENTS

We thank the support of US National Institutes of Health grants GM108972 and GM114343 (to Y.-M.H.) and European Molecular Biology Organization (EMBO) installation grant 2057 and National Science Center grant Sonata BIS 2012/07/E/NZ1/01900 (to J.I.S). I.M. is a Japan Society for the Promotion of Science (JSPS) Overseas Research Fellow. The content is solely the responsibility of the authors and does not represent the official views of the granting agencies.

REFERENCES

[1] L.C. Keffer-Wilkes, G.R. Veerareddygari, U. Kothe, RNA modification enzyme TruB is a tRNA chaperone, Proc. Natl. Acad. Sci. U.S.A. 113 (2016) 14306–14311. PubMed, 27849601.

[2] G.L. Winsor, D.K. Lam, L. Fleming, R. Lo, M.D. Whiteside, N.Y. Yu, R.E. Hancock, F.S. Brinkman, Pseudomonas Genome Database: improved comparative analysis and population genomics capability for Pseudomonas genomes, Nucleic Acids Res. 39 (2011) D596–D600. PubMed, 20929876.

[3] V. de Berardinis, D. Vallenet, V. Castelli, M. Besnard, A. Pinet, C. Cruaud, S. Samair, C. Lechaplais, G. Gyapay, C. Richez, M. Durot, A. Kreimeyer, et al., A complete collection of single-gene deletion mutants of Acinetobacter baylyi ADP1, Mol. Syst. Biol. 4 (2008) 174. PubMed, 18319726.

[4] R.A. Forsyth, R.J. Haselbeck, K.L. Ohlsen, R.T. Yamamoto, H. Xu, J.D. Trawick, D. Wall, L. Wang, V. Brown-Driver, J.M. Froelich, K.G. C, P. King, et al., A genome-wide strategy for the identification of essential genes in *Staphylococcus aureus*, Mol. Microbiol. 43 (2002) 1387–1400. PubMed, 11952893.

[5] T. Baba, T. Ara, M. Hasegawa, Y. Takai, Y. Okumura, M. Baba, K.A. Datsenko, M. Tomita, B.L. Wanner, H. Mori, Construction of *Escherichia coli* K-12 in-frame, single-gene knockout mutants: the Keio collection, Mol. Syst. Biol. 2 (2006) 2006.0008. PubMed, 16738554.
[6] C.M. Sassetti, D.H. Boyd, E.J. Rubin, Genes required for mycobacterial growth defined by high density mutagenesis, Mol. Microbiol. 48 (2003) 77–84. PubMed, 12657046.
[7] J.A. Thanassi, S.L. Hartman-Neumann, T.J. Dougherty, B.A. Dougherty, M.J. Pucci, Identification of 113 conserved essential genes using a high-throughput gene disruption system in Streptococcus pneumoniae, Nucleic Acids Res. 30 (2002) 3152–3162. PubMed, 12136097.
[8] S.Y. Gerdes, M.D. Scholle, J.W. Campbell, G. Balazsi, E. Ravasz, M.D. Daugherty, A.L. Somera, N.C. Kyrpides, I. Anderson, M.S. Gelfand, A. Bhattacharya, V. Kapatral, et al., Experimental determination and system level analysis of essential genes in *Escherichia coli* MG1655, J. Bacteriol. 185 (2003) 5673–5684. PubMed, 13129938.
[9] A.S. Bystrom, G.R. Bjork, The structural gene (trmD) for the tRNA(m1G)methyltransferase is part of a four polypeptide operon in *Escherichia coli* K-12, Mol. Gen. Genet. 188 (1982) 447–454. PubMed, 6298574.
[10] H.B. Gamper, I. Masuda, M. Frenkel-Morgenstern, Y.M. Hou, Maintenance of protein synthesis reading frame by EF-P and m(1)G37-tRNA, Nat. Commun. 6 (2015) 7226. PubMed, 26009254.
[11] H.B. Gamper, I. Masuda, M. Frenkel-Morgenstern, Y.M. Hou, The UGG isoacceptor of tRNAPro is naturally prone to frameshifts, Int. J. Mol. Sci. 16 (2015) 14866–14883. PubMed, 26140378.
[12] G.R. Bjork, P.M. Wikstrom, A.S. Bystrom, Prevention of translational frameshifting by the modified nucleoside 1-methylguanosine, Science 244 (1989) 986–989. PubMed, 2471265.
[13] G.R. Bjork, K. Jacobsson, K. Nilsson, M.J. Johansson, A.S. Bystrom, O.P. Persson, A primordial tRNA modification required for the evolution of life? EMBO J. 20 (2001) 231–239. PubMed, 11226173.
[14] T. Christian, C. Evilia, S. Williams, Y.M. Hou, Distinct origins of tRNA(m1G37) methyltransferase, J. Mol. Biol. 339 (2004) 707–719. PubMed, 15165845.
[15] H.J. Ahn, H.W. Kim, H.J. Yoon, B.I. Lee, S.W. Suh, J.K. Yang, Crystal structure of tRNA(m1G37)methyltransferase: insights into tRNA recognition, EMBO J. 22 (2003) 2593–2603. PubMed, 12773376.
[16] P.A. Elkins, J.M. Watts, M. Zalacain, A. van Thiel, P.R. Vitazka, M. Redlak, C. Andraos-Selim, F. Rastinejad, W.M. Holmes, Insights into catalysis by a knotted TrmD tRNA methyltransferase, J. Mol. Biol. 333 (2003) 931–949. PubMed, 14583191.
[17] T. Ito, I. Masuda, K. Yoshida, S. Goto-Ito, S. Sekine, S.W. Suh, Y.M. Hou, S. Yokoyama, Structural basis for methyl-donor-dependent and sequence-specific binding to tRNA substrates by knotted methyltransferase TrmD, Proc. Natl. Acad. Sci. U.S.A. 112 (2015) E4197–E4205. PubMed, 26183229.
[18] S. Goto-Ito, T. Ito, R. Ishii, Y. Muto, Y. Bessho, S. Yokoyama, Crystal structure of archaeal tRNA(m(1)G37)methyltransferase aTrm5, Proteins 72 (2008) 1274–1289. PubMed, 18384044.
[19] S. Goto-Ito, T. Ito, M. Kuratani, Y. Bessho, S. Yokoyama, Tertiary structure checkpoint at anticodon loop modification in tRNA functional maturation, Nat. Struct. Mol. Biol. 16 (2009) 1109–1115. PubMed, 19749755.
[20] G. Lahoud, S. Goto-Ito, K. Yoshida, T. Ito, S. Yokoyama, Y.M. Hou, Differentiating analogous tRNA methyltransferases by fragments of the methyl donor, RNA 17 (2011) 1236–1246. PubMed, 21602303.

[21] T. Christian, Y.M. Hou, Distinct determinants of tRNA recognition by the TrmD and Trm5 methyl transferases, J. Mol. Biol. 373 (2007) 623–632. PubMed, 17868690.
[22] R. Sakaguchi, A. Giessing, Q. Dai, G. Lahoud, Z. Liutkeviciute, S. Klimasauskas, J. Piccirilli, F. Kirpekar, Y.M. Hou, Recognition of guanosine by dissimilar tRNA methyltransferases, RNA 18 (2012) 1687–1701. PubMed, 22847817.
[23] T. Christian, G. Lahoud, C. Liu, Y.M. Hou, Control of catalytic cycle by a pair of analogous tRNA modification enzymes, J. Mol. Biol. 400 (2010) 204–217. PubMed, 20452364.
[24] T. Christian, C. Evilia, Y.M. Hou, Catalysis by the second class of tRNA(m1G37) methyl transferase requires a conserved proline, Biochemistry 45 (2006) 7463–7473. PubMed, 16768442.
[25] T. Christian, G. Lahoud, C. Liu, K. Hoffmann, J.J. Perona, Y.M. Hou, Mechanism of N-methylation by the tRNA m1G37 methyltransferase Trm5, RNA 16 (2010) 2484–2492. PubMed, 20980671.
[26] R. Sakaguchi, G. Lahoud, T. Christian, H. Gamper, Y.M. Hou, A divalent metal ion-dependent N(1)-methyl transfer to G37-tRNA, Chem. Biol. 21 (2014) 1351–1360. PubMed, 25219964.
[27] T. Christian, H. Gamper, Y.M. Hou, Conservation of structure and mechanism by Trm5 enzymes, RNA 19 (2013) 1192–1199. PubMed, 23887145.
[28] M.Y. Galperin, D.R. Walker, E.V. Koonin, Analogous enzymes: independent inventions in enzyme evolution, Genome Res. 8 (1998) 779–790. PubMed, 9724324.
[29] T.A. White, D.B. Kell, Comparative genomic assessment of novel broad-spectrum targets for antibacterial drugs, Comp. Funct. Genomics 5 (2004) 304–327. PubMed, 18629165.
[30] M. Redlak, C. Andraos-Selim, R. Giege, C. Florentz, W.M. Holmes, Interaction of tRNA with tRNA (guanosine-1)methyltransferase: binding specificity determinants involve the dinucleotide G36pG37 and tertiary structure, Biochemistry 36 (1997) 8699–8709. PubMed, 9220956.
[31] T. Christian, R. Sakaguchi, A.P. Perlinska, G. Lahoud, T. Ito, E.A. Taylor, S. Yokoyama, J.I. Sulkowska, Y.M. Hou, Methyl transfer by substrate signaling from a knotted protein fold, Nat. Struct. Mol. Biol. 23 (2016) 941–948. PubMed, 27571175.
[32] S.I. Hauenstein, Y.M. Hou, J.J. Perona, The homotetrameric phosphoseryl-tRNA synthetase from Methanosarcina mazei exhibits half-of-the-sites activity, J. Biol. Chem. 283 (2008) 21997–22006. PubMed, 18559342.
[33] S.H. Ha, J.E. Ferrell Jr., Thresholds and ultrasensitivity from negative cooperativity, Science 352 (2016) 990–993. PubMed, 27174675.
[34] J.I. Sulkowska, E.J. Rawdon, K.C. Millett, J.N. Onuchic, A. Stasiak, Conservation of complex knotting and slipknotting patterns in proteins, Proc. Natl. Acad. Sci. U.S.A. 109 (2012) E1715–E1723. PubMed, 22685208.
[35] J.I. Sulkowska, P. Sulkowski, P. Szymczak, M. Cieplak, Stabilizing effect of knots on proteins, Proc. Natl. Acad. Sci. U.S.A. 105 (2008) 19714–19719. PubMed, 19064918.
[36] J. Liu, W. Wang, D.H. Shin, H. Yokota, R. Kim, S.H. Kim, Crystal structure of tRNA (m1G37) methyltransferase from Aquifex aeolicus at 2.6 A resolution: a novel methyltransferase fold, Proteins 53 (2003) 326–328. PubMed, 14517984.
[37] G.R. Bjork, K. Nilsson, 1-methylguanosine-deficient tRNA of Salmonella enterica serovar Typhimurium affects thiamine metabolism, J. Bacteriol. 185 (2003) 750–759. PubMed, 12533450.
[38] I. Masuda, R. Sakaguchi, C. Liu, H. Gamper, Y.M. Hou, The temperature sensitivity of a mutation in the essential tRNA modification enzyme tRNA methyltransferase D (TrmD), J. Biol. Chem. 288 (2013) 28987–28996. PubMed, 23986443.
[39] J. Kim, H. Xiao, J.B. Bonanno, C. Kalyanaraman, S. Brown, X. Tang, N.F. Al-Obaidi, Y. Patskovsky, P.C. Babbitt, M.P. Jacobson, Y.S. Lee, S.C. Almo, Structure-guided

discovery of the metabolite carboxy-SAM that modulates tRNA function, Nature 498 (2013) 123–126. PubMed, 23676670.

[40] Y. Sakai, K. Miyauchi, S. Kimura, T. Suzuki, Biogenesis and growth phase-dependent alteration of 5-methoxycarbonylmethoxyuridine in tRNA anticodons, Nucleic Acids Res. 44 (2016) 509–523. PubMed, 26681692.

[41] J.J. Perona, Y.M. Hou, Indirect readout of tRNA for aminoacylation, Biochemistry 46 (2007) 10419–10432. PubMed, 17718520.

[42] T. Maehigashi, J.A. Dunkle, S.J. Miles, C.M. Dunham, Structural insights into +1 frameshifting promoted by expanded or modification-deficient anticodon stem loops, Proc. Natl. Acad. Sci. U.S.A. 111 (2014) 12740–12745. PubMed, 25128388.

[43] W.J. Craigen, C.T. Caskey, Expression of peptide chain release factor 2 requires high-efficiency frameshift, Nature 322 (1986) 273–275. PubMed, 3736654.

[44] W.J. Craigen, R.G. Cook, W.P. Tate, C.T. Caskey, Bacterial peptide chain release factors: conserved primary structure and possible frameshift regulation of release factor 2, Proc. Natl. Acad. Sci. U.S.A. 82 (1985) 3616–3620. PubMed, 3889910.

[45] J. Chen, A. Petrov, M. Johansson, A. Tsai, S.E. O'Leary, J.D. Puglisi, Dynamic pathways of −1 translational frameshifting, Nature 512 (2014) 328–332. PubMed, 24919156.

[46] S. Yan, J.D. Wen, C. Bustamante, I. Tinoco Jr., Ribosome excursions during mRNA translocation mediate broad branching of frameshift pathways, Cell 160 (2015) 870–881. PubMed, 25703095.

[47] F. Jorgensen, C.G. Kurland, Processivity errors of gene expression in *Escherichia coli*, J. Mol. Biol. 215 (1990) 511–521. PubMed, 2121997.

[48] S.J. Nasvall, P. Chen, G.R. Bjork, The modified wobble nucleoside uridine-5-oxyacetic acid in tRNAPro(cmo5UGG) promotes reading of all four proline codons in vivo, RNA 10 (2004) 1662–1673. PubMed, 15383682.

[49] M.C. Ganoza, H. Aoki, Peptide bond synthesis: function of the efp gene product, Biol. Chem. 381 (2000) 553–559. PubMed, 10987361.

[50] L.K. Doerfel, I. Wohlgemuth, C. Kothe, F. Peske, H. Urlaub, M.V. Rodnina, EF-P is essential for rapid synthesis of proteins containing consecutive proline residues, Science 339 (2013) 85–88. PubMed, 23239624.

[51] S. Ude, J. Lassak, A.L. Starosta, T. Kraxenberger, D.N. Wilson, K. Jung, Translation elongation factor EF-P alleviates ribosome stalling at polyproline stretches, Science 339 (2013) 82–85. PubMed, 23239623.

[52] V. Marquez, D.N. Wilson, W.P. Tate, F. Triana-Alonso, K.H. Nierhaus, Maintaining the ribosomal reading frame: the influence of the E site during translational regulation of release factor 2, Cell 118 (2004) 45–55. PubMed, 15242643.

[53] A. Devaraj, S. Shoji, E.D. Holbrook, K. Fredrick, A role for the 30S subunit E site in maintenance of the translational reading frame, RNA 15 (2009) 255–265. PubMed, 19095617.

[54] G. Yusupova, L. Jenner, B. Rees, D. Moras, M. Yusupov, Structural basis for messenger RNA movement on the ribosome, Nature 444 (2006) 391–394. PubMed, 17051149.

[55] J. Fei, A.C. Richard, J.E. Bronson, R.L. Gonzalez Jr., Transfer RNA-mediated regulation of ribosome dynamics during protein synthesis, Nat. Struct. Mol. Biol. 18 (2011) 1043–1051. PubMed, 21857664.

[56] S.B. Zou, S.J. Hersch, H. Roy, J.B. Wiggers, A.S. Leung, S. Buranyi, J.L. Xie, K. Dare, M. Ibba, W.W. Navarre, Loss of elongation factor P disrupts bacterial outer membrane integrity, J. Bacteriol. 194 (2012) 413–425. PubMed, 22081389.

[57] J.L. Ferrer, C. Zubieta, R.A. Dixon, J.P. Noel, Crystal structures of alfalfa caffeoyl coenzyme A 3-O-methyltransferase, Plant Physiol. 137 (2005) 1009–1017. PubMed, 15734921.

[58] J.G. Kopycki, D. Rauh, A.A. Chumanevich, P. Neumann, T. Vogt, M.T. Stubbs, Biochemical and structural analysis of substrate promiscuity in plant Mg2+-dependent O-methyltransferases, J. Mol. Biol. 378 (2008) 154–164. PubMed, 18342334.

[59] R. Lukacin, U. Matern, S. Specker, T. Vogt, Cations modulate the substrate specificity of bifunctional class I O-methyltransferase from *Ammi majus*, FEBS Lett. 577 (2004) 367–370. PubMed, 15556611.
[60] P. Bist, D.N. Rao, Identification and mutational analysis of Mg2+ binding site in EcoP15I DNA methyltransferase: involvement in target base eversion, J. Biol. Chem. 278 (2003) 41837–41848. PubMed, 12917398.
[61] L.I. Fedoreyeva, B.F. Vanyushin, N(6)-Adenine DNA-methyltransferase in wheat seedlings, FEBS Lett. 514 (2002) 305–308. PubMed, 11943171.
[62] S.R. Das, J.A. Piccirilli, General acid catalysis by the hepatitis delta virus ribozyme, Nat. Chem. Biol. 1 (2005) 45–52. PubMed, 16407993.
[63] A.M. Pyle, Metal ions in the structure and function of RNA, J. Biol. Inorg. Chem. 7 (2002) 679–690. PubMed, 12203005.
[64] P. Gangola, B.P. Rosen, Maintenance of intracellular calcium in *Escherichia coli*, J. Biol. Chem. 262 (1987) 12570–12574. PubMed, 2442165.
[65] T.H. Hohle, M.R. O'Brian, The mntH gene encodes the major Mn(2+) transporter in Bradyrhizobium japonicum and is regulated by manganese via the Fur protein, Mol. Microbiol. 72 (2009) 399–409. PubMed, 19298371.
[66] A.R. Gall, K.A. Datsenko, N. Figueroa-Bossi, L. Bossi, I. Masuda, Y.M. Hou, L.N. Csonka, Mg2+ regulates transcription of mgtA in Salmonella Typhimurium via translation of proline codons during synthesis of the MgtL peptide, Proc. Natl. Acad. Sci. U.S.A. 113 (2016) 15096–15101. PubMed, 27849575.
[67] E.A. Groisman, The ins and outs of virulence gene expression: Mg2+ as a regulatory signal, Bioessays 20 (1998) 96–101. PubMed, 9504051.
[68] L.R. Prost, S.I. Miller, The Salmonellae PhoQ sensor: mechanisms of detection of phagosome signals, Cell. Microbiol. 10 (2008) 576–582. PubMed, 18182085.
[69] E.A. Groisman, The pleiotropic two-component regulatory system PhoP-PhoQ, J. Bacteriol. 183 (2001) 1835–1842. PubMed, 11222580.
[70] S.I. Miller, A.M. Kukral, J.J. Mekalanos, A two-component regulatory system (phoP phoQ) controls Salmonella typhimurium virulence, Proc. Natl. Acad. Sci. U.S.A. 86 (1989) 5054–5058. PubMed, 2544889.
[71] M.W. Bader, S. Sanowar, M.E. Daley, A.R. Schneider, U. Cho, W. Xu, R.E. Klevit, H. Le Moual, S.I. Miller, Recognition of antimicrobial peptides by a bacterial sensor kinase, Cell 122 (2005) 461–472. PubMed, 16096064.
[72] S.Y. Park, M.J. Cromie, E.J. Lee, E.A. Groisman, A bacterial mRNA leader that employs different mechanisms to sense disparate intracellular signals, Cell 142 (2010) 737–748. PubMed, 20813261.
[73] G. Zhao, W. Kong, N. Weatherspoon-Griffin, J. Clark-Curtiss, Y. Shi, Mg2+ facilitates leader peptide translation to induce riboswitch-mediated transcription termination, EMBO J. 30 (2011) 1485–1496. PubMed, 21399613.
[74] E.J. Lee, E.A. Groisman, Control of a Salmonella virulence locus by an ATP-sensing leader messenger RNA, Nature 486 (2012) 271–275. PubMed, 22699622.
[75] E.J. Lee, J. Choi, E.A. Groisman, Control of a Salmonella virulence operon by proline-charged tRNA(Pro), Proc. Natl. Acad. Sci. U.S.A. 111 (2014) 3140–3145. PubMed, 24516160.
[76] K. Hollands, S. Proshkin, S. Sklyarova, V. Epshtein, A. Mironov, E. Nudler, E.A. Groisman, Riboswitch control of Rho-dependent transcription termination, Proc. Natl. Acad. Sci. U.S.A. 109 (2012) 5376–5381. PubMed, 22431636.
[77] K. Hollands, A. Sevostiyanova, E.A. Groisman, Unusually long-lived pause required for regulation of a Rho-dependent transcription terminator, Proc. Natl. Acad. Sci. U.S.A. 111 (2014) E1999–E2007. PubMed, 24778260.
[78] M.J. Cromie, Y. Shi, T. Latifi, E.A. Groisman, An RNA sensor for intracellular Mg(2+), Cell 125 (2006) 71–84. PubMed, 16615891.

[79] E. Merino, C. Yanofsky, Transcription attenuation: a highly conserved regulatory strategy used by bacteria, Trends Genet. 21 (2005) 260–264. PubMed, 15851059.
[80] M.H. Pontes, J. Yeom, E.A. Groisman, Reducing ribosome biosynthesis promotes translation during low Mg2+ stress, Mol. Cell 64 (2016) 480–492. PubMed, 27746019.
[81] H. Schramke, F. Tostevin, R. Heermann, U. Gerland, K. Jung, A dual-sensing receptor confers robust cellular homeostasis, Cell Rep. 16 (2016) 213–221. PubMed, 27320909.
[82] K.M. Papp-Wallace, M.E. Maguire, Magnesium transport and magnesium homeostasis, EcoSal Plus 3 (2008) PubMed, 26443723.
[83] E.A. Groisman, K. Hollands, M.A. Kriner, E.J. Lee, S.Y. Park, M.H. Pontes, Bacterial Mg2+ homeostasis, transport, and virulence, Annu. Rev. Genet. 47 (2013) 625–646. PubMed, 24079267.
[84] P.J. Hill, A. Abibi, R. Albert, B. Andrews, M.M. Gagnon, N. Gao, T. Grebe, L.I. Hajec, J. Huang, S. Livchak, S.D. Lahiri, D.C. McKinney, et al., Selective inhibitors of bacterial t-RNA-(N(1)G37) methyltransferase (TrmD) that demonstrate novel ordering of the lid domain, J. Med. Chem. 56 (2013) 7278–7288. PubMed, 23981144.
[85] T. Kawasuji, M. Fuji, T. Yoshinaga, A. Sato, T. Fujiwara, R. Kiyama, A platform for designing HIV integrase inhibitors. Part 2: a two-metal binding model as a potential mechanism of HIV integrase inhibitors, Bioorg. Med. Chem. 14 (2006) 8420–8429. PubMed, 17005407.
[86] Y.Q. Long, X.H. Jiang, R. Dayam, T. Sanchez, R. Shoemaker, S. Sei, N. Neamati, Rational design and synthesis of novel dimeric diketoacid-containing inhibitors of HIV-1 integrase: implication for binding to two metal ions on the active site of integrase, J. Med. Chem. 47 (2004) 2561–2573. PubMed, 15115398.
[87] E.A. Semenova, A.A. Johnson, C. Marchand, D.A. Davis, R. Yarchoan, Y. Pommier, Preferential inhibition of the magnesium-dependent strand transfer reaction of HIV-1 integrase by alpha-hydroxytropolones, Mol. Pharmacol. 69 (2006) 1454–1460. PubMed, 16418335.
[88] S.H. Kim, F.L. Suddath, G.J. Quigley, A. McPherson, J.L. Sussman, A.H. Wang, N.C. Seeman, A. Rich, Three-dimensional tertiary structure of yeast phenylalanine transfer RNA, Science 185 (1974) 435–440.

CHAPTER FOUR

Structures and Activities of the Elongator Complex and Its Cofactors

Olga Kolaj-Robin[*,†,‡,§], **Bertrand Séraphin**[*,†,‡,§,1]
[*]Institut de Génétique et de Biologie Moléculaire et Cellulaire (IGBMC), Illkirch, France
[†]Centre National de Recherche Scientifique (CNRS) UMR 7104, Illkirch, France
[‡]Institut National de Santé et de Recherche Médicale (INSERM) U964, Illkirch, France
[§]Université de Strasbourg, Illkirch, France
[1]Corresponding author: e-mail address: bertrand.seraphin@igbmc.fr

Contents

Abstract

Elongator is a highly conserved eukaryotic protein complex consisting of two sets of six Elp proteins, while homologues of its catalytic subunit Elp3 are found in all the kingdoms of life. Although it was originally described as a transcription elongation factor, cumulating evidence suggests that its primary function is catalyzing tRNA modifications. In humans, defects in Elongator subunits are associated with neurological disorders and cancer. Although further studies are still required, a clearer picture of the molecular mechanism of action of Elongator and its cofactors has started to emerge within recent years that have witnessed significant development in the field. In this review we summarize recent Elongator-related findings provided largely by crystal structures of several subunits of the complex, the electron microscopy structure of the entire yeast holoenzyme, as well as the structure of the Elongator cofactor complex Kti11/Kti13.

The Enzymes, Volume 41
ISSN 1874-6047
http://dx.doi.org/10.1016/bs.enz.2017.03.001

1. YEAST ELONGATOR: tRNA-MODIFYING COMPLEX

The Elongator complex was first identified in the yeast *Saccharomyces cerevisiae* as copurifying three-subunit (Elp1, Elp2, and Elp3) assembly interacting with an elongating form of yeast RNA polymerase II (Pol.II) and hence its name [1,2]. Shortly after, the Elongator was shown to contain six subunits forming two stable subcomplexes, Elp1–Elp2–Elp3 and Elp4–Elp5–Elp6 [3–5]. The Elp3 subunit harbors an *N*-acetyltransferase domain originally named HAT for histone acetyltransferase [2] and now named KAT for lysine acetyltransferase. Accordingly, partially purified Elp3 was described as a modifier of all four core histone proteins using an in-gel assay [2]. It is, however, noteworthy that this assay could not discriminate between a modification of histone by acetylation and formation of a covalent acetyl-Elp3 intermediate occurring in a histone-dependent manner. Later tests using all six subunits revealed that the Elongator complex acetylates only the tails of histone H3 and H4 in vitro, while in vivo acetylation of histones was reported to be reduced in Elongator mutants [6]. The observation of synthetic genetic interaction between Elongator mutants and mutations altering histones, other HATs, Pol.II subunits, and other factors modulating the activity of transcription factors was taken as further indications of the role of Elongator in chromatin modification and transcription and, hence, its nuclear localization [7–11]. Consistently, Elongator was reported to interact with nascent RNA [12] and Elongator mutants were reported to alter the kinetic of transcription activation in response to changes in growth conditions [1,13].

The yeast Elongator's implication in other nuclear functions, namely local transcriptional repression and genome stability maintenance, was proposed based on Elongator mutants displaying partial loss of telomeric gene silencing and increased sensitivity to DNA-damaging agents, further supported by the observation of its interaction with proliferating cell nuclear antigen [14].

The nuclear localization of the complex and its implication in elongation of transcription and histone acetylation were almost immediately questioned as the complex was not detected on transcribing open reading frames by chromatin immunoprecipitation experiments, and Elp1 and Elp3 proteins were found to be localized in the cytosol [15]. Later unbiased large-scale localization experiments clearly indicated that Elongator was mostly localized in the cytoplasm [16]. In line with its cytoplasmic localization, the yeast complex was also shown to interact through its Elp1 subunit with Sec2p which led to conclusion of its implications in exocytosis [17].

The first data linking Elongator to tRNA modification were generated 15 years before the official identification of the complex. A screen for mutants abolishing the activity of a stop codon tRNA suppressor in *Schizosaccharomyces pombe* identified the *sin3*$^{+}$ gene [18], which was shown 21 years later to encode a protein of 77% identity to *S. cerevisiae* Elp3 [19]. Interestingly, this mutation prevents the formation of a modified 5-methoxycarbonylmethyl-2-thiouridine (mcm^5s^2U, Fig. 1) at the wobble position of the anticodon of tRNAs [18,20]. Most eukaryotic cytoplasmic tRNAs carrying a U at their wobble position are indeed modified in a manner that enhances decoding efficiency and to allow wobble interactions [20,21]. The nature of this modification varies according to the tRNA. In *S. cerevisiae*, tRNAs carrying a U at the wobble position are modified by addition of related groups (Fig. 1): a mcm^5 (5-methoxycarbonylmethyl) group for two anticodons, a mcm^5s^2 (5-methoxycarbonylmethyl and 2-thio) groups for three anticodons, a ncm^5 (5-carbamoylmethyl) group for five anticodons, and a ncm^5 and a ribose methyl group for one anticodon [20]. Interestingly, with the exception of tRNAs carrying a UAG anticodon corresponding to leucine, all *S. cerevisiae* tRNAs with a U at the wobble position allowing A/G wobble interactions are modified. Moreover, the SUP4

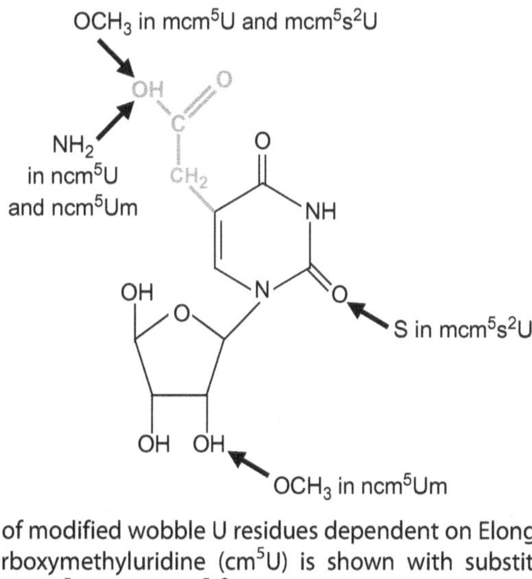

Fig. 1 Structures of modified wobble U residues dependent on Elongator function. The structure of 5-carboxymethyluridine (cm^5U) is shown with substitutions present in ncm^5U, ncm^5Um, mcm^5U, and mcm^5s^2U indicated. The carboxymethyl group derived from acetyl-CoA is indicated in *gray*.

nonsense suppressor tRNA resulting from a mutation introducing a U in the wobble position of a tyrosine-inserting tRNA is also modified to mcm^{5}U [19]. Given that the corresponding wild-type tRNA is unmodified due to the presence of a wobble G, this observation indicates that the presence of a U at the wobble position is a major determinant to trigger the base modification.

Identification of the gene affecting 5-(methoxycarbonylmethyl)-2-thiouridine formation in *S. pombe* as a homolog of Elp3 led Huang and coworkers to demonstrate that inactivation of all other subunits of Elongator in *S. cerevisiae* results in the same phenotype [19]. Interestingly, inactivation of Elongator prevented the formation of both 5-methoxycarbonylmethyl and 5-carbamoylmethyl groups, suggesting that these related modifications arise in part through a common pathway (Fig. 2). Moreover, inactivation of Elongator in *S. cerevisiae* also abolished simultaneously the modification

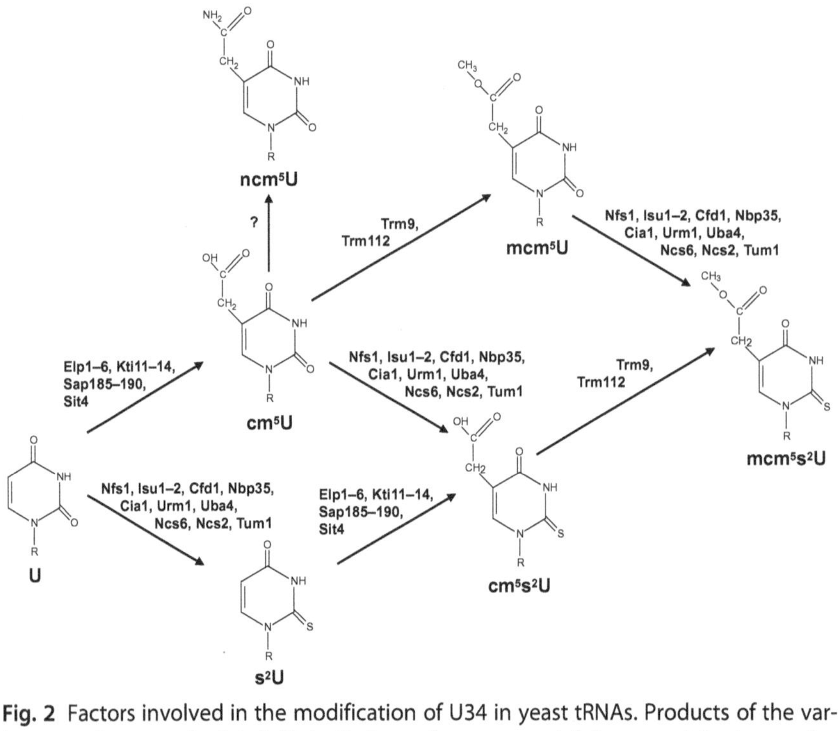

Fig. 2 Factors involved in the modification of U34 in yeast tRNAs. Products of the various reactions are depicted. Note that reactions may not follow an obligatory order. ncm^{5}U is likely to derive from mcm^{5}U, but this has not been formally demonstrated and alternative pathways have been proposed [22]. Factors specifically involved in the proposed conversion of mcm^{5}U in ncm^{5}U have not been identified.

and the activity of the SUP4 tRNA suppressor [20]. Additional modifications of wobble U residues through thiolation and ribose methylation in some tRNAs were shown to occur independently of Elongator through dedicated biosynthetic pathways (Fig. 2). Biochemical evidence supporting Elongator's function in tRNA modification was recently delivered by Selvadurai and colleagues who reported that the Elp3 homologue from archaea *Methanocaldococcus infernus* was able to modify U34 tRNA to cm^5U in vitro in the presence of SAM, acetyl-CoA and $Na_2S_2O_4$ [23]. It should be mentioned, however, that this observation could not be reproduced in an independent study [24].

In summary, Elongator was reported to be implicated in at least four very distinct processes in yeast (histone modification, transcriptional repression and genome stability, exocytosis, and tRNA modification, see earlier) and in several additional ones in other eukaryotic organisms (see later). It is a highly conserved protein complex and the fact that human Elp3 and Elp4 are able to complement corresponding yeast mutants [25,26] argues for the conservation of its activity between eukaryotic species. These seemingly opposite conclusions can be reunited through the hypothesis that altering one central function directly mediated by Elongator resulted in the alteration of distant cellular processes. This point was elegantly addressed by a genetic approach in yeast. Reported data indicated that Elongator impacts tRNA modification and, hence, translation, while other functions were indirectly altered. Indeed, overexpression of various combinations of $tRNA^{Lys}{}_{UUU}$, $tRNA^{Gln}{}_{UUG}$, and $tRNA^{Glu}{}_{UUC}$ (carrying the S^2 group on the wobble U but lacking the mcm^5 substitution in the absence of Elongator) led to suppression of defect in transcription response, histone acetylation, genetic interaction between acetylases, sensitivity to DNA-damaging agents, and telomere silencing in Elongator mutants [27,28]. Moreover, expression of the aforementioned tRNAs also suppressed the defects observed in the *ncs2* mutant which is unable to form 2-thio group (but not mcm^5) at wobble uridine position [27,28]. Elongator mutants were also suppressed by tRNA overexpression in *S. pombe*, and specific phenotypes resulting from Elongator mutation in this organism could be attributed to poor translation of a key factor [29]. Recently, yeast Elongator was also shown to be implicated in mitochondrial function under stress conditions. The observed respiratory deficiency was linked to the impaired mitochondrial protein synthesis that could be selectively rescued by overexpression of cytoplasmic $tRNA^{Lys}{}_{UUU}$ as well as two genes with a strong bias for the AAA codon read by this tRNA [30]. Overall, these results argued that the primary defect of

Elongator inactivation is the absence of tRNA modifications, which leads to altered translation itself resulting in distinct phenotypes.

2. ELONGATOR IN OTHER EUKARYOTES

2.1 Elongator in Nonhuman Species

Elongator mutants in other eukaryotes show an extremely diverse range of phenotypes, some of which are described below. In *Drosophila*, mutation of the Elp6 homologue affected viability and impacted growth, the TOR pathway and insulin signaling [31], while mutation of the Elp3 homologue was reported to induce death at the larval stage [32], affected development of and functioning of the nervous system and also immunity [33–35]. The neurodevelopmental defects in the fruit fly were suggested to be caused by disruption of the Elongator-mediated acetylation of Bruchpilot, a protein important for neuronal differentiation [34]. Mouse mutants demonstrated that Elp1/IKAP inactivation causes embryonic lethality [36,37]. Reduced expression of Elp1/IKAP leads to peripheral neuropathies [37], while its conditional inactivation leads to meiotic defects during spermatogenesis [38]. Knockdown of mouse homologues of Elp1, Elp3, or Elp4 was reported to impair zygotic paternal genome demethylation [39]. Independently, analysis of mouse brain demonstrated expression of Elongator in cortical neurons during development and the inhibition of Elongator with shRNA reduced the neurons migration rate. Elongator was found to coprecipitate with tubulin and was reported to acetylate tubulin [40] in a manner consistent with results obtained in the nematode *Caenorhabditis elegans* [41]. In the latter organism mutation of Elp1 and Elp3 homologues was shown to lead to neurological dysfunctions and in combination with a mutation of a protein TUC1 to cause developmental defects [42]. These phenotypes have also been directly associated with defects in Elongator-mediated tRNA modifications as well as with the TUC1-mediated thiolation step of the uridine at position 34 already modified by the Elongator [42]. Defective tubulin acetylation in *C. elegans* is also likely to be explained by poor translation of (a) key factor(s) [43]. Similarly, in mice, the role of Elp3 in embryonic development was recently reported to be associated with regulation of translation [36]. Furthermore, recent studies also showed that loss of Elp3-impaired codon translation speed in mouse cortical stem cells triggers unfolded protein response that contributes to the control of cortical neurogenesis [44]. However, another recent study defined the function of Elongator as regulating the distribution and turnover of actin microfilaments during cell migration and, hence, controlling cortical interneuron migration in mouse embryos [45]. Another argument in support of Elongator importance in cell migration was also recently delivered in *Xenopus* embryos. Elp3 mutants showed inhibited neural crest cell migration due to disrupted Elongator-mediated stabilization of a master regulator of cell motility protein, Snail1, that was reported to bind to Elp3 through its zinc finger domain [46].

Disruption of the Elongator complex in *Arabidopsis* leads to aberrant growth and development, resistance to abiotic stresses, and susceptibility to plant pathogens and has been associated with reduced histone acetylation and/or DNA methylation ([47] and references therein). However, the direct role of plant Elongator in histone acetylation, DNA demethylation, or methylation was not established to date. The complex was also demonstrated to be involved in microRNA biogenesis in *Arabidopsis* [48]. Limited studies suggest a possible contribution of Elongator in tRNA modification process. The *elp3* mutant was shown to have defect in tRNA wobble uridine modification, while coexpression of plant *elp1* and *elp3* genes restored Elongator's tRNA modification function in vivo in *elp1 elp3* yeast mutant [49]. In addition, the gene encoding a homologue of Elp1 was independently identified as being involved in tRNA modification [50]. Considering increasing amount of evidences demonstrating the implication of Elongator in tRNA modification, it is likely that with time a direct link between Elongator function and tRNA modification in plants will also be provided.

This long and still incomplete catalog suggests either that Elongator is extremely pleiotropic or that it impacts one or a limited number of molecular function(s), the disruption of which results in numerous phenotypes because of its(their) central role(s) for the cell activity. It is indeed difficult to imagine that a single complex would be endowed with so many enzymatic activities. Moreover, some of the proposed functions of Elongator are difficult to reconcile with its phylogenetic distribution. This pertains, for example, to the proposed function of Elongator in DNA demethylation in mouse [39] and plants [47], which would question the function and conservation of Elongator in *S. cerevisiae*, an organism that is known not to carry methylated bases in its genome. It is noteworthy that in most cases, analyses of the phenotypic consequences of Elongator inactivation were not complemented by biochemical assays demonstrating its ability to catalyze the corresponding reaction. In the light of the studies cited earlier, diverse and unrelated functions of Elongator in the nucleus and cytoplasm [51,52] cannot be excluded. Yet, it appears that in animals, plants, and fungi, defects in Elongator result in reduced modification of wobble uridines in tRNAs (reviewed in Ref. [53]). The consequences of altered translation that would ensue may be responsible for a wide variety of indirect phenotypes. Moreover, a role of Elongator in tRNA modification could easily explain the presence of Elp3 homologs in archaea and some eubacteria.

2.2 Elongator and Human Diseases

Mutations in genes encoding Elongator subunits were shown to be associated to human diseases. Two independent studies identified two mutations in the human gene encoding IKAP (for IκB kinase (IKK) complex-associated protein), a human homologue of Elp1, as causative of familial dysautonomia (FD, initially Riley–Day syndrome) [54,55]. This autosomal recessive neurodegenerative disease primarily found among the Ashkenazi Jewish population [56,57] is characterized by impaired development and progressive degeneration of the sensory and autonomic nerves. Consistently, mouse mutants demonstrated that inactivation of the gene encoding IKAP/Elp1 is lethal, while reduced expression of this protein leads to peripheral neuropathies [37]. In 99.5% of the cases, FD is caused by homozygous mutations in intron 20 of the gene encoding the IKK complex-associated protein IKAP [54]. (Note, however, that IKAP/Elp1 was later shown not to associate with IKK [58].) The intron 20 mutation results in aberrant splicing leading to skipping of exon 20, a frameshift and the generation of a premature termination codon in exon 21 of IKAP-encoding mRNA. Consequently the levels of the full-length IKAP protein are reduced and a truncated version is expressed (residues 1–914) [54,59]. It is possible that exon skipping constitutes the molecular mechanism for the tissue-specific expression of the disease as various cells from patients were found to contain normal as well as truncated IKAP proteins at variable levels [54]. Elongator was shown to support neuronal migration target tissue innervation and survival of sensory and sympathetic neurons [60–62]. Relatively recently, evidence was provided to link FD with defects in tRNA modification. Yoshida and coworkers showed that the levels of modified wobble uridine are reduced in cells from FD patients. They also identified a small compound, rectifier of aberrant splicing (RECTAS) that amends the aberrant splicing of the pre-mRNA encoding IKAP in cells from patients with FD and also rescues the levels of tRNA modifications [63]. Simultaneously and independently, another group also showed reduced levels of the wobble uridine nucleoside mcm^5s^2U in brain tissue and fibroblast cell line from FD patients [64]. These findings suggest that FD is likely to be the result of an inefficient translation caused by lower levels of wobble uridine nucleosides modifications.

Mutations in the gene encoding IKAP were also shown to increase risk for bronchial asthma in children [65]. This phenomenon can most likely also be explained by defective Elongator activity as it was shown that the mutations in question (C1072S and P1158L) altered the interactions with Elp2 Elongator subunit and, hence, the integrity of the complex [66].

Variants of the catalytic subunit of the Elongator complex, Elp3, detected in three human populations were associated with amyotrophic lateral sclerosis (ALS) [33], a spontaneous, relentlessly progressive motor neuron disease, usually resulting in death from respiratory failure within 3 years. The same study provided also additional evidences of the importance of Elp3 in neuronal communication and survival, as mutagenesis screens in *Drosophila* for genes important in these processes identified two loss-of-function mutations in Elp3, while its knockdown in zebrafish led to motor axonal abnormalities [33]. Genome-wide association screen of sporadic ALS also provided some evidences of Elp3 implication in this disease, although genomic variations responsible for these associations remain to be clarified [67]. It was postulated that given the proposed function for Elongator in α-tubulin acetylation in neurons, it might play a role in other neurological disorders such as Alzheimer's, Parkinson's, and Huntington's diseases in addition to ALS [68]. However, the role of α-tubulin acetylation in neurological diseases has not been directly demonstrated to date.

Allelic variants of remaining subunits of the complex have been associated with neurological diseases. Elp4 was shown to be associated with the centrotemporal sharp wave electroencephalogram trait in Rolandic epilepsy, the most common form of human epilepsy characterized by classic focal seizures often preceded by several developmental defects, and in speech sound disorder [69,70]. Homozygous modifications of Elp2 were associated with intellectual disability (ID) in a large-scale sequencing study [71], while two cases of severe ID related to heterozygous missense variant of this subunit were also reported in an independent study [72].

Elongator complex activity is also associated with cancer. Elongator depletion through Elp1, Elp3, Elp5, and Elp6 was reported to restrict cell migration and reduce tumorigenicity in melanoma cells [73]. More recently, two reports showed that the role of Elongator in cancer is strictly related to the regulation of the levels of key proteins via tRNA modification. In the first one, Elp3 was shown to be implicated in Wnt-dependent tumor initiation and regeneration in the intestine via regulation of the levels of Sox9 needed to maintain the subpool of relevant cancer stem cells [52]. The second showed the importance of Elp3, together with cytosolic thiouridylase (CTU) 1/2, in human breast cancer and metastasis via regulation of translation of the oncoprotein DEK and hence of the proinvasive transcription factor LEF1 [74].

Examples described herein demonstrate that the role of Elongator in neurological diseases is undeniable and its contribution to cell growth and cancer is significant. Although these roles might be linked to other enzymatic functions, increasing lines of evidence seem to relate them to the Elongator-mediated tRNA modifications. This is further supported by two recent studies. In the first one, ribosome profiling experiments performed in *S. cerevisiae* and *C. elegans* showed that loss of Elongator function and, hence, anticodon wobble uridine (U34) modification in a subset of tRNAs leads to ribosome pausing affecting global translation efficiency and causing proteotoxic stress [75]. The second study, performed in yeast, demonstrated that loss of Elongator-mediated tRNA modifications results in an altered metabolic profile that could be only partially suppressed by overexpression of hypomodified $tRNA^{Lys}{}_{UUU}$, $tRNA^{Gln}{}_{UUG}$, and $tRNA^{Glu}{}_{UUC}$ species carrying the wobble U S^2 group but not the mcm^5 substitution [76].

3. PROTEINS REQUIRED FOR WOBBLE URIDINE MODIFICATION

Elongator complex was independently functionally identified by screening for yeast mutants insensitive to zymocin [77,78], the product of a gene encoded by an optional natural plasmid of the yeast *Kluyveromyces lactis* that is secreted and allows expressing strains to outcompete nonresistant yeasts present in the same environment [79]. These genes were originally named *KTI* for killer toxin insensitive. These data suggested that Elongator activity resulted in the production of a compound targeted by zymocin. The later observation that Elongator mutations prevent tRNA modifications subsequently suggested that modified tRNAs could be the targets of zymocin. The zymocin ribonuclease activity was shown to split specifically tRNAs containing a mcm^5s^2-modified wobble uridine [80,81]. Overall, a number of genes related to the zymocin resistance and formation of the mcm^5 and s^2 modifications of wobble uridine was identified [19,82–84]. Two methyltransferase subunits Trm9 and Trm112 are required for the final step of mcm^5 formation [22,85–89], and 11 additional proteins are necessary for formation of s^2 moiety in mcm^5s^2U (Fig. 2). For the early step of ncm^5/mcm^5 side-chain formation, in addition to the six Elongator subunits, killer toxin insensitive (Kti) proteins 11–14, Sit4, as well as Sap185 and Sap190 were shown to be required (Fig. 2). If Trm9/Trm112 needs to act after cm^5U synthesis, formation of the cm^5 and s^2 groups does not appear to occur in a defined order. ncm^5U is likely to be derived from cm^5U, but alternative pathways have not yet been ruled out and factor(s) involved in this process have not yet been identified. The structure–function aspects of Elongator complex and its cofactors are discussed below.

3.1 Elp1 Is a Phosphorylated Scaffold/Docking Platform of the Elp Complex

Elp1 is the largest (~150 kDa) of the six Elongator complex subunits and contains two WD40 β-propeller domains in its N-terminus and a tetratricopeptide repeat domain (TPR) in its C-terminus [90] (Fig. 3).

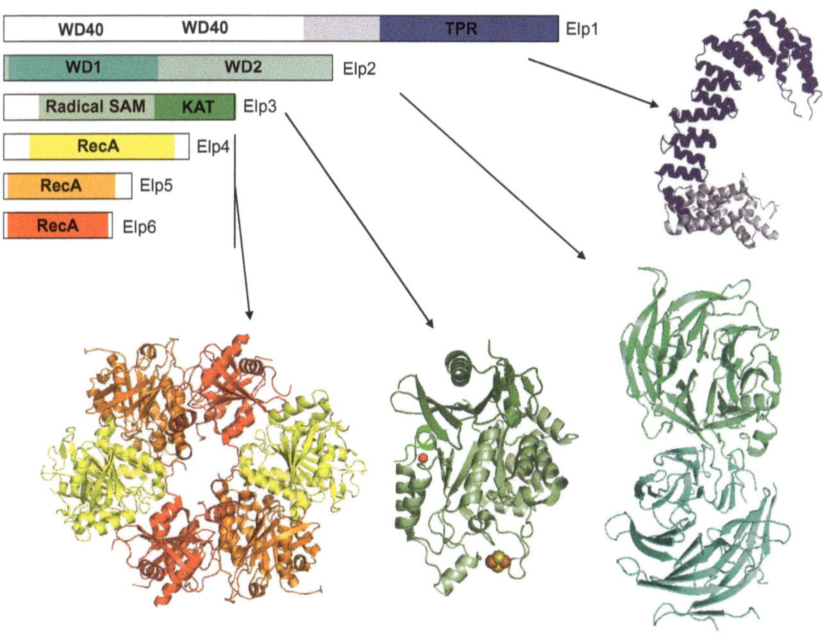

Fig. 3 Structural information available on Elongator subunits. Domains present in Elongator subunits are depicted. Structures available for human Elp1, *Saccharomyces cerevisiae* Elp2, *Dehalococcoides mccartyi* Elp3, and *S. cerevisiae* Elp4–6 are presented. The [Fe–S] cluster and Zn ion bound to *D. mccartyi* Elp3 are shown in *orange/brown* and *red*, respectively. The Elp1 horseshoe dimerization domain is shown at the *top* of the structure. Protein Data Bank entries 5cqr, 4xfv, 5l7j, and 4a8j were used to prepare this figure.

Elp1 was suggested to act as a scaffold for the complex assembly [91] and was shown to interact strongly with two other subunits of the complex, Elp2 and Elp3 to form a subcomplex stable in high salt concentrations under which the second Elongator subcomplex, Elp456, dissociates [4]. A cross-linking/ MS study performed recently showed strong interactions of Elp1 also with Elp4 but not with Elp5 or Elp6 [92], while coprecipitation with a second Elp1 subunit carrying a different tag suggested its dimerization [93]. The largest subunit of the complex was reported to interact also with Sec2 [17], α-tubulin [41], cell migration proteins [60], JNK-associated protein [94], as well as with protein found necessary for tRNA modification, casein kinase Kti14/Hrr25 [95], and with the tRNA itself [96]. Interestingly, interaction with tRNAs was shown to be mediated by the highly conserved arginine/lysine-rich C-terminal region of Elp1 [94] that resembles a nuclear localization sequence [97], somewhat in contrary to the mainly cytoplasmic localization of the complex. Two crystal structures of the C-terminal domains of human Elp1 (residues 715–1332) and yeast Elp1 (residues 919–1349) were solved to date at 3.0 and 2.7 Å, respectively [66]. Despite the low sequence identity, both C-terminal domains of Elp1 exhibited high structural similarity. The horseshoe shaped was shown to be responsible and sufficient for Elp1 dimerization in vivo, required for the assembly and stability of the Elongator complex but not for tRNA binding [66]. Defects in Elp1 dimerization, however, cannot be considered as universal explanation of Elp1-associated diseases as, unlike the major mutation causing FD, minor FD- and bronchial asthma-associated mutations in Elp1 did not alter Elp1 dimer formation [66].

The yeast Elp1 is phosphorylated and its dephosphorylation was shown to be dependent on the phosphatase Sit4 and its associated partners, Sap185 and Sap190 [98]. The function of Elongator was suggested to be regulated by the antagonistic activities of Sit4, and the kinase Kti14 as Elp1 was hypo- and hyperphosphorylated in *kti14* and *sit4* null mutants, respectively [95]. The kinase Kti14 binds to the Elongator through another cofactor protein required for tRNA modifications, Kti12, overexpression of which leads to hyperphosphorylation of Elp1 [95]. Kti12 and Sit4 were also shown to be involved in response to ceramide stress in yeast, although in Elongator-unrelated process [99]. Of the nine in vivo phosphorylation sites identified in Elp1, the four tightly clustered at the Elp1 C-terminus nearby the tRNA-binding region were shown to be important for the Elongator's tRNA modification activity. Their mutations decreased the level of mcm^5U and ncm^5U modifications to different extent and revealed the phosphorylation of Ser-1209 playing a major, positive role in Elongator-dependent tRNA modification [100]. It was shown that defects in phosphorylation do not affect the assembly of the complex. Ser-1209 was suggested to modulate interaction of Elongator with its cofactors Kti12 and Kti14. The latter kinase directly phosphorylates two functionally important sites but not the key residue Ser-1209. This observation suggests the involvement of at least one other kinase modulating the function of Elongator [100].

In addition to phosphorylation and dephosphorylation, the activity of the Elongator complex was suggested to be regulated also by proteolysis as purification of the complex revealed two major and a minor form of Elp1 and the increasing levels of the N-terminally truncated form in mutants lacking Urm1 and Kti11 [3,97]. Recently, a vacuolar protease Prb1 was identified as mediating cleavage of Elp1 between Lys203 and Ala204. However, as removal of Prb1 did not affect the levels of, and the N-truncated form was shown not to be active in Elongator-mediated tRNA modifications, the proteolytic regulation of Elongator was questioned and truncation of Elp1 was finally shown to be a purification artifact [101].

3.2 Elp2: WD β-Propeller

The second largest subunit of Elongator complex, Elp2 (~90 kDa), was predicted to contain WD repeats [13] that were shown to bind to various targets [102,103]. Recently, the crystal structure of Elp2 from yeast was solved at 3.2 Å [104] and independently at 2.8 Å resolution [92] confirming the presence of two seven-bladed β-propellers slightly twisted with respect to each other (Fig. 3). As the Elongator complex was reported to acetylate histones and α-tubulin (see earlier), Dong and colleagues investigated interactions between those proteins and Elp2. Elp2 did not interact with histone octamers or nucleosomes in vitro [104], a result that agrees with the observation that Elongator lacking the Elp2 subunit can still mediate protein acetylation [105]. Because deletion of Elp2 leads to phenotypes identical to the deletion of other subunits of the Elongator complex [1,19,77], this result suggests that there is no direct connection between the HAT activity and the observed *elp* phenotypes in vivo. Interestingly, Elp2, but not its variant where four highly conserved arginine residues clustering at the top of β-propeller II were mutated, was shown to interact with α-tubulin in the context of microtubules (MTs) in vitro and in vivo [104]. By contrast, loss of Elongator function did not affect levels of acetylated α-tubulin in

HEK293T cells and mutants unable to interact with MTs showed the same level of H3K14 acetylation. Interestingly, this mutant does not affect the assembly of the Elp123 subcomplex in vitro but in vivo displays *elp* phenotypes [104]. These contradictory results together with indications that α-tubulin acetylation might be controlled by Elongator indirectly through tRNA modification [29] as well as the information obtained from the recently reported structure of Elp3 [24] (see later) question the relevance of the observed interactions of Elp2 with MTs in Elongator activity. However, it is clear that the integrity of Elp2 plays an important role in the complex function as *elp2* mutation results in the same phenotypes than deletion of any other subunit of the complex. The N-terminal β-propeller domain I of Elp2 consists of a continuous string of residues, while the very first 10 residues constitute the final β-strand of the last blade of propeller II (Fig. 3), a so-called molecular Velcro observed in other WD40 and β-propeller proteins. Removal of the first 14 residues from the N-terminus of Elp2, which affects the integrity of its WD40 fold, completely abolished interactions with Elp1 and Elp3 in vitro while exhibiting typical *elp* phenotypes in vivo [104]. Based on these observations Elp2 was suggested to be important for complex assembly [104]. However, Elongator complexes comprising only the five remaining subunits were reported to form in vivo and in vitro [92,105]. Moreover, unlike Elp1, Elp2 was reported to exist in solution as monomer [104]. Therefore, it is Elp1 rather than Elp2 that ensures complex assembly and through its dimerization, Elp1 recruits two copies of Elp2 as well as of Elp3 in the complex [66].

In human, Elp2 is also known as Stat3-interacting protein 1 (StlP1) and it was shown to interact preferentially with nonphosphorylated signal transducer and activator of transcription (STAT) proteins Stat1, Stat3, and Stat5 as well as with Janus kinases (JAKs) [106,107]. Since the JAK/STAT signaling pathway is present only in higher eukaryotes, it was suggested that Elp2 might have evolved in mammals to provide regulation of the Elongator's activity through its physical connection with various signaling cascades [90]. The importance of interactions of Elp2 with external factors was also suggested for yeast Elongator as the conserved basic region of Elp2 was shown to be required for tRNA modification in vivo but not for protein–protein interactions within the complex [92]. Finally, Elp2 function might be also regulated by phosphorylation as a phosphorylation site at Ser-492 of yeast Elp2 was recently identified in vivo [100]. However, the functional significance of this phosphorylation, if any, remains to be elucidated.

3.3 Elp3: The Catalytic Heart of the Enzyme

Since the identification of Elongator, Elp3 (~56 kDa) is considered as the catalytic subunit of the complex since it contains two distinctive domains: an N-terminal radical *S*-adenosylmethionine (SAM)-binding domain and a C-terminal KAT, previously referred to as HAT domain, both of which were shown to be crucial for Elongator activity in vivo [2] (Fig. 3). This interpretation is further supported by the fact that the conservation of Elp3 extends beyond eukaryotes, and almost all archaea, some eubacteria and two viruses also possess homologues of Elp3 [23] but not of all the remaining five Elongator subunits. Interestingly, the presence of Elp3 coincides with the lack of genes encoding for MnmE and MnmG that are responsible for or implicated in tRNA wobble uridine modification in bacteria and archaea [108,109]. This phylogenetic correlation is a strong argument further supporting the idea that tRNA modification is the principal molecular function of Elongator. As mentioned earlier, this hypothesis was recently supported by the observation of tRNA modification by an archaeal Elp3 reconstituted in vitro in the presence of SAM, acetyl-CoA, and $Na_2S_2O_4$ in anaerobic conditions [23].

In agreement with its association with the family of radical SAM enzymes, Elp3 was predicted to contain a [4Fe–4S] cluster based on sequence comparison [110]. This was subsequently supported by detection of unstable [4Fe–4S] cluster in Elp3 homologues from archaea *Methanocaldococcus jannaschii* that were also shown to bind SAM [111], and more recently in *M. infernus* [23]. Bacterial Elp3 from *Dehalococcoides mccartyi* was also shown to contain a fragile [4Fe–4S] cluster that degraded to [2Fe–2S] cluster upon reduction with $Na_2S_2O_4$ or in the crystal structure [24]. Initial analysis of the yeast [Fe–S] cluster in Elp3 concluded that it was not required for the HAT or RNA-binding activity of Elongator. Instead, the authors proposed that the [Fe–S] cluster played a structural role required for Elongator integrity as well as for the association of the complex with its accessory factors Kti11 and Kti12 [110]. More recent studies, however, have demonstrated the role of [Fe–S] cluster in Elongator-mediated tRNA modification. An archaeal Elp3 C95A C98S double mutant was not able to modify tRNA in vitro [23], while equivalent mutations in yeast Elp3 abolished U34 modification in vivo [27]. The corresponding residues were shown to coordinate the [Fe–S] cluster in the recently solved crystal structure of Elp3 from *D. mccartyi* [24]. Interestingly, the latter study also confirms to some extent the structural role of the [Fe–S] cluster that is located at the interface of two

Elp3 molecules. Moreover, although an Elp3 with a mutated [Fe–S]-coordinating cysteine residues was still capable of dimerization, the presence of the cluster increased dimerization, and both dimerization and coordination of [Fe–S] were concluded to be essential for the catalytic activity of Elongator in vivo [24]. In the view of these findings, the detection of Elp3 monomers in solution could be interpreted as a purification artifact. Yet, the presence of these monomers allowed the demonstration that dimerization of Elp3 was not necessary for tRNA binding as both monomers and dimers bind in vitro-transcribed tRNA with similar affinities. tRNAs are proposed to bind to the basic cavity of Elp3 located on the interface between the KAT and SAM domains through the anticodon and D-loop regions. This binding was confirmed to be independent from the coordination of [Fe–S] clusters or from the binding of SAM or acetyl-CoA [24].

In the crystal structure of bacterial Elp3 solved at 2.15 Å resolution, the SAM domain completely blocks the canonical substrate-binding site of the KAT domain practically ruling out the acetylation of lysine residues of histone tails or other substrates as a primary and conserved function of Elp3 and hence of the Elongator complex [24]. Interestingly, replacing the blocking loop with a short linker increased affinity to tRNA fourfold and led to a proposal of loop displacement upon tRNA binding. This step would allow the wobble base to be positioned in proximity to the center of the KAT domain. tRNA binding would promote in turn binding of acetyl-CoA and its hydrolysis [24]. This observation is in agreement with the observed increased turnover of acetyl-CoA in the presence of tRNA and questions the alternative mechanism of the Elp3-catalyzed reaction, where a stable acetyl-Elp3 covalent intermediate has been proposed to be formed in the first step followed by tRNA binding [23]. It remains to be established whether this mechanism is conserved in eukaryotes where a higher level of sequence conservation is observed within the loop in question.

The structural analysis of bacterial Elp3 demonstrated that it is not a classical member of the already remarkably diverse family of radical SAM enzymes. The coordination and location of the [Fe–S] cluster are different from those of other known radical SAM enzymes [112]. If Elp3 binds SAM like other enzymes of this family, the iron-cluster coordination loop would have to undergo structural arrangements. Indeed, inspection of the Elp3 structure shows that relatively small movements within Elp3 would place 5′-dA•, acetyl-CoA, and U34 in vicinity enabling the proposed reaction (Fig. 4). These rearrangements are likely to be triggered by tRNA binding

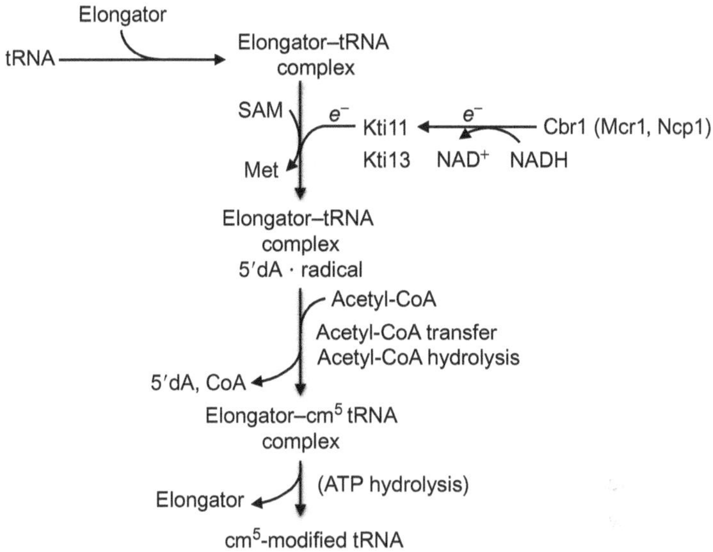

Fig. 4 Proposed pathway for cm^{5}U biosynthesis by Elongator. The various steps that would lead to the formation of cm^{5}U at the wobble position of tRNA by Elongator and associated factors are depicted. If parts of this scheme are well established (e.g., requirement of electron transfer mediated by Cbr1 and Kti11/Kti13), it is noteworthy that this pathway remains hypothetical and that additional studies are required to elucidate details including exact order of substrates binding and transfer (e.g., is the acetyl-CoA hydrolyzed before the transfer of the acetyl group to the recipient U base or is acetyl-CoA transferred to the wobble base and hydrolyzed in a second step? Is acetyl-CoA bound before or after SAM?). Similarly the role of ATP in tRNA release (or binding) needs to be ascertained. These analyses await the development of a reconstituted in vitro system for tRNA modification.

as no major conformational changes were detected by small-angle X-ray scattering in bacterial Elp3 upon addition of SAM or acetyl-CoA [24]. This hypothesis is consistent with the phenomenon recently reported for another member of radical SAM family, RlmN, catalyzing methylation of C2 of adenosine residues in 23 rRNA and A37 in several *Escherichia coli* tRNAs. The structure of this enzyme cross-linked to a $tRNA_{Glu}$ substrate was recently solved [113]. Upon tRNA binding, the terminal methylene unit of Cys355 residing otherwise near the SAM methyl group but far (~6 Å) from the site of 5′-dA• formation is brought within 4.1 Å of SAM cleavage product following a 5-Å shift in the position of the backbone loop [113].

The first structural information about the catalytic domain of Elongator in the form of crystal structure of Elp3 from *D. mccartyi* constitutes undoubtedly a great step toward understanding of the complicated action mechanism of the complex. However, more analyses are necessary to understand each step of this reaction as well as to understand the function of the zinc-binding loop located between the SAM and KAT domain, which is conserved in bacterial and archaeal but not eukaryotic Elp3 homologues [24].

3.4 Elp456: Heterohexameric Ring ATPase

Discovered as a part of the Elongator complex upon purification in mild conditions [4,5], the Elp456 subcomplex provided the first crystallographic insight into the complex structure [93]. The electron microscopy and crystal structure of the yeast Elp456 solved at 2.1 Å revealed that all the three component proteins, despite the lack of previously recognized sequence homology between them, adopted almost identical RecA folds. Furthermore, the Elp456 formed a heterohexameric, ring-like structure containing two copies of each subunit. This structure, characteristic of homohexameric RecA-like ATPases, was also independently confirmed in another study [114].

Elp456 was shown to specifically bind tRNA but not single-stranded DNA or oligo (U) RNA [93]. Despite the lack of the key signature of ATPases in Elp4, Elp5, and Elp6, the formed Elongator subcomplex exhibited intrinsic basal ATPase activity that could not be stimulated by tRNA or the anticodon loop, unlike other RecA ATPases whose activity increases in the presence of their substrates [93]. Nevertheless, the functional connection between the ATPase activity of the Elp456 and Elongator's tRNA modification activity in vivo was demonstrated through experiments where mutants with compromised ATPase activity only partially complemented corresponding yeast deletion mutants. Based on the observed lower affinity to tRNA in the presence of ATP and other hydrolyzable nucleoside triphosphates, it was concluded that binding and/or release of tRNA to the subcomplex is controlled in an ATP-dependent manner [93] (Fig. 4). The tRNA molecule could be successfully docked to the Elp456 hexamer with the anticodon loop pointing into the central cavity, in the same arrangement observed for hexameric ATPases binding nucleic acids. The binding of tRNA to Elp456 is mediated through its anticodon loop but was not to limited to modifiable tRNAs [93]. This suggested that substrate specificity might be achieved directly during the modification reaction [93].

Similar to Elp1 and Elp2, subunits Elp4 and Elp5 were shown to contain two and one in vivo phosphorylation sites, respectively [100,115]. As in the case of Elp2, their functional importance remains to be determined.

3.5 The HoloElongator Complex

The assembly of the Elongator itself seems a rather complex process and was suggested to be dependent on posttranslational modifications or chaperones [100,116]. Although a complex lacking the Elp2 subunit could be formed in *elp2* yeast mutants [105], it could not be assembled in vitro from individually purified proteins expressed in *E. coli* [92]. Based on coimmunoprecipitation analyses, it was previously concluded that two copies of Elp123 are associated with the hexameric ring of Elp456 [93]. In accordance with the twofold symmetry of the Elp456 subcomplex, a model was proposed in which each Elp123 copy is peripherally attached to the Elp456 ring [93]. However, this model contradicts the finding that dimerization of the C-terminal domain of Elp1 facilitates subsequent dimerization of Elongator [66]. Recently, two electron microscopy structures of the yeast Elongator holoenzyme were simultaneously reported by two independent groups [92,117] (Fig. 5). They revealed that Elongator consists of two symmetric triangular-like wings formed by two copies of Elp123 linked together by an arch formed by the C-terminal domains of the two Elp1 subunits, disproving the previously proposed model [93] and in agreement with the importance of dimerization of Elp1 in complex assembly [66]. Surprisingly, in both of the recently reported holoenzyme structures, the heterohexameric Elp456 ring was bound asymmetrically to one Elp123 wing of the heterotrimeric Elp123 subcomplex [92,117] (Fig. 5). The rationale for this asymmetric binding as well as its mechanism remain unclear. As conformational changes were observed in the structure of the Elp123 subcomplex upon binding of the Elp456 [92,117], it was hypothesized that binding of one hexameric ring

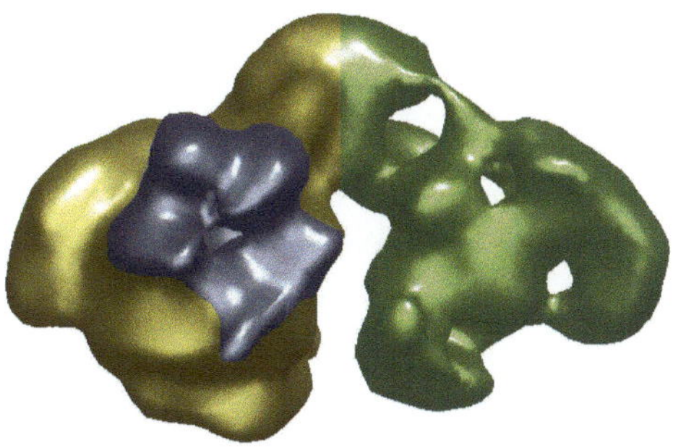

Fig. 5 Envelope model of the holoElongator complex. Approximate positions of the two pseudosymmetric Elp123 subcomplexes are shown in *yellow* and *green*, respectively, while the asymmetrically located Elp456 hexameric ring is in *blue*. Coordinates from EM Data Bank entry EMD-8239 were used to prepare this figure.

to one Elp123 wing would hinder binding of another Elp456 ring to the second Elp123 subcomplex [92,117]. However, when purification was performed in mild conditions (75 mM NaCl), the presence of a minor population of particles containing two densities on each Elp123 wing was detected, suggesting that Elongator may be able to accommodate two bound Elp456 hexameric rings [117]. Furthermore, it was shown that purified yeast Elongator supplemented with recombinant Elp456 can bind two hexameric rings even in the salt conditions originally used for Elongator purification (150 mM NaCl) [117]. In the light of these results it was concluded that no mechanism preventing loading of the second hexameric ring to Elp123 existed [117]. However, in both studies, Elongator purified from yeast contained only one copy of Elp456 [92,117]. The mechanism ensuring binding of Elp456 to only one Elp123 wing remains therefore to be elucidated. A possible regulation by levels of expression was suggested by Setiaputra and colleagues [117]. However, this model could provide only a partial explanation as it does not explain why the naked, singly, and doubly loaded Elp123 were not observed in equilibrium. It remains to be verified whether double loading of the Elp456 ring exists in vivo. Both studies confirmed the scaffolding function of Elp1 and the peripheral location of Elp2 interacting with the C-terminal domain of Elp1 and Elp3 [92,117], in agreement with the previous observations of easy dissociation of Elp2 from the complex during its purification.

Considering the structural information available about bacterial Elp3, it is somewhat surprising that no direct interactions between the Elp3 molecules were observed in the yeast holoElongator model. It is possible that dimerization of Elp3 could be triggered via conformational changes of the complex upon binding cofactors that were not present in the structure, as previously suggested [92]. It is, however, equally possible that eukaryotic Elp3 does not require dimerization, while this is prerequisite for its function in bacteria where no other components of the complex exist. Moreover, the fact that two Elp456 subcomplexes can be loaded onto the Elp123 subcomplex suggests that this spatial arrangement of Elp3 could perhaps ensure double rates of the tRNA modification, at least in some conditions. Indeed, the two splayed-apart copies of Elp123 form a central cleft containing Elp3 active site exactly on the top of where the Elp456 hexamer is placed [92]. This architecture also suggests a direct role of the Elp456 in tRNA modification. tRNA was reported to bind through its anticodon and D-loop to the basic cavity of bacterial Elp3 that is conserved between species [24]. Binding of tRNA to Elp456 was also proposed to be mediated through the anticodon loop [93]. Therefore, current models of Elongator activity have proposed a role for Elp456 in the recruitment of tRNA through initial binding to Elp456, followed by transfer to the Elp123 complex in an ATP-dependent manner where the wobble uridine modification would take place [92].

3.6 The Kti11/Kti13 Elongator Cofactors

Of all the Kti proteins, the smallest one, Kti11 is the most extensively characterized. It is a highly conserved protein with homologues found in most eukaryotes. Loss of Kti11 was shown to cause embryonic lethality in mice [118], suggesting an important cellular function. In addition to the *elp* phenotypes exhibited by the *kti11* mutant, copurification of Kti11 with the Elp123 subcomplex [97] as well as coprecipitation with Elp2 and Elp5 [119] further demonstrated the functional relationship of yeast Kti11 with Elongator. Yeast Kti11 was independently found to be identical to Dph3. Kti11/Dph3, together with diphthamide synthesis factors Dph1 and Dph2 with which it interacts, are required for the first step of biosynthesis of diphthamide [97,120,121], a posttranslationally modified histidine residue found in eukaryotic translation elongation factor 2 (eEF2), which interacts with Kti11. As diphthamide modification creates a target for the diphtheria toxin, deletion of *kti11* confers resistance to this agent [119]. In addition to other binding partners such as ribosomal proteins, Kti11 was reported to interact with the Ran guanine exchange (GEF) factor-like protein Kti13, also known as α-tubulin suppressor 1 (Ats1) [122]. Kti11 and Kti13 also seem to work in tandem, similar to Kti14 and Kti12. As found for other *kti* genes, deletion of *kti13* in yeast also protects against zymocin [123,124]. However, modification of wobble uridine nucleosides in this mutant is not completely abolished but significantly reduced [19,83], suggesting an important but not critical role of Kti13 in Elongator function. Yeast Kti13 was also shown to interact with cell polarity factor Nap1 [125,126], but the molecular role of this interaction remains to be elucidated. In humans, a putative GEF homologous to Kti13, DelGEF (also known as SERGEF) was shown to be critical for two forms of hereditary deafness [127]. DelGEF acts together with the Kti11 homologue DESR1 (also named DelGIP1), which was reported to affect proteoglycan secretion [128]. As the effect of various secretory mutations that could be suppressed by *kti* or *elp* deletions [126] was reversed by increased levels of Elongator-dependent tRNAs in yeast [28], it is possible that the reported proteoglycan secretion defects are caused by altered translation, which would further demonstrate a functional relationship between Kti11/Kti13 and Elongator. Recently, the involvement of Kti13 in the diphthamide modification pathway was reported [129] further demonstrating the close functional relationship of Kti11 and Kti13. Those two proteins were shown to coimmunoprecipitate, even in the absence of a properly assembled Elongator complex [126]. The recent purification of the yeast Kti11/Kti13 complex expressed in *E. coli* further demonstrates that a stable heterodimer is formed in the absence of any other yeast proteins [130]. Following the resolution of the atomic structure of Kti11 by nuclear magnetic resonance [131], crystal structures of the Kti11/Kti13 complex purified from *E. coli* were solved in two independent studies at 1.45 and 2.9 Å resolution, respectively [129,130] (Fig. 6A and B). These structures were solved using either a C-terminally truncated Kti11 or the full-length proteins connected with a 10-amino acid-long linker. In addition, the structure of Kti13 was also solved on its own at 2.4 Å resolution [129]. Containing one canonical RCC1 repeat, Kti13 adopts a typical arrangement for this superfamily with a seven-bladed β-propeller fold and a characteristic Velcro-like arrangement of the first/last β-sheet formed by two strands from each N-terminal and C-terminal half repeats. With a unique mixture of four- and three-stranded blades characteristic for other eukaryotic RCC1-like proteins and bacterial structural homologues, respectively, yeast Kti13 represents an intermediate form of seven-bladed propellers between

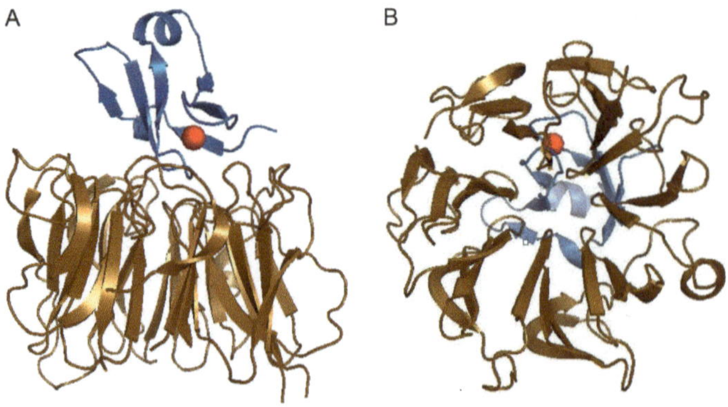

Fig. 6 Structural information for the Kti11/Kti13 Elongator cofactor. The *Saccharomyces cerevisiae* Kti11–Kti13 complex is depicted. Kti11 is in *blue*, while Kti13 is in *light brown*. The redox-active iron atom is shown as a *red sphere*. (A) *Side* and (B) *bottom views* of the complex are presented. The latter highlights the seven-bladed propeller structure of Kti13, while the former emphasizes the shielding of the Kti11-bound iron atom by Kti13. Protein Data Bank entry 4x33 was used to prepare this figure.

bacteria and higher eukaryotes [130] (Fig. 6B). Although initially described as a zinc-binding protein [131], Kti11 was also shown to bind iron [132]. Most importantly, in a high-resolution structural analysis, only iron was found to be associated with Kti11 when in complex with Kti13 [130] (Fig. 6A), indicating that Kti13 may act as a chaperone ensuring that only biologically relevant and active iron is loaded on Kti11. The presence of both zinc and iron in the complex in a second structural study may result from Kti11 binding to a neighboring Kti13 molecule and not to the one to which it was covalently linked to, and might thus not reflect bona fide complex formation in vivo [129]. Consistent with the binding of iron, Kti11 was confirmed to be a redox-active protein [130,132] and was recently shown to deliver electrons to the [4Fe–4S] cluster-containing Dph2 [133] triggering the synthesis of diphthamide via an unusual SAM radical reaction [134]. Given certain similarities of Dph2 and Elp3, it is very likely that Kti11 also acts as electron donor to [Fe–S]-containing Elongator subunit Elp3 (Fig. 4). Consistently, the iron bound to Kti11 was shown to be essential to support both Elongator-mediated tRNA modification and diphthamide biosynthesis, although its coordination seems to be important only for interactions with Elp3 and not Dph1 [129]. Formation of the Kti11/Kti13 complex was also shown to be important for both Elongator-mediated tRNA modification and diphthamide biosynthesis [129]. Some hypotheses regarding the role of Kti13 in this process were proposed based on electron transfer analysis [130]. Asymmetric complexing with Kti13, which is largely mediated via hydrophobic interactions and does not trigger major structural rearrangements in the complex components, sterically restricts the access to the iron atom of Kti11 (Fig. 6A). As this affects the electron transfer abilities of Kti11, it was proposed that Kti13 orients and shields, at least partially, the otherwise exposed iron atom in Kti11 preventing electron transfer to random redox-active molecules in the cell and, hence, ensures biologically relevant electron flow [130]. Given the involvement of Kti13 in the biosynthesis of diphthamide, the hypothesis that Kti13 lowers the midpoint potential of Kti11 upon complex formation to become a more efficient electron donor [130] has to be extended also to the reaction involving Dph1–2. Indeed, although functional separation of tRNA modification- and diphthamide biosynthesis-related activities of Kti11/Kti13 could be achieved [129], the highly conserved His297 of Kti13 directly faces the iron atom of Kti11 and was proposed to influence the redox potential of Kti11 [130]. This residue was shown to be involved in both processes as its mutation conferred resistance to zymocin as well as to diphtheria toxins [129].

Recently, Cbr1, an NADH-specific reductase, was shown act as the major endogenous reductase of Kti11 using NADH as substrate [135]. However, a Cbr1 mutation was not sufficient to block tRNA modification and diphthamide biosynthesis in vivo. This led to the identification of Mcr1, and possibly Ncp1, as redundant reductases acting in the same pathway [135] (Fig. 4). Interestingly, Cbr1 is a transmembrane protein embedded in the endoplasmic reticulum and mitochondrion outer membrane with its catalytic domain residing on the cytosolic side, suggesting that at least part of the Kti11-mediated processes might be performed in association with cell membranes. Somewhat in agreement with the localization of the Kti11 reductase, Kti13 was shown to specifically bind phospholipids that are enriched in cytosolic side of cell membrane, suggesting its possible cellular localization or enrichment at (but not in) the cell membrane [129]. The colocalization of the Kti11 reductase and Kti13 is in agreement with the hypothesis of the role of Kti13 in shielding of the reduced iron atom and suggests that Kti11/Kti13 complex formation might be coupled to the reduction of Kti11 that occurs at the cytosolic side of the cell membrane. It is noteworthy that Elp3 from *Toxoplasma gondii* was previously shown to contain a C-terminal transmembrane domain anchoring the protein at the mitochondrial outer membrane [136]. In the absence of clear homologues of Kti11, Kti13, as well as other subunits of Elongator beside Elp3, it is likely that electron transfer necessary for Elp3-mediated tRNA modification is ensured differently in this protozoan and therefore requires the direct localization of Elp3 in a membrane.

4. CONCLUDING REMARKS

The highly conserved eukaryotic Elongator complex is composed of dimeric heterotrimer Elp123 complex with an asymmetrically attached heterohexameric Elp456 ring. Although originally identified as a transcription elongation factor and despite being associated with many other cellular functions over the years, the vast majority of studies performed within last decade suggest that tRNA modification is the predominant function of Elongator. Therefore, the name of this complex, Elongator, which was given because it was thought to interact with an elongating form of yeast Pol.II now seems rather unfortunate and inadequate, unless it is interpreted as reflecting its involvement in elongation of translation.

Recent high-resolution structural studies provided information about all six Elp proteins and/or its homologues and about the complex cofactor

Kti11/Kti13, while the architecture of the entire complex was elucidated at lower resolution by electron microscopy analyses. Due to those studies our understanding about Elongator and its function has dramatically improved. Additional functional and structural information about the formation of modified wobble U residues in eukaryotes recently emerged from the study of the Trm9–Trm112 dimer [137]. Yet further high-resolution structural analyses coupled to biochemical assays are required to reveal the mode of action and biological function of Elongator and its functional partners. Detailed elucidation of the molecular mechanism underlying the activity of Elongator and its regulation will in turn allow us to define how alterations of this complex contribute to neurological disorders. Further characterization of the emerging role of Elongator in the context of these diseases should provide the basis for development of treatment for a variety of devastating neuropathies.

ACKNOWLEDGMENTS

We thank our colleagues and team members for discussions and IGBMC services for support. We apologize for publications that could not be quoted. This work was supported by Ligue Contre le Cancer (Equipe Labellisée 2014), Centre National pour la Recherche Scientifique, CERBM IGBMC, the project Elongator (grant ANR-13-BSV8-0005-01). This study was also supported by the grant ANR-10-LABX-0030-INRT, a French State fund managed by the Agence Nationale de la Recherche under the frame program Investissements d'Avenir ANR-10-IDEX-0002-02.

REFERENCES

[1] G. Otero, J. Fellows, Y. Li, T. de Bizemont, A.M. Dirac, C.M. Gustafsson, H. Erdjument-Bromage, P. Tempst, J.Q. Svejstrup, Elongator, a multisubunit component of a novel RNA polymerase II holoenzyme for transcriptional elongation, Mol. Cell 3 (1999) 109–118.

[2] B.O. Wittschieben, G. Otero, T. de Bizemont, J. Fellows, H. Erdjument-Bromage, R. Ohba, Y. Li, C.D. Allis, P. Tempst, J.Q. Svejstrup, A novel histone acetyltransferase is an integral subunit of elongating RNA polymerase II holoenzyme, Mol. Cell 4 (1) (1999) 123–128.

[3] N.J. Krogan, J.F. Greenblatt, Characterization of a six-subunit holo-elongator complex required for the regulated expression of a group of genes in Saccharomyces cerevisiae, Mol. Cell. Biol. 21 (23) (2001) 8203–8212.

[4] G.S. Winkler, T.G. Petrakis, S. Ethelberg, M. Tokunaga, H. Erdjument-Bromage, P. Tempst, J.Q. Svejstrup, RNA polymerase II elongator holoenzyme is composed of two discrete subcomplexes, J. Biol. Chem. 276 (35) (2001) 32743–32749.

[5] Y. Li, Y. Takagi, Y. Jiang, M. Tokunaga, H. Erdjument-Bromage, P. Tempst, R.D. Kornberg, A multiprotein complex that interacts with RNA polymerase II elongator, J. Biol. Chem. 276 (32) (2001) 29628–29631.

[6] G.S. Winkler, A. Kristjuhan, H. Erdjument-Bromage, P. Tempst, J.Q. Svejstrup, Elongator is a histone H3 and H4 acetyltransferase important for normal histone acetylation levels in vivo, Proc. Natl. Acad. Sci. U.S.A. 99 (6) (2002) 3517–3522.

[7] B.O. Wittschieben, J. Fellows, W. Du, D.J. Stillman, J.Q. Svejstrup, Overlapping roles for the histone acetyltransferase activities of SAGA and elongator in vivo, EMBO J. 19 (12) (2000) 3060–3068.

[8] V. Van Mullem, M. Wery, M. Werner, J. Vandenhaute, P. Thuriaux, The Rpb9 subunit of RNA polymerase II binds transcription factor TFIIE and interferes with the SAGA and elongator histone acetyltransferases, J. Biol. Chem. 277 (12) (2002) 10220–10225.

[9] S.E. Kong, M.S. Kobor, N.J. Krogan, B.P. Somesh, T.M. Sogaard, J.F. Greenblatt, J.Q. Svejstrup, Interaction of Fcp1 phosphatase with elongating RNA polymerase II holoenzyme, enzymatic mechanism of action, and genetic interaction with elongator, J. Biol. Chem. 280 (6) (2005) 4299–4306.

[10] G. Jona, B.O. Wittschieben, J.Q. Svejstrup, O. Gileadi, Involvement of yeast carboxy-terminal domain kinase I (CTDK-I) in transcription elongation in vivo, Gene 267 (1) (2001) 31–36.

[11] T. Formosa, P. Eriksson, J. Wittmeyer, J. Ginn, Y. Yu, D.J. Stillman, Spt16-Pob3 and the HMG protein Nhp6 combine to form the nucleosome-binding factor SPN, EMBO J. 20 (13) (2001) 3506–3517.

[12] C. Gilbert, A. Kristjuhan, G.S. Winkler, J.Q. Svejstrup, Elongator interactions with nascent mRNA revealed by RNA immunoprecipitation, Mol. Cell 14 (4) (2004) 457–464.

[13] J. Fellows, H. Erdjument-Bromage, P. Tempst, J.Q. Svejstrup, The Elp2 subunit of elongator and elongating RNA polymerase II holoenzyme is a WD40 repeat protein, J. Biol. Chem. 275 (17) (2000) 12896–12899.

[14] Q. Li, A.M. Fazly, H. Zhou, S. Huang, Z. Zhang, B. Stillman, The elongator complex interacts with PCNA and modulates transcriptional silencing and sensitivity to DNA damage agents, PLoS Genet. 5 (10) (2009) e1000684.

[15] D.K. Pokholok, N.M. Hannett, R.A. Young, Exchange of RNA polymerase II initiation and elongation factors during gene expression in vivo, Mol. Cell 9 (4) (2002) 799–809.

[16] W.K. Huh, J.V. Falvo, L.C. Gerke, A.S. Carroll, R.W. Howson, J.S. Weissman, E.K. O'Shea, Global analysis of protein localization in budding yeast, Nature 425 (6959) (2003) 686–691.

[17] P.B. Rahl, C.Z. Chen, R.N. Collins, Elp1p, the yeast homolog of the FD disease syndrome protein, negatively regulates exocytosis independently of transcriptional elongation, Mol. Cell 17 (6) (2005) 841–853.

[18] W.D. Heyer, P. Thuriaux, J. Kohli, P. Ebert, H. Kersten, C. Gehrke, K.C. Kuo, P.F. Agris, An antisuppressor mutation of Schizosaccharomyces pombe affects the post-transcriptional modification of the "wobble" base in the anticodon of tRNAs, J. Biol. Chem. 259 (5) (1984) 2856–2862.

[19] B. Huang, M.J. Johansson, A.S. Bystrom, An early step in wobble uridine tRNA modification requires the Elongator complex, RNA 11 (4) (2005) 424–436.

[20] M.J. Johansson, A. Esberg, B. Huang, G.R. Bjork, A.S. Bystrom, Eukaryotic wobble uridine modifications promote a functionally redundant decoding system, Mol. Cell. Biol. 28 (10) (2008) 3301–3312.

[21] F.A. Vendeix, F.V. Murphy IV, W.A. Cantara, G. Leszczynska, E.M. Gustilo, B. Sproat, A. Malkiewicz, P.F. Agris, Human tRNA(Lys3)(UUU) is pre-structured by natural modifications for cognate and wobble codon binding through keto-enol tautomerism, J. Mol. Biol. 416 (4) (2012) 467–485.

[22] C. Chen, B. Huang, J.T. Anderson, A.S. Bystrom, Unexpected accumulation of ncm (5)U and ncm(5)S(2) (U) in a trm9 mutant suggests an additional step in the synthesis of mcm(5)U and mcm(5)S(2)U, PLoS One 6 (6) (2011) e20783.

[23] K. Selvadurai, P. Wang, J. Seimetz, R.H. Huang, Archaeal Elp3 catalyzes tRNA wobble uridine modification at C5 via a radical mechanism, Nat. Chem. Biol. 10 (10) (2014) 810–812.

[24] S. Glatt, R. Zabel, O. Kolaj-Robin, O.F. Onuma, F. Baudin, A. Graziadei, V. Taverniti, T.Y. Lin, F. Baymann, B. Seraphin, K.D. Breunig, C.W. Muller, Structural basis for tRNA modification by Elp3 from Dehalococcoides mccartyi, Nat. Struct. Mol. Biol. 23 (9) (2016) 794–802.

[25] F. Li, J. Lu, Q. Han, G. Zhang, B. Huang, The Elp3 subunit of human Elongator complex is functionally similar to its counterpart in yeast, Mol. Genet. Genomics 273 (3) (2005) 264–272.

[26] F. Li, Q.J. Han, D.H. Luo, J. Lu, B.Q. Huang, The Elp4 subunit of human Elongator complex partially complements the growth defects of yeast ELP4 deletion strain, Yi Chuan Xue Bao 31 (7) (2004) 668–674.

[27] C. Chen, B. Huang, M. Eliasson, P. Ryden, A.S. Bystrom, Elongator complex influences telomeric gene silencing and DNA damage response by its role in wobble uridine tRNA modification, PLoS Genet. 7 (9) (2011) e1002258.

[28] A. Esberg, B. Huang, M.J. Johansson, A.S. Bystrom, Elevated levels of two tRNA species bypass the requirement for elongator complex in transcription and exocytosis, Mol. Cell 24 (1) (2006) 139–148.

[29] F. Bauer, A. Matsuyama, J. Candiracci, M. Dieu, J. Scheliga, D.A. Wolf, M. Yoshida, D. Hermand, Translational control of cell division by Elongator, Cell Rep. 1 (5) (2012) 424–433.

[30] M. Tigano, R. Ruotolo, C. Dallabona, F. Fontanesi, A. Barrientos, C. Donnini, S. Ottonello, Elongator-dependent modification of cytoplasmic tRNALysUUU is required for mitochondrial function under stress conditions, Nucleic Acids Res. 43 (17) (2015) 8368–8380.

[31] E. Bolukbasi, S. Vass, N. Cobbe, B. Nelson, V. Simossis, D.R. Dunbar, M.M. Heck, Drosophila poly suggests a novel role for the Elongator complex in insulin receptor-target of rapamycin signalling, Open Biol. 2 (1) (2012) 110031.

[32] J. Walker, S.Y. Kwon, P. Badenhorst, P. East, H. McNeill, J.Q. Svejstrup, Role of elongator subunit Elp3 in Drosophila melanogaster larval development and immunity, Genetics 187 (4) (2011) 1067–1075.

[33] C.L. Simpson, et al., Variants of the elongator protein 3 (ELP3) gene are associated with motor neuron degeneration, Hum. Mol. Genet. 18 (3) (2009) 472–481.

[34] K. Miskiewicz, L.E. Jose, A. Bento-Abreu, M. Fislage, I. Taes, J. Kasprowicz, J. Swerts, S. Sigrist, W. Versees, W. Robberecht, P. Verstreken, ELP3 controls active zone morphology by acetylating the ELKS family member Bruchpilot, Neuron 72 (5) (2011) 776–788.

[35] N. Singh, M.T. Lorbeck, A. Zervos, J. Zimmerman, F. Elefant, The histone acetyltransferase Elp3 plays in active role in the control of synaptic bouton expansion and sleep in Drosophila, J. Neurochem. 115 (2) (2010) 493–504.

[36] H. Yoo, D. Son, Y.J. Jang, K. Hong, Indispensable role for mouse ELP3 in embryonic stem cell maintenance and early development, Biochem. Biophys. Res. Commun. 478 (2) (2016) 631–636.

[37] Y.T. Chen, M.M. Hims, R.S. Shetty, J. Mull, L. Liu, M. Leyne, S.A. Slaugenhaupt, Loss of mouse Ikbkap, a subunit of elongator, leads to transcriptional deficits and embryonic lethality that can be rescued by human IKBKAP, Mol. Cell. Biol. 29 (3) (2009) 736–744.

[38] F.J. Lin, L. Shen, C.W. Jang, P.O. Falnes, Y. Zhang, Ikbkap/Elp1 deficiency causes male infertility by disrupting meiotic progression, PLoS Genet. 9 (5) (2013) e1003516.

[39] Y. Okada, K. Yamagata, K. Hong, T. Wakayama, Y. Zhang, A role for the elongator complex in zygotic paternal genome demethylation, Nature 463 (7280) (2010) 554–558.
[40] C. Creppe, L. Malinouskaya, M.L. Volvert, M. Gillard, P. Close, O. Malaise, S. Laguesse, I. Cornez, S. Rahmouni, S. Ormenese, S. Belachew, B. Malgrange, J.P. Chapelle, U. Siebenlist, G. Moonen, A. Chariot, L. Nguyen, Elongator controls the migration and differentiation of cortical neurons through acetylation of alpha-tubulin, Cell 136 (3) (2009) 551–564.
[41] J.A. Solinger, R. Paolinelli, H. Kloss, F.B. Scorza, S. Marchesi, U. Sauder, D. Mitsushima, F. Capuani, S.R. Sturzenbaum, G. Cassata, The Caenorhabditis elegans Elongator complex regulates neuronal alpha-tubulin acetylation, PLoS Genet. 6 (1) (2010) e1000820.
[42] C. Chen, S. Tuck, A.S. Bystrom, Defects in tRNA modification associated with neurological and developmental dysfunctions in Caenorhabditis elegans elongator mutants, PLoS Genet. 5 (7) (2009) e1000561.
[43] F. Bauer, D. Hermand, A coordinated codon-dependent regulation of translation by Elongator, Cell Cycle 11 (24) (2012) 4524–4529.
[44] S. Laguesse, C. Creppe, D.D. Nedialkova, P.P. Prevot, L. Borgs, S. Huysseune, B. Franco, G. Duysens, N. Krusy, G. Lee, N. Thelen, M. Thiry, P. Close, A. Chariot, B. Malgrange, S.A. Leidel, J.D. Godin, L. Nguyen, A dynamic unfolded protein response contributes to the control of cortical neurogenesis, Dev. Cell 35 (5) (2015) 553–567.
[45] S. Tielens, S. Huysseune, J.D. Godin, A. Chariot, B. Malgrange, L. Nguyen, Elongator controls cortical interneuron migration by regulating actomyosin dynamics, Cell Res. 26 (10) (2016) 1131–1148.
[46] X. Yang, J. Li, W. Zeng, C. Li, B. Mao, Elongator protein 3 (Elp3) stabilizes Snail1 and regulates neural crest migration in Xenopus, Sci. Rep. 6 (2016) 26238.
[47] Y. Ding, Z. Mou, Elongator and its epigenetic role in plant development and responses to abiotic and biotic stresses, Front. Plant Sci. 6 (2015) 296.
[48] X. Fang, Y. Cui, Y. Li, Y. Qi, Transcription and processing of primary microRNAs are coupled by Elongator complex in Arabidopsis, Nat. Plants 1 (2015) 15075.
[49] C. Mehlgarten, D. Jablonowski, U. Wrackmeyer, S. Tschitschmann, D. Sondermann, G. Jager, Z. Gong, A.S. Bystrom, R. Schaffrath, K.D. Breunig, Elongator function in tRNA wobble uridine modification is conserved between yeast and plants, Mol. Microbiol. 76 (5) (2010) 1082–1094.
[50] P. Chen, G. Jager, B. Zheng, Transfer RNA modifications and genes for modifying enzymes in Arabidopsis thaliana, BMC Plant Biol. 10 (2010) 201.
[51] B.J. Landgraf, E.L. McCarthy, S.J. Booker, Radical S-adenosylmethionine enzymes in human health and disease, Annu. Rev. Biochem. 85 (2016) 485–514.
[52] A. Ladang, et al., Elp3 drives Wnt-dependent tumor initiation and regeneration in the intestine, J. Exp. Med. 212 (12) (2015) 2057–2075.
[53] T. Karlsborn, H. Tukenmez, A.K. Mahmud, F. Xu, H. Xu, A.S. Bystrom, Elongator, a conserved complex required for wobble uridine modifications in eukaryotes, RNA Biol. 11 (12) (2014) 1519–1528.
[54] S.A. Slaugenhaupt, A. Blumenfeld, S.P. Gill, M. Leyne, J. Mull, M.P. Cuajungco, C.B. Liebert, B. Chadwick, M. Idelson, L. Reznik, C. Robbins, I. Makalowska, M. Brownstein, D. Krappmann, C. Scheidereit, C. Maayan, F.B. Axelrod, J.F. Gusella, Tissue-specific expression of a splicing mutation in the IKBKAP gene causes familial dysautonomia, Am. J. Hum. Genet. 68 (3) (2001) 598–605.
[55] S.L. Anderson, R. Coli, I.W. Daly, E.A. Kichula, M.J. Rork, S.A. Volpi, J. Ekstein, B.Y. Rubin, Familial dysautonomia is caused by mutations of the IKAP gene, Am. J. Hum. Genet. 68 (3) (2001) 753–758.

[56] C. Maayan, E. Kaplan, S. Shachar, O. Peleg, S. Godfrey, Incidence of familial dysautonomia in Israel 1977–1981, Clin. Genet. 32 (2) (1987) 106–108.
[57] C.M. Riley, R.L. Day, et al., Central autonomic dysfunction with defective lacrimation; report of five cases, Pediatrics 3 (4) (1949) 468–478.
[58] L. Cohen, W.J. Henzel, P.A. Baeuerle, IKAP is a scaffold protein of the IkappaB kinase complex, Nature 395 (6699) (1998) 292–296.
[59] E.C. Ibrahim, M.M. Hims, N. Shomron, C.B. Burge, S.A. Slaugenhaupt, R. Reed, Weak definition of IKBKAP exon 20 leads to aberrant splicing in familial dysautonomia, Hum. Mutat. 28 (1) (2007) 41–53.
[60] L.D. Johansen, T. Naumanen, A. Knudsen, N. Westerlund, I. Gromova, M. Junttila, C. Nielsen, T. Bottzauw, A. Tolkovsky, J. Westermarck, E.T. Coffey, M. Jaattela, T. Kallunki, IKAP localizes to membrane ruffles with filamin A and regulates actin cytoskeleton organization and cell migration, J. Cell Sci. 121 (Pt. 6) (2008) 854–864.
[61] L. George, M. Chaverra, L. Wolfe, J. Thorne, M. Close-Davis, A. Eibs, V. Riojas, A. Grindeland, M. Orr, G.A. Carlson, F. Lefcort, Familial dysautonomia model reveals Ikbkap deletion causes apoptosis of Pax3+ progenitors and peripheral neurons, Proc. Natl. Acad. Sci. U.S.A. 110 (46) (2013) 18698–18703.
[62] M.Z. Jackson, K.A. Gruner, C. Qin, W.G. Tourtellotte, A neuron autonomous role for the familial dysautonomia gene ELP1 in sympathetic and sensory target tissue innervation, Development 141 (12) (2014) 2452–2461.
[63] M. Yoshida, N. Kataoka, K. Miyauchi, K. Ohe, K. Iida, S. Yoshida, T. Nojima, Y. Okuno, H. Onogi, T. Usui, A. Takeuchi, T. Hosoya, T. Suzuki, M. Hagiwara, Rectifier of aberrant mRNA splicing recovers tRNA modification in familial dysautonomia, Proc. Natl. Acad. Sci. U.S.A. 112 (9) (2015) 2764–2769.
[64] T. Karlsborn, H. Tukenmez, C. Chen, A.S. Bystrom, Familial dysautonomia (FD) patients have reduced levels of the modified wobble nucleoside mcm(5)s(2)U in tRNA, Biochem. Biophys. Res. Commun. 454 (3) (2014) 441–445.
[65] S. Takeoka, M. Unoki, Y. Onouchi, S. Doi, H. Fujiwara, A. Miyatake, K. Fujita, I. Inoue, Y. Nakamura, M. Tamari, Amino-acid substitutions in the IKAP gene product significantly increase risk for bronchial asthma in children, J. Hum. Genet. 46 (2) (2001) 57–63.
[66] H. Xu, Z. Lin, F. Li, W. Diao, C. Dong, H. Zhou, X. Xie, Z. Wang, Y. Shen, J. Long, Dimerization of elongator protein 1 is essential for Elongator complex assembly, Proc. Natl. Acad. Sci. U.S.A. 112 (34) (2015) 10697–10702.
[67] L.C. Kwee, Y. Liu, C. Haynes, J.R. Gibson, A. Stone, S.A. Schichman, F. Kamel, L.M. Nelson, B. Topol, S.K. Van den Eeden, C.M. Tanner, M.E. Cudkowicz, D.L. Grasso, R. Lawson, S. Muralidhar, E.Z. Oddone, S. Schmidt, M.A. Hauser, A high-density genome-wide association screen of sporadic ALS in US veterans, PLoS One 7 (3) (2012) e32768.
[68] L. Nguyen, S. Humbert, F. Saudou, A. Chariot, Elongator—an emerging role in neurological disorders, Trends Mol. Med. 16 (1) (2010) 1–6.
[69] L.J. Strug, T. Clarke, T. Chiang, M. Chien, Z. Baskurt, W. Li, R. Dorfman, B. Bali, E. Wirrell, S.L. Kugler, D.E. Mandelbaum, S.M. Wolf, P. McGoldrick, H. Hardison, E.J. Novotny, J. Ju, D.A. Greenberg, J.J. Russo, D.K. Pal, Centrotemporal sharp wave EEG trait in rolandic epilepsy maps to Elongator Protein Complex 4 (ELP4), Eur. J. Hum. Genet. 17 (9) (2009) 1171–1181.
[70] D.K. Pal, W. Li, T. Clarke, P. Lieberman, L.J. Strug, Pleiotropic effects of the 11p13 locus on developmental verbal dyspraxia and EEG centrotemporal sharp waves, Genes Brain Behav. 9 (8) (2010) 1004–1012.
[71] H. Najmabadi, et al., Deep sequencing reveals 50 novel genes for recessive cognitive disorders, Nature 478 (7367) (2011) 57–63.

[72] J.S. Cohen, S. Srivastava, K.D. Farwell, H.M. Lu, W. Zeng, H. Lu, E.C. Chao, A. Fatemi, ELP2 is a novel gene implicated in neurodevelopmental disabilities, Am. J. Med. Genet. A 167 (6) (2015) 1391–1395.
[73] P. Close, M. Gillard, A. Ladang, Z. Jiang, J. Papuga, N. Hawkes, L. Nguyen, J.P. Chapelle, F. Bouillenne, J. Svejstrup, M. Fillet, A. Chariot, DERP6 (ELP5) and C3ORF75 (ELP6) regulate tumorigenicity and migration of melanoma cells as subunits of Elongator, J. Biol. Chem. 287 (39) (2012) 32535–32545.
[74] S. Delaunay, F. Rapino, L. Tharun, Z. Zhou, L. Heukamp, M. Termathe, K. Shostak, I. Klevernic, A. Florin, H. Desmecht, C.J. Desmet, L. Nguyen, S.A. Leidel, A.E. Willis, R. Buttner, A. Chariot, P. Close, Elp3 links tRNA modification to IRES-dependent translation of LEF1 to sustain metastasis in breast cancer, J. Exp. Med. 213 (11) (2016) 2503–2523.
[75] D.D. Nedialkova, S.A. Leidel, Optimization of Codon translation rates via tRNA modifications maintains proteome integrity, Cell 161 (7) (2015) 1606–1618.
[76] T. Karlsborn, A.K. Mahmud, H. Tukenmez, A.S. Bystrom, Loss of ncm5 and mcm5 wobble uridine side chains results in an altered metabolic profile, Metabolomics 12 (12) (2016) 177.
[77] F. Frohloff, L. Fichtner, D. Jablonowski, K.D. Breunig, R. Schaffrath, Saccharomyces cerevisiae Elongator mutations confer resistance to the Kluyveromyces lactis zymocin, EMBO J. 20 (8) (2001) 1993–2003.
[78] D. Jablonowski, F. Frohloff, L. Fichtner, M.J. Stark, R. Schaffrath, Kluyveromyces lactis zymocin mode of action is linked to RNA polymerase II function via Elongator, Mol. Microbiol. 42 (4) (2001) 1095–1105.
[79] M.J. Stark, A. Boyd, The killer toxin of Kluyveromyces lactis: characterization of the toxin subunits and identification of the genes which encode them, EMBO J. 5 (8) (1986) 1995–2002.
[80] J. Lu, B. Huang, A. Esberg, M.J. Johansson, A.S. Bystrom, The Kluyveromyces lactis gamma-toxin targets tRNA anticodons, RNA 11 (11) (2005) 1648–1654.
[81] J. Lu, A. Esberg, B. Huang, A.S. Bystrom, Kluyveromyces lactis gamma-toxin, a ribonuclease that recognizes the anticodon stem loop of tRNA, Nucleic Acids Res. 36 (4) (2008) 1072–1080.
[82] D. Jablonowski, S. Zink, C. Mehlgarten, G. Daum, R. Schaffrath, tRNAGlu wobble uridine methylation by Trm9 identifies Elongator's key role for zymocin-induced cell death in yeast, Mol. Microbiol. 59 (2) (2006) 677–688.
[83] B. Huang, J. Lu, A.S. Bystrom, A genome-wide screen identifies genes required for formation of the wobble nucleoside 5-methoxycarbonylmethyl-2-thiouridine in Saccharomyces cerevisiae, RNA 14 (10) (2008) 2183–2194.
[84] S. Leidel, P.G. Pedrioli, T. Bucher, R. Brost, M. Costanzo, A. Schmidt, R. Aebersold, C. Boone, K. Hofmann, M. Peter, Ubiquitin-related modifier Urm1 acts as a sulphur carrier in thiolation of eukaryotic transfer RNA, Nature 458 (7235) (2009) 228–232.
[85] H.R. Kalhor, S. Clarke, Novel methyltransferase for modified uridine residues at the wobble position of tRNA, Mol. Cell. Biol. 23 (24) (2003) 9283–9292.
[86] M.H. Mazauric, L. Dirick, S.K. Purushothaman, G.R. Bjork, B. Lapeyre, Trm112p is a 15-kDa zinc finger protein essential for the activity of two tRNA and one protein methyltransferases in yeast, J. Biol. Chem. 285 (24) (2010) 18505–18515.
[87] L. Songe-Moller, E. van den Born, V. Leihne, C.B. Vagbo, T. Kristoffersen, H.E. Krokan, F. Kirpekar, P.O. Falnes, A. Klungland, Mammalian ALKBH8 possesses tRNA methyltransferase activity required for the biogenesis of multiple wobble uridine modifications implicated in translational decoding, Mol. Cell. Biol. 30 (7) (2010) 1814–1827.
[88] S.K. Purushothaman, J.M. Bujnicki, H. Grosjean, B. Lapeyre, Trm11p and Trm112p are both required for the formation of 2-methylguanosine at position 10 in yeast tRNA, Mol. Cell. Biol. 25 (11) (2005) 4359–4370.

[89] V. Leihne, F. Kirpekar, C.B. Vagbo, E. van den Born, H.E. Krokan, P.E. Grini, T.J. Meza, P.O. Falnes, Roles of Trm9- and ALKBH8-like proteins in the formation of modified wobble uridines in Arabidopsis tRNA, Nucleic Acids Res. 39 (17) (2011) 7688–7701.
[90] S. Glatt, C.W. Muller, Structural insights into Elongator function, Curr. Opin. Struct. Biol. 23 (2) (2013) 235–242.
[91] F. Frohloff, D. Jablonowski, L. Fichtner, R. Schaffrath, Subunit communications crucial for the functional integrity of the yeast RNA polymerase II elongator (gamma-toxin target (TOT)) complex, J. Biol. Chem. 278 (2) (2003) 956–961.
[92] M.I. Dauden, J. Kosinski, O. Kolaj-Robin, A. Desfosses, A. Ori, C. Faux, N.A. Hoffmann, O.F. Onuma, K.D. Breunig, M. Beck, C. Sachse, B. Seraphin, S. Glatt, C.W. Müller, Architecture of the yeast Elongator complex. EMBO Rep. 18 (2017) 264–279, http://dx.doi.org/10.15252/embr.201643353.
[93] S. Glatt, J. Letoquart, C. Faux, N.M. Taylor, B. Seraphin, C.W. Muller, The Elongator subcomplex Elp456 is a hexameric RecA-like ATPase, Nat. Struct. Mol. Biol. 19 (3) (2012) 314–320.
[94] C. Holmberg, S. Katz, M. Lerdrup, T. Herdegen, M. Jaattela, A. Aronheim, T. Kallunki, A novel specific role for I kappa B kinase complex-associated protein in cytosolic stress signaling, J. Biol. Chem. 277 (35) (2002) 31918–31928.
[95] C. Mehlgarten, D. Jablonowski, K.D. Breunig, M.J. Stark, R. Schaffrath, Elongator function depends on antagonistic regulation by casein kinase Hrr25 and protein phosphatase Sit4, Mol. Microbiol. 73 (5) (2009) 869–881.
[96] R. Di Santo, S. Bandau, M.J. Stark, A conserved and essential basic region mediates tRNA binding to the Elp1 subunit of the Saccharomyces cerevisiae Elongator complex, Mol. Microbiol. 92 (2014) 1227–1242.
[97] L. Fichtner, D. Jablonowski, A. Schierhorn, H.K. Kitamoto, M.J. Stark, R. Schaffrath, Elongator's toxin-target (TOT) function is nuclear localization sequence dependent and suppressed by post-translational modification, Mol. Microbiol. 49 (5) (2003) 1297–1307.
[98] D. Jablonowski, L. Fichtner, M.J. Stark, R. Schaffrath, The yeast elongator histone acetylase requires Sit4-dependent dephosphorylation for toxin-target capacity, Mol. Biol. Cell 15 (3) (2004) 1459–1469.
[99] A. Woodacre, M.A. Lone, D. Jablonowski, R. Schneiter, F. Giorgini, R. Schaffrath, A novel Sit4 phosphatase complex is involved in the response to ceramide stress in yeast, Oxid. Med. Cell. Longev. 2013 (2013) 129645.
[100] W. Abdel-Fattah, D. Jablonowski, R. Di Santo, K.L. Thuring, V. Scheidt, A. Hammermeister, S. Ten Have, M. Helm, R. Schaffrath, M.J. Stark, Phosphorylation of Elp1 by Hrr25 is required for elongator-dependent tRNA modification in yeast, PLoS Genet. 11 (1) (2015) e1004931.
[101] H. Xu, J. Bygdell, G. Wingsle, A.S. Bystrom, Yeast Elongator protein Elp1p does not undergo proteolytic processing in exponentially growing cells, Microbiologyopen 4 (6) (2015) 867–878.
[102] C.U. Stirnimann, E. Petsalaki, R.B. Russell, C.W. Muller, WD40 proteins propel cellular networks, Trends Biochem. Sci. 35 (10) (2010) 565–574.
[103] C. Xu, J. Min, Structure and function of WD40 domain proteins, Protein Cell 2 (3) (2011) 202–214.
[104] C. Dong, Z. Lin, W. Diao, D. Li, X. Chu, Z. Wang, H. Zhou, Z. Xie, Y. Shen, J. Long, The Elp2 subunit is essential for elongator complex assembly and functional regulation, Structure 23 (6) (2015) 1078–1086.
[105] T.G. Petrakis, B.O. Wittschieben, J.Q. Svejstrup, Molecular architecture, structure-function relationship, and importance of the Elp3 subunit for the RNA binding of holo-elongator, J. Biol. Chem. 279 (31) (2004) 32087–32092.

[106] N.A. Hawkes, G. Otero, G.S. Winkler, N. Marshall, M.E. Dahmus, D. Krappmann, C. Scheidereit, C.L. Thomas, G. Schiavo, H. Erdjument-Bromage, P. Tempst, J.Q. Svejstrup, Purification and characterization of the human elongator complex, J. Biol. Chem. 277 (4) (2002) 3047–3052.

[107] R.G. Collum, S. Brutsaert, G. Lee, C. Schindler, A Stat3-interacting protein (StIP1) regulates cytokine signal transduction, Proc. Natl. Acad. Sci. U.S.A. 97 (18) (2000) 10120–10125.

[108] I. Moukadiri, S. Prado, J. Piera, A. Velazquez-Campoy, G.R. Bjork, M.E. Armengod, Evolutionarily conserved proteins MnmE and GidA catalyze the formation of two methyluridine derivatives at tRNA wobble positions, Nucleic Acids Res. 37 (21) (2009) 7177–7193.

[109] J.A. McCloskey, D.E. Graham, S. Zhou, P.F. Crain, M. Ibba, J. Konisky, D. Soll, G.J. Olsen, Post-transcriptional modification in archaeal tRNAs: identities and phylogenetic relations of nucleotides from mesophilic and hyperthermophilic Methanococcales, Nucleic Acids Res. 29 (22) (2001) 4699–4706.

[110] C. Greenwood, L.A. Selth, A.B. Dirac-Svejstrup, J.Q. Svejstrup, An iron-sulfur cluster domain in Elp3 important for the structural integrity of elongator, J. Biol. Chem. 284 (1) (2009) 141–149.

[111] C. Paraskevopoulou, S.A. Fairhurst, D.J. Lowe, P. Brick, S. Onesti, The Elongator subunit Elp3 contains a Fe4S4 cluster and binds S-adenosylmethionine, Mol. Microbiol. 59 (3) (2006) 795–806.

[112] J. Wang, R.P. Woldring, G.D. Roman-Melendez, A.M. McClain, B.R. Alzua, E.N. Marsh, Recent advances in radical SAM enzymology: new structures and mechanisms, ACS Chem. Biol. 9 (9) (2014) 1929–1938.

[113] E.L. Schwalm, T.L. Grove, S.J. Booker, A.K. Boal, Crystallographic capture of a radical S-adenosylmethionine enzyme in the act of modifying tRNA, Science 352 (6283) (2016) 309–312.

[114] Z. Lin, W. Zhao, W. Diao, X. Xie, Z. Wang, J. Zhang, Y. Shen, J. Long, Crystal structure of elongator subcomplex Elp4-6, J. Biol. Chem. 287 (25) (2012) 21501–21508.

[115] A. Soulard, A. Cremonesi, S. Moes, F. Schutz, P. Jeno, M.N. Hall, The rapamycin-sensitive phosphoproteome reveals that TOR controls protein kinase A toward some but not all substrates, Mol. Biol. Cell 21 (19) (2010) 3475–3486.

[116] A.G. Van der Veen, K. Schorpp, C. Schlieker, L. Buti, J.R. Damon, E. Spooner, H.L. Ploegh, S. Jentsch, Role of the ubiquitin-like protein Urm1 as a noncanonical lysine-directed protein modifier, Proc. Natl. Acad. Sci. U.S.A. 108 (5) (2011) 1763–1770.

[117] D.T. Setiaputra, D.T.H. Cheng, S. Lu, J.M. Hansen, U. Dalwadi, C.H.Y. Lam, J.T. To, M.-Q. Dong, C.K. Yip, Molecular architecture of the yeast Elongator complex reveals an unexpected asymmetric subunit arrangement, EMBO Rep. 18 (2017) 280–291, http://dx.doi.org/10.15252/embr.201642548.

[118] S. Liu, J.F. Wiggins, T. Sreenath, A.B. Kulkarni, J.M. Ward, S.H. Leppla, Dph3, a small protein required for diphthamide biosynthesis, is essential in mouse development, Mol. Cell. Biol. 26 (10) (2006) 3835–3841.

[119] C. Bar, R. Zabel, S. Liu, M.J. Stark, R. Schaffrath, A versatile partner of eukaryotic protein complexes that is involved in multiple biological processes: Kti11/Dph3, Mol. Microbiol. 69 (5) (2008) 1221–1233.

[120] S. Liu, G.T. Milne, J.G. Kuremsky, G.R. Fink, S.H. Leppla, Identification of the proteins required for biosynthesis of diphthamide, the target of bacterial ADP-ribosylating toxins on translation elongation factor 2, Mol. Cell. Biol. 24 (21) (2004) 9487–9497.

[121] S. Liu, S.H. Leppla, Retroviral insertional mutagenesis identifies a small protein required for synthesis of diphthamide, the target of bacterial ADP-ribosylating toxins, Mol. Cell 12 (3) (2003) 603–613.

[122] D. Kirkpatrick, F. Solomon, Overexpression of yeast homologs of the mammalian checkpoint gene RCC1 suppresses the class of alpha-tubulin mutations that arrest with excess microtubules, Genetics 137 (2) (1994) 381–392.

[123] L. Fichtner, R. Schaffrath, KTI11 and KTI13, Saccharomyces cerevisiae genes controlling sensitivity to G1 arrest induced by Kluyveromyces lactis zymocin, Mol. Microbiol. 44 (3) (2002) 865–875.

[124] A.R. Butler, J.H. White, Y. Folawiyo, A. Edlin, D. Gardiner, M.J. Stark, Two Saccharomyces cerevisiae genes which control sensitivity to G1 arrest induced by Kluyveromyces lactis toxin, Mol. Cell. Biol. 14 (9) (1994) 6306–6316.

[125] C.M. Shields, R. Taylor, T. Nazarenus, J. Cheatle, A. Hou, A. Tapprich, A. Haifley, A.L. Atkin, Saccharomyces cerevisiae Ats1p interacts with Nap1p, a cytoplasmic protein that controls bud morphogenesis, Curr. Genet. 44 (4) (2003) 184–194.

[126] R. Zabel, C. Bar, C. Mehlgarten, R. Schaffrath, Yeast alpha-tubulin suppressor Ats1/Kti13 relates to the Elongator complex and interacts with Elongator partner protein Kti11, Mol. Microbiol. 69 (1) (2008) 175–187.

[127] J. Uhlmann, S. Wiemann, H. Ponstingl, DelGEF, an RCC1-related protein encoded by a gene on chromosome 11p14 critical for two forms of hereditary deafness, FEBS Lett. 460 (1) (1999) 153–160.

[128] M. Sjolinder, J. Uhlmann, H. Ponstingl, Characterisation of an evolutionary conserved protein interacting with the putative guanine nucleotide exchange factor DelGEF and modulating secretion, Exp. Cell Res. 294 (1) (2004) 68–76.

[129] S. Glatt, R. Zabel, I. Vonkova, A. Kumar, D.J. Netz, A.J. Pierik, V. Rybin, R. Lill, A.C. Gavin, J. Balbach, K.D. Breunig, C.W. Muller, Structure of the Kti11/Kti13 heterodimer and its double role in modifications of tRNA and eukaryotic elongation factor 2, Structure 23 (1) (2015) 149–160.

[130] O. Kolaj-Robin, A.G. McEwen, J. Cavarelli, B. Seraphin, Structure of the Elongator cofactor complex Kti11/Kti13 provides insight into the role of Kti13 in Elongator-dependent tRNA modification, FEBS J. 282 (5) (2015) 819–833.

[131] J. Sun, J. Zhang, F. Wu, C. Xu, S. Li, W. Zhao, Z. Wu, J. Wu, C.Z. Zhou, Y. Shi, Solution structure of Kti11p from Saccharomyces cerevisiae reveals a novel zinc-binding module, Biochemistry 44 (24) (2005) 8801–8809.

[132] M. Proudfoot, S.A. Sanders, A. Singer, R. Zhang, G. Brown, A. Binkowski, L. Xu, J.A. Lukin, A.G. Murzin, A. Joachimiak, C.H. Arrowsmith, A.M. Edwards, A.V. Savchenko, A.F. Yakunin, Biochemical and structural characterization of a novel family of cystathionine beta-synthase domain proteins fused to a Zn ribbon-like domain, J. Mol. Biol. 375 (1) (2008) 301–315.

[133] M. Dong, X. Su, B. Dzikovski, E.E. Dando, X. Zhu, J. Du, J.H. Freed, H. Lin, Dph3 is an electron donor for dph1-dph2 in the first step of eukaryotic diphthamide biosynthesis, J. Am. Chem. Soc. 136 (5) (2014) 1754–1757.

[134] Y. Zhang, X. Zhu, A.T. Torelli, M. Lee, B. Dzikovski, R.M. Koralewski, E. Wang, J. Freed, C. Krebs, S.E. Ealick, H. Lin, Diphthamide biosynthesis requires an organic radical generated by an iron-sulphur enzyme, Nature 465 (7300) (2010) 891–896.

[135] Z. Lin, M. Dong, Y. Zhang, E.A. Lee, H. Lin, Cbr1 is a Dph3 reductase required for the tRNA wobble uridine modification, Nat. Chem. Biol. 12 (2016) 995–997.

[136] K.L. Stilger, W.J. Sullivan Jr., Elongator protein 3 (Elp3) lysine acetyltransferase is a tail-anchored mitochondrial protein in Toxoplasma gondii, J. Biol. Chem. 288 (35) (2013) 25318–25329.

[137] J. Letoquart, N. van Tran, V. Caroline, A. Aleksandrov, N. Lazar, H. van Tilbeurgh, D. Liger, M. Graille, Insights into molecular plasticity in protein complexes from Trm9-Trm112 tRNA modifying enzyme crystal structure, Nucleic Acids Res. 43 (22) (2015) 10989–11002.

CHAPTER FIVE

Posttranscriptional RNA Pseudouridylation

Meemanage D. De Zoysa, Yi-Tao Yu[1]
University of Rochester Medical Center, Center for RNA Biology, Rochester, NY, United States
[1]Corresponding author: e-mail address: yitao_yu@urmc.rochester.edu

Contents

Abstract

Pseudouridine (Ψ) is the most abundant posttranscriptional modification in noncoding RNAs. Pseudouridines are often clustered in important regions of rRNAs (ribosomal RNAs), snRNAs (small nuclear RNAs), and tRNAs (transfer RNAs), contributing to RNA function. Pseudouridylation is governed by two independent mechanisms. The first involves single protein enzymes called pseudouridine synthases (PUSs) that alone recognize the substrate and catalyze the isomerization of uridine to pseudouridine (RNA-independent pseudouridylation). The second is an RNA-guided pseudouridylation by a family of box H/ACA RNPs (ribonucleoproteins), each of which consists of a unique RNA (box H/ACA RNA) and four common core proteins (Cbf5/NAP57/Dyskerin, Nhp2/L7Ae, Nop10, and Gar1). The RNA component serves as a guide that base pairs with the substrate RNA and directs the enzyme (Cbf5) to carry out the pseudouridylation reaction at a specific site. The crystal structures of many PUSs have been solved in numerous organisms including *E. coli* and human. Several partial and complete crystal structures of archaea and yeast box H/ACA RNPs are available, providing a rich source of information

The Enzymes, Volume 41
ISSN 1874-6047
http://dx.doi.org/10.1016/bs.enz.2017.02.001

regarding the molecular interactions between protein components and box H/ACA RNA. Over the years, several experimental systems have been developed to study the mechanism and function of pseudouridylation. Apart from noncoding RNA pseudouridylation, recent experiments have provided evidence of mRNA pseudouridylation as well. Despite remarkable progress, there is a need to accelerate efforts in order to understand the detailed mechanisms and functions of RNA pseudouridylation.

1. INTRODUCTION

Ribonucleic acids (RNAs) undergo over 140 distinct types of post-transcriptional modifications [1]. Among them, pseudouridylation is the most abundant modification. Pseudouridine was initially found in total RNA hydrolysates of calf liver and considered as the fifth ribonucleoside due to its high abundance. Later it was renamed as pseudouridine, which is derived from the Greek letter psi (Ψ) [2,3]. In the formation of pseudouridine from uridine (pseudouridylation), the nitrogen–carbon (N1–C1′) bond, which links the uracil base to the ribose sugar, is first broken and the liberated uracil base is rotated 180 degree along the N3–C6 axis. The rotated base then establishes a new carbon–carbon (C5–C1′) bond between the base and the sugar (Fig. 1). Pseudouridine gains two new features that differentiate it from uridine. First, the canonical C–N glycosidic bond is

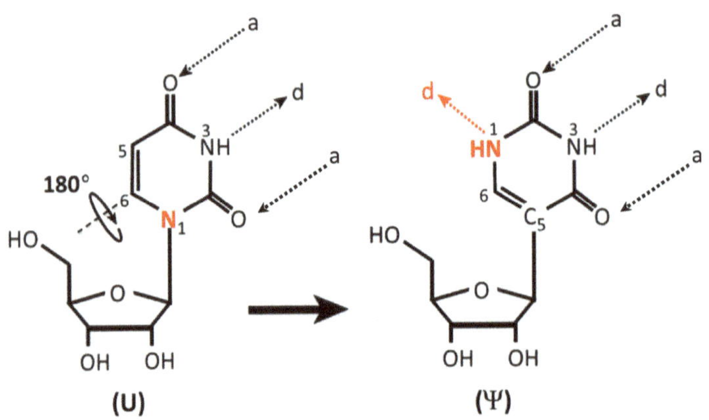

Fig. 1 The isomerization reaction of uridine to pseudouridine. The bond between the N1 of uridine base (marked with *red*) and the C1 of ribose sugar, is first broken. The liberated uridine base rotates 180 degree around the N3–C6 axis and attaches to the ribose sugar via a new bond between the C5 of the base and C1 of the ribose sugar. "a" and "d" stand for H-bonds acceptors and donors, respectively.

changed to a more inert C–C bond [3]. Second, there is an extra hydrogen bond donor at the N1 of the pseudouridine base. These distinctions cause efficient base stacking and water coordination of pseudouridine, thereby increasing the rigidity of the phosohodiester backbone and thermodynamic stability of the Ψ-A base pair compared to U-A base pair [4].

Pseudouridylation mainly occurs in noncoding RNAs, such as snRNAs (small nuclear RNAs), rRNAs (ribosomal RNAs), and tRNAs (transfer RNAs). For example, there are a total of 24 pseudouridines in the 5 vertebrate spliceosomal snRNAs (U1, U2, U4, U5, and U6), and 13 of them are concentrated in U2 snRNA [5]. In *Saccharomyces cerevisiae* (yeast), a total of six pseudouridines have been identified in snRNAs, including three in U2, two in U1, and one in U5 [6] (Fig. 2). Significantly, pseudouridine residues are concentrated in evolutionarily conserved and functionally important regions of snRNAs. For instance, the two pseudouridines in both vertebrate and yeast U1 snRNA are in the 5′ end region

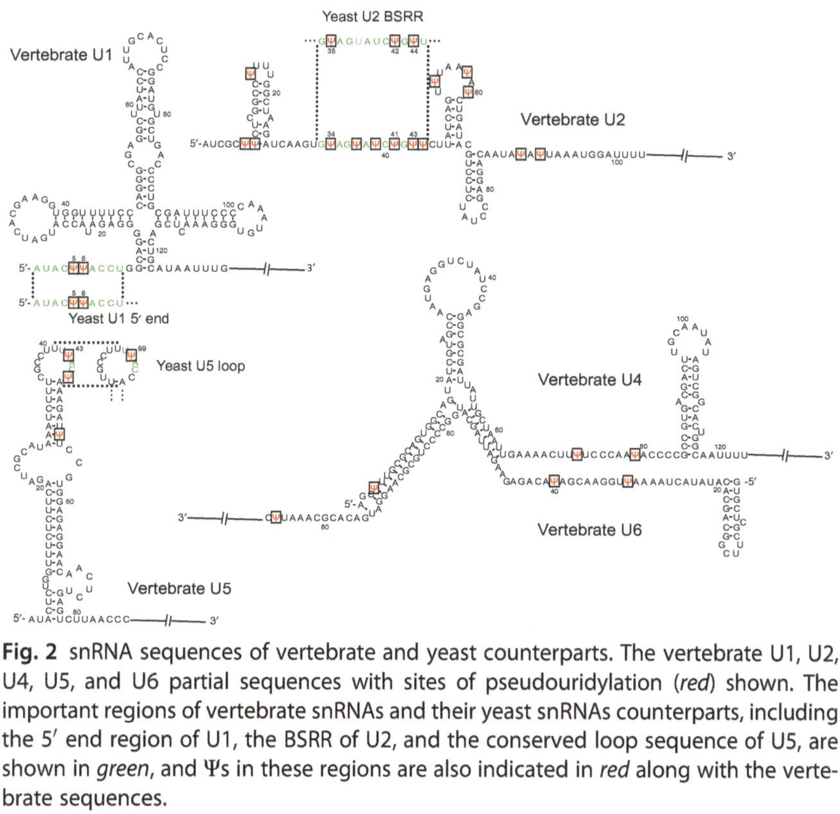

Fig. 2 snRNA sequences of vertebrate and yeast counterparts. The vertebrate U1, U2, U4, U5, and U6 partial sequences with sites of pseudouridylation (*red*) shown. The important regions of vertebrate snRNAs and their yeast snRNAs counterparts, including the 5′ end region of U1, the BSRR of U2, and the conserved loop sequence of U5, are shown in *green*, and Ψs in these regions are also indicated in *red* along with the vertebrate sequences.

(Ψ5 and Ψ6), which forms base-pairing interactions with the 5′ splice site during spliceosome assembly. There are multiple pseudouridines in the branch site recognition region (BSRR) of yeast and vertebrate U2 snRNA that are involved in base pairing with the pre-mRNA branch site during splicing. There is a Ψ (Ψ43) in the conserved loop of vertebrate U5 snRNA that interacts with the 5′ and 3′ exon sequences. Ψ is also present in yeast U5 snRNA at the equivalent position (Ψ99) within the conserved loop (Fig. 2) [7]. Furthermore, there is experimental evidence available to show that pseudouridines contribute significantly to RNA function. For instance, the loss of pseudouridines at the BSRR of U2 impacts pre-mRNA splicing [8]. In a more detailed study, it has been demonstrated that yeast U2 lacking Ψ42 and Ψ44 shows reduced binding affinity with Prp5, an ATPase required for spliceosome assembly, thus resulting in inefficient spliceosome assembly and splicing [9]. In an exciting observation, pseudouridylation of U6 snRNA at U28 (catalyzed by Pus1p) is important to initiate the filamentous growth in yeast, demonstrating the impact of Ψ on cell growth program and development [10].

In rRNA, there are a large number of pseudouridylation sites as well. For instance, there are approximately 97 pseudouridines in mammalian rRNAs and 46 in yeast rRNAs [11]. As in snRNAs, the pseudouridine sites are clustered in functionally important regions of rRNAs. For instance, multiple Ψs are present in the ribosome peptidyl transferase center and blocking pseudouridylation in this region results in defects in protein translation and cell growth [12]. Furthermore, pseudouridines are abundant in the ribosome decoding center of 18S rRNA and A-site finger region of 25S rRNA; importantly, elimination of Ψs in these regions leads to functional defects [13]. Hypopseudouridylated rRNAs cause impaired ribosome–ligand interactions, resulting in decreased affinity for tRNAs and poor translational fidelity in yeast and mammalian cells [14]. Therefore, the presence of pseudouridine at strategic locations of snRNAs and rRNAs together with the relative high degree of conservation clearly suggests that pseudouridylation plays an important role in pre-mRNA splicing and protein synthesis.

2. MECHANISM OF PSEUDOURIDYLATION

The pseudouridylation of RNA substrates occurs via two main mechanisms. One is an RNA-independent mechanism that involves only a single protein enzyme called pseudouridine synthase (PUS). The other mechanism is an RNA-dependent mechanism involving an RNA–protein complex known as box H/ACA RNP. In prokaryotes, the reaction of pseudouridylation is catalyzed solely by the protein-only mechanism, while both mechanisms appear to coexist in eukaryotic organisms.

2.1 RNA-Independent Mechanism

PUSs carry out both substrate recognition and catalysis of the isomerization reaction of uridine to pseudouridine without using any cofactors (Fig. 3A). Depending on the sequence and structure conservation, PUS enzymes are divided into six distinct families, TruA, RluA, RsuA, PUS10, TruB, and TruD, that are named after the founding protein member [15]. Despite the low conservation of structure and sequence, all of these PUS families catalyze the same reaction by using aspartate (Asp) as the nucleophile. The (crystal) structures of some PUS enzymes have been solved (see "The evolution of multi-substrate specificity by RNA modification

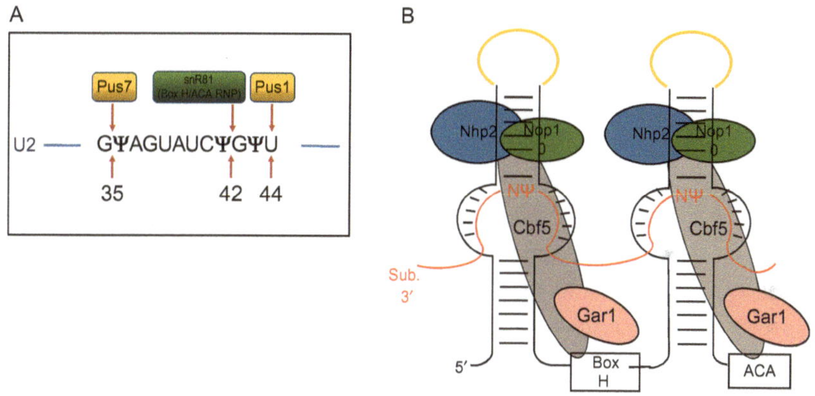

Fig. 3 Schematic of uridine-to-pseudouridine conversion catalyzed by two mechanisms in *Saccharomyces cerevisiae*. (A). The pseudouridylation catalyzed by a standalone protein pseudouridine synthase 1 (Pus1p) and pseudouridine synthase 7 (Pus7p) in conversion of U44 to Ψ44 and U35 to Ψ35, respectively, in U2 snRNA. Pseudouridylation at position 42 by snR81 box H/ACA RNP is also indicated. (For the structure of a pseudouridine synthase, refer to chapter "The evolution of multi-substrate specificity by RNA modification enzymes" by Alfonzo) (B). The box H/ACA RNP, consisting of a small noncoding box H/ACA guide RNA that folds into a typical hairpin–hinge–hairpin–tail secondary structure forming two hairpins and four common core proteins. The core proteins are centromere-binding factor 5 (Cbf5; known as dyskerin/NAP57 in mammals), glycine–arginine-rich protein 1 (Gar1), nonhistone protein 2 (Nhp2), and nucleolar protein 10 (Nop10). The substrate RNA (indicated in *red lines*) drapes around the box H/ACA RNP complex via complementary base-pairing interactions with the guiding pockets of box H/ACA RNA, positioning the target uridine, and its 3′ adjacent nucleotide (N) at the base of the upper stem and leaving them unpaired. The apical loops of box H/ACA RNA that hold the CAB boxes are indicated in *yellow*.

enzymes" by Alfonzo). However, the exact mechanism of isomerization is poorly understood. It is known that pseudouridylation of tRNA in both eukaryotes and prokaryotes is catalyzed by PUS enzymes. Interestingly, pseudouridylation of some of the yeast snRNAs is catalyzed by both RNA-independent and RNA-dependent mechanisms [16,17]. For example, the Branlant group found that a pseudouridylase called Pus1p, which was earlier identified as a tRNA pseudouridylase, can also catalyze the pseudouridylation of yeast U2 snRNA at position 44 [18]. Later, using a GST-ORF fusion protein library, Ma *et al.* discovered that the position 35 (U35) of yeast U2 snRNA is also pseudouridylated by a PUS enzyme, Pus7p [19].

2.2 RNA-Dependent Mechanism

The RNA-dependent pseudouridylation is catalyzed by box H/ACA RNPs, each of which contains an RNA component called a box H/ACA RNA and a set of common core proteins (Fig. 3B). Box H/ACA RNAs are noncoding RNAs that fold into a hairpin–hinge–hairpin–tail secondary structure. The hinge region and the tail region contain evolutionary conserved box H with the consensus sequence "ANANNA" and the trinucleotide Box "ACA," respectively. The two hairpins each contain an internal loop called pseudouridylation guide pocket, which has a short specific sequence complementary to the substrate RNA. The guiding pockets recognize the sites of modifications through Watson–Crick base-pairing interactions with substrate RNAs, thereby positioning the uridine to be modified at the base of the upper stem and leaving it unpaired. This brings the target uridine 13–16 nucleotides upstream of either box H or box ACA [20]. Although the typical box H/ACA guide RNA consists of two hairpins, it can vary in some organisms. For example, in archaea, the number of hairpins can vary from one to three [21] while in humans there is a box H/ACA RNA that contains a structure of four hairpins [22].

All box H/ACA RNPs examined to date each also contain an evolutionarily conserved set of four core proteins [23]. These proteins are essential for the stability of each other and the stability of box H/ACA RNA. They are also essential for catalysis of pseudouridylation. In humans, the four core proteins are dyskerin (Cbf5p in *S. cerevisiae*, NAP57 in rats, and Nop60B in *Drosophila*), Gar1p, Nhp2p (L7Ae in archaea), and Nop10p. All of these core proteins are essential for cell viability and have been characterized in the context of box H/ACA RNA-guided pseudouridylation [24–28]. Among the four proteins, dyskerin (Cbf5p, NAP57, or Nop60B) is the pseudouridylation enzyme that catalyzes the reaction. The pseudouridylase activity of *S. cerevisiae* Cbf5p was initially identified by mutational analysis [29], and it was later realized that the rat NAP57 and the yeast Cbf5p were homologs with 71% sequence identity [30,31]. Interestingly, Cbf5p is also homologous to the bacterial PUS TruB [29]. Detailed analysis of Cbf5p revealed that it contained a catalytic domain, which is common to all known PUSs, as well as a carboxyl-terminal PUA domain (PUS and archaeosine transglycosylase). Both domains show a high degree of structural conservation among bacterial, archaeal, and yeast pseudouridylases [32,33].

While Cbf5p is the most well-studied protein component, the other three proteins are also important for RNA-guided pseudouridylation. Gar1p is a small protein consisting of two glycine–arginine-rich (GAR) domains that flank either side of the central core domain [34]. Structural analysis shows that Gar1p does not directly interact with box H/ACA RNA but is essential for substrate turnover in the reaction of pseudouridylation [32]. Nhp2p (nonhistone protein) is a small basic protein [23]. Recent fluorescent data on the archaeal homologue of Nhp2p, L7Ae, suggests that it might be involved in the correct positioning of the substrate uridine at the active site of the box H/ACA RNA-guiding pocket [35]. The smallest protein component, Nop10p consists of only 64 amino acids. To date, no recognizable or known motifs have been reported in Nop10p, but it has been shown that only Nop10p and Cbf5p proteins alone display a very low level of pseudouridylation activity when incubated with a box H/ACA guide RNA and a complementary substrate RNA. This low level of pseudouridylation was significantly increased when L7Ae protein was present [36].

Box H/ACA RNPs that are involved in the pseudouridylation of rRNAs are localized to the nucleolus (snoRNPs; small nucleolar RNP), whereas the box H/ACA RNPs that guide the pseudouridylation of spliceosomal snRNAs are localized to the Cajal bodies (scaRNPs; small Cajal body-specific RNP) [37]. The nucleolar targeting of box H/ACA snoRNPs requires an intact box H and box ACA together called the box H/ACA motif [38]. The localization of scaRNPs into Cajal bodies involves an additional element, called the CAB box, in the 5′ and 3′ apical loops of box H/ACA RNA, and a CAB-binding protein [39,40].

In catalyzing pseudouridylation, all pseudouridylation enzymes, including RNA dependent (box H/ACA RNPs) and independent (PUS), use a universally conserved aspartic acid at the active site to nucleophilically attack the target uridine. The site of attack, whether at the base (C6) or at the ribose (C1′), was under debate [41]. However, the structural analysis of archaeal box H/ACA RNP revealed a closer proximity of attacking aspartate to the C6 of the uridine base [42,43]. On the other hand, how the separation happens between the base and the sugar and how the rotation of the base occurs remain known. A thumb-loop element in the catalytic domain appears to play an important role in both substrate RNA binding and turnover [44,45].

3. STRUCTURE OF BOX H/ACA RNP

The crystal structure of many PUSs, derived from bacteria, archaea, and yeast, has been solved. The bacterial TruB, complexed with a part of tRNA, was the first to be crystallized [46]. Then, several partial and complete structures of archaea and yeast box H/ACA RNP with or without substrate RNA were solved [32,35,44,47–51]. For instance, the crystal structure of *Pyrococcus furiosus* (Pf) H/ACA RNP, assembled with a single-hairpin H/ACA RNA, was solved. The upper stem of guide RNA binds to Cbf5p, Nop10p, and L7Ae (Nhp2p) while the lower stem and ACA motif are bound to the PUA domain of Cbf5p. Gar1p directly contacts Cbf5p and is important for substrate turnover (loading and release) through controlling the confirmation of thumb-loop of Cbf5p. These interactions between the protein components and the lower stem, upper stem, or the H/ACA motif, specifically position the catalytic domain of Cbf5p over the pseudouridylation guide pocket, forming a molecular bracket [32,44,50,51].

The structure of eukaryotic box H/ACA RNP was poorly understood. However, recently the crystal structure of yeast box H/ACA RNPs with two-hairpin H/ACA RNA and recombinant proteins Cbf5p, Nop10p, Gar1p, and Nhp2p has been reconstituted, displaying a general similarity to that of archaea [33]. However, there are some major differences between the two structures, specifically in the region around the upper stem of the box H/ACA RNA. In archaea, there is a K-turn motif bound by L7Ae, which contacts Nop10p, but does not stably associate with Cbf5p–Nop10p in the absence of RNA [21,32,36,52]. In contrast, no K-turn is present in the upper stem of H/ACA RNA in eukaryotic complex and Nhp2p stably interacts with Cbf5p–Nop10p protein complex [53]. Removal of archaeal L7Ae affects the basic catalytic activity and the substrate turnover, but Nhp2p in yeast seems less important compared to L7Ae because yeast box H/ACA RNP can assemble without Nhp2p and the complex lacking

Nhp2 displayed noticeable but reduced activity in multiple-turnover reactions [33].

Even though the mammalian box H/ACA RNP structure is not yet available, a lot of information can be deduced from the available structures of other species, due to high degree of evolutional similarity. Based on the information, the structure of mammalian box H/ACA RNP has been modeled. Accordingly, NAP57 (or Dyskerin) forms the core of the RNP particle with NOP10 and GAR1 binding independently to the catalytic domain of NAP57. NHP2 appears to dock on NOP10 without contacting NAP57. The guide RNA is draped over the core trimer, which consists of NAP57, NOP10, and NHP2 and does not seem to contact GAR1 [6].

4. THREE ELEMENTS REQUIRED FOR BOX H/ACA RNP ACTIVITY

In order to understand the mechanism of uridine-to-pseudouridine conversion, a great deal of effort has been made to dissect the box H/ACA RNP system in detail. Based on the initial work, the conserved box H (in the hinge region) and box ACA (at the tail) appear to be important for proper nucleolar localization as well as for pseudouridylation [38,54]. Later, several reconstitution systems were developed to dissect the minute details in the function of box H/ACA RNP [36,52,55,56]. All these studies identified three important sequence and structural elements that are critical for RNA-guided pseudouridylation by box H/ACA RNA (Fig. 3B). The first required element is the stability of the hairpin structure harboring the pseudouridylation pocket. This includes the upper and lower stems in between the guide pocket (pseudouridylation pocket). The stable base-pairing interactions of these stems are important to maintain a proper guide pocket. The second element is the stability of base-pairing between the guide sequence and the substrate RNA. Stable interactions are necessary to position the target uridine at the catalytic center of pseudouridylase Cbf5p (Dyskerin/NAP57). The third element is the distance between the target uridine and box H or box ACA. Usually this distance is approximately 13–16 nucleotides [57]. This distance not only allows the upper stem of box H/ACA RNA to bind with Cbf5p, Nop10p and Nhp2, and the lower stem and box H or ACA to bind with the PUA domain of Cbf5p, but also establishes direct contact between the catalytic center of Cbf5p and the target uridine [44,55,58].

5. EXPERIMENTAL SYSTEMS TO STUDY PSEUDOURIDYLATION

Owing to the importance and abundance of the spliceosomal snRNA pseudouridylation, a great effort has also been made to elucidate the function of snRNA pseudouridylation. As a result, several reconstitution systems have been developed, including the mammalian in vitro system, the *Xenopus* oocyte microinjection system, and the yeast system.

5.1 The Mammalian in vitro System

In the early 1990s, Jeffery Patton carried out the first functional analysis of U2 snRNA modification using nuclear extracts prepared from HeLa cells. In these experiments, he showed that in vitro synthesized spliceosomal snRNA, when incubated with nuclear extracts under appropriate conditions, could be efficiently pseudouridylated. Interestingly, he demonstrated that 5-fluorouridine can block pseudouridylation when it is incorporated into U2 snRNA at the sites of pseudouridylation. Using this knowledge, he carried out a series of in vitro experiments and showed that the inhibition of pseudouridylation of spliceosomal snRNA impaired snRNP assembly, suggesting an important role of pseudouridylation in snRNP biogenesis and pre-mRNA splicing [59,60]. Later, in 2004, using functional reconstitution, the Luhrmann group provided experimental evidence suggesting that U2 snRNA pseudouridylation might indeed be important for pre-mRNA splicing. In this study, the endogenous U2 snRNP was depleted from HeLa cell nuclear extracts using antisense oligonucleotides and reconstituted using in vitro synthesized or cellularly derived U2 snRNA [61]. Using this strategy, they later identified Ψ6, Ψ7, and Ψ15 as important pseudouridines for pre-mRNA splicing [62].

5.2 The *Xenopus* Oocyte Microinjection System

The *Xenopus* oocyte microinjection system specifically relies on the fact that an endogenous spliceosomal snRNA can be depleted from oocytes upon injection of an antisense oligonucleotide that is complementary to the target snRNA, thereby directing an endogenous RNase H to degrade the RNA strand (snRNA) of the RNA–DNA hybrids. The antisense oligonucleotide itself is later degraded by an endogenous DNase activity. The depleted snRNA can be restored by injecting the respective in vitro synthesized snRNA. After a short reconstitution period pre-mRNA splicing can be assayed. Using this system, Yu *et al.* demonstrated that in vitro transcribed (and thus unmodified) U2 snRNA did not rescue pre-mRNA splicing in U2-depleted *Xenopus* oocytes, while the cellularly derived (and therefore modified) U2 effectively restored the splicing activity in the same oocytes [63]. Furthermore, using a series of chimeric U2 snRNAs that have both modified and unmodified parts, they identified many pseudouridines in the 5′ end region and in the BSRR of U2 snRNA to be important for splicing [63,64].

5.3 The Yeast System

In *S. cerevisiae* U2 snRNA, there are three pseudouridines at the BSRR. Ψ35 and Ψ44 are modified by the standalone protein enzymes Pus7p and Pus1p, respectively. The third site, Ψ42, is modified by snR81, a box H/ACA RNP; Ψ42 is the only RNA-guided modification site in the yeast U2 snRNA [18,19,56]. The identification of three enzymes responsible for pseudouridylation at three sites in U2 snRNA made it possible to carry out a genetic analysis to dissect the function of U2 pseudouridylation in yeast pre-mRNA splicing. The experiments showed that removal of pseudouridines, either individually or in combination, resulted in splicing defects (to various extents), which, in turn, led to growth defects [9,65].

6. INDUCIBLE PSEUDOURIDYLATION

Apart from the constitutive pseudouridylation discussed earlier, it has been shown that pseudouridylation can be induced as well. The first line of evidence on inducible pseudouridylation came from Wu *et al.*, who showed that U2 snRNA in yeast can be pseudouridylated at novel sites upon changes in growth conditions [66]. In this study, they subjected yeast cells to nutrient deprivation stress by growing cells to saturation or using nutrient-depleted media. Total RNA was isolated from stressed cells and U2 pseudouridylation was assayed, and they detected novel modification sites, positions 56 and 93, which were previously identified as unmodified uridines. Further analysis showed that position 56 was pseudouridylated by the standalone protein enzyme Pus7p, which was previously known to constitutively pseudouridylate position 35 (Ψ35) in U2 snRNA. Position 93 was found to be pseudouridylated by the 3′ guide pocket of box H/ACA RNA snR81, which is normally responsible for the constitutive pseudouridylation of U1051 in 25S rRNA [*Note*: The 5′ guide pocket of snR81 is responsible for the constitutive pseudouridylation of U2 at position 42 (Ψ42)]. Interestingly, the sequences surrounding positions 56 and 93 were not identical but similar to the sequences surrounding the constitutively pseudouridylated sites, positions 35 of U2 and 1051 of 25S rRNA, respectively. Wu *et al.* further showed that the imperfect base pairing between the guide pocket of snR81 and substrate was necessary for induction of pseudouridylation at position 93 in U2 snRNA under stress conditions [66]. Interestingly, the Pus1p-catalyzed U6 snRNA pseudouridylation at position 28 (U28-to-Ψ28 conversion), which plays an important role in regulating the yeast filamentous growth program, appears to be developmentally induced as well [10].

7. mRNA PSEUDOURIDYLATION

Even though pseudouridylation was previously considered as a post-transcriptional modification that occurs exclusively in noncoding RNAs, Karijolich *et al.* demonstrated that mRNA can also be pseudouridylated by designer box H/ACA RNPs [67]. They inserted a premature termination codon (PTC) into *CUP1* gene in the ACT1–CUP1 reporter mRNA [68] and engineered the yeast snR81 box H/ACA RNA (by changing its guide sequence) to base pair with the ACT1–CUP1 mRNA, thereby specifically targeting the uridine in the PTC for pseudouridylation. They showed that the PTC within the ACT1–CUP1 mRNA (and other mRNAs) can indeed be pseudouridylated [67]. Remarkably, pseudouridylation of PTC resulted in stop codon readthrough or nonsense suppression.

It had been reported that there are numerous putative box H/ACA RNAs with typical hairpin–hinge–hairpin–tail structure that fail to match any known pseudouridine sites in stable noncoding RNAs. These RNAs have thus been dubbed as “orphan” box H/ACA guide RNAs. However, the fact that designer box H/ACA RNAs can guide mRNA pseudouridylation at specific sites [67] raises a possibility that some of the naturally occurring “orphan” box H/ACA might target mRNA for pseudouridylation [69–71].

Excitingly, several groups recently presented evidence of naturally occurring mRNA pseudouridylation in *S. cerevisiae* and mammalian cells [72–75]. They developed pseudouridine-seq methods by coupling the conventional CMCT (*N*-cyclohexyl-*N*′-(2-morpholinoethyl)-carbodiimide metho-*p*-toluenesulfonate) modification–primer-extension, which is often used to detect pseudouridines in noncoding RNA, with deep sequencing and bioinformatics analysis. Using these methods, a large number of pseudouridines, ranging from 50–100 in yeast to 100–400 in human mRNAs, were identified. There is no positional bias of mRNA pseudouridylation and the pseudouridine residues were detected within the 5′ and 3′ untranslated regions (UTRs) and coding sequences. Interestingly, however, all the studies were in agreement with the fact that most of the mRNA pseudouridylation is catalyzed by the standalone PUS enzymes and only a few were catalyzed by box H/ACA RNPs. In addition, mRNA pseudouridylation was found to be highly inducible, as evidenced by the fact that the number of pseudouridines in yeast and human mRNAs increases upon nutritional stress and serum starvation, respectively.

8. CONCLUDING REMARKS

In this chapter, we have discussed one of the most abundant posttranscriptional RNA modifications, pseudouridylation. Since the discovery of the box H/ACA RNA family, RNA pseudouridylation has attracted much attention and has been studied extensively. However, even though we have come a long way in the field of RNA pseudouridylation, much still needs to be done to fully understand the mechanism and function of this modification. For instance, it is completely unclear as to whether mRNA pseudouridylation plays a role in mRNA processing and/or protein coding. At the same time, it has been reported that pseudouridylation (or pseudouridylation enzymes) can be linked to a range of diseases such as mitochondrial myopathy and sideroblastic anemia [76], and yet, the detailed roles of pseudouridylation in human diseases remain elusive. However, with recent rapid advances in the field of RNA modification, we are hopeful that some of the major questions related to RNA pseudouridylation will soon be addressed.

ACKNOWLEDGMENTS

We thank the members of the Yu laboratory for critical reading of and helpful comments on the manuscript. The work carried out in the Yu laboratory was supported by grant GM104077 from NIH (to Y.-T.Y.).

REFERENCES

[1] M.A. Machnicka, et al., MODOMICS: a database of RNA modification pathways—2013 update, Nucleic Acids Res. 41 (Database issue) (2013) D262–D267.
[2] F.F. Davis, F.W. Allen, Ribonucleic acids from yeast which contain a fifth nucleotide, J. Biol. Chem. 227 (2) (1957) 907–915.

[3] W.E. Cohn, 5-Ribosyl uracil, a carbon-carbon ribofuranosyl nucleoside in ribonucleic acids, Biochim. Biophys. Acta 32 (1959) 569–571.
[4] M. Charette, M.W. Gray, Pseudouridine in RNA: what, where, how, and why, IUBMB Life 49 (2000) 341051.
[5] R. Reddy, H. Busch, Small nuclear RNAs: RNA sequences, structure, and modifications, in: M.L. Birnstiel (Ed.), Structure and Function of Major and Minor Small Nuclear Ribonucleoprotein Particles, Springer Berlin Heidelberg, Berlin, Heidelberg, 1988, pp. 1–37.
[6] Y.-T. Yu, U.T. Meier, RNA-guided isomerization of uridine to pseudouridine—pseudouridylation, RNA Biol. 11 (12) (2014) 1483–1494.
[7] H. Adachi, Y.-T. Yu, Insight into the mechanisms and functions of spliceosomal snRNA pseudouridylation, World J. Biol. Chem. 5 (4) (2014) 398–408.
[8] X. Zhao, et al., An H/ACA guide RNA directs U2 pseudouridylation at two different sites in the branchpoint recognition region in Xenopus oocytes, RNA 8 (12) (2002) 1515–1525.
[9] G. Wu, et al., Pseudouridines in U2 snRNA stimulate the ATPase activity of Prp5 during spliceosome assembly, EMBO J. 35 (6) (2016) 654–667.
[10] A. Basak, C.C. Query, A pseudouridine residue in the spliceosome core is part of the filamentous growth program in yeast, Cell Rep. 8 (4) (2014) 966–973.
[11] P. Schattner, S. Barberan-Soler, T.M. Lowe, A computational screen for mammalian pseudouridylation guide H/ACA RNAs, RNA 12 (1) (2006) 15–25.
[12] T.H. King, et al., Ribosome structure and activity Are altered in cells lacking snoRNPs that form pseudouridines in the peptidyl transferase center, Mol. Cell 11 (2) (2003) 425–435.
[13] X.-H. Liang, Q. Liu, M.J. Fournier, rRNA modifications in an intersubunit bridge of the ribosome strongly affect both ribosome biogenesis and activity, Mol. Cell 28 (6) (2007) 965–977.
[14] K. Jack, et al., rRNA pseudouridylation defects affect ribosomal ligand binding and translational fidelity from yeast to human cells, Mol. Cell 44 (4) (2011) 660–666.
[15] F. Spenkuch, Y. Motorin, M. Helm, Pseudouridine: still mysterious, but never a fake (uridine)! RNA Biol. 11 (12) (2014) 1540–1554.
[16] H. Grosjean, Modification and editing of RNA: historical overview and important facts to remember, in: H. Grosjean (Ed.), Fine-Tuning of RNA Functions by Modification and Editing, Springer Berlin Heidelberg, Berlin, Heidelberg, 2005, pp. 1–22.
[17] W.A. Decatur, M.N. Schnare, Different mechanisms for pseudouridine formation in yeast 5S and 5.8S rRNAs, Mol. Cell. Biol. 28 (10) (2008) 3089–3100.
[18] S. Massenet, et al., Pseudouridine mapping in the Saccharomyces cerevisiae spliceosomal U small nuclear RNAs (snRNAs) reveals that pseudouridine synthase Pus1p exhibits a dual substrate specificity for U2 snRNA and tRNA, Mol. Cell. Biol. 19 (3) (1999) 2142–2154.
[19] X. Ma, X. Zhao, Y.-T. Yu, Pseudouridylation (Ψ) of U2 snRNA in S. cerevisiae is catalyzed by an RNA-independent mechanism, EMBO J. 22 (8) (2003) 1889–1897.
[20] Y.-T. Yu, R.M. Terns, M.P. Terns, Mechanisms and functions of RNA-guided RNA modification, in: H. Grosjean (Ed.), Fine-Tuning of RNA Functions by Modification and Editing, Springer Berlin Heidelberg, Berlin, Heidelberg, 2005, pp. 223–262.
[21] T.S. Rozhdestvensky, et al., Binding of L7Ae protein to the K-turn of archaeal snoRNAs: a shared RNA binding motif for C/D and H/ACA box snoRNAs in Archaea, Nucleic Acids Res. 31 (3) (2003) 869–877.
[22] A.M. Kiss, et al., A Cajal body-specific pseudouridylation guide RNA is composed of two box H/ACA snoRNA-like domains, Nucleic Acids Res. 30 (21) (2002) 4643–4649.

[23] M. Terns, R. Terns, Noncoding RNAs of the H/ACA family, Cold Spring Harb. Symp. Quant. Biol. 71 (2006) 395–405.
[24] A. Henras, et al., Nhp2p and Nop10p are essential for the function of H/ACA snoRNPs, EMBO J. 17 (23) (1998) 7078–7090.
[25] N.J. Watkins, et al., Cbf5p, a potential pseudouridine synthase, and Nhp2p, a putative RNA-binding protein, are present together with Gar1p in all H BOX/ACA-motif snoRNPs and constitute a common bipartite structure, RNA 4 (12) (1998) 1549–1568.
[26] F. Dragon, V. Pogačić, W. Filipowicz, In vitro assembly of human H/ACA small nucleolar RNPs reveals unique features of U17 and telomerase RNAs, Mol. Cell. Biol. 20 (9) (2000) 3037–3048.
[27] V. Pogacic, F. Dragon, W. Filipowicz, Human H/ACA small nucleolar RNPs and telomerase share evolutionarily conserved proteins NHP2 and NOP10, Mol. Cell. Biol. 20 (23) (2000) 9028–9040.
[28] Y.-I. Watanabe, M.W. Gray, Evolutionary appearance of genes encoding proteins associated with box H/ACA snoRNAs: Cbf5p in Euglena gracilis, an early diverging eukaryote, and candidate Gar1p and Nop10p homologs in archaebacteria, Nucleic Acids Res. 28 (12) (2000) 2342–2352.
[29] Y. Zebarjadian, et al., Point mutations in yeast CBF5 can abolish in vivo pseudouridylation of rRNA, Mol. Cell. Biol. 19 (1999) 7461–7472.
[30] W. Jiang, et al., An essential yeast protein, CBF5p, binds in vitro to centromeres and microtubules, Mol. Cell. Biol. 13 (8) (1993) 4884–4893.
[31] U.T. Meier, G. Blobel, NAP57, a mammalian nucleolar protein with a putative homolog in yeast and bacteria, J. Cell Biol. 127 (6) (1994) 1505–1514 (published erratum appears in J. Cell Biol. 140 (2) (1998) 447).
[32] L. Li, K. Ye, Crystal structure of an H/ACA box ribonucleoprotein particle, Nature 443 (7109) (2006) 302–307.
[33] S. Li, et al., Reconstitution and structural analysis of the yeast box H/ACA RNA-guided pseudouridine synthase, Genes Dev. 25 (22) (2011) 2409–2421.
[34] J.P. Girard, et al., GAR1 is an essential small nucleolar RNP protein required for pre-rRNA processing in yeast, EMBO J. 11 (2) (1992) 673–682.
[35] B. Liang, et al., Substrate RNA positioning in the archaeal H/ACA ribonucleoprotein complex, Nat. Struct. Mol. Biol. 14 (12) (2007) 1189–1195.
[36] B. Charpentier, S. Muller, C. Branlant, Reconstitution of archaeal H/ACA small ribonucleoprotein complexes active in pseudouridylation, Nucleic Acids Res. 33 (10) (2005) 3133–3144.
[37] T. Kiss, Small nucleolar RNAs: an abundant group of noncoding RNAs with diverse cellular functions, Cell 109 (2) (2002) 145–148.
[38] A. Narayanan, et al., Nucleolar localization signals of Box H/ACA small nucleolar RNAs, EMBO J. 18 (18) (1999) 5120–5130.
[39] P. Richard, et al., A common sequence motif determines the Cajal body-specific localization of box H/ACA scaRNAs, EMBO J. 22 (16) (2003) 4283–4293.
[40] K.T. Tycowski, et al., A conserved WD40 protein binds the Cajal body localization signal of scaRNP particles, Mol. Cell 34 (1) (2009) 47–57.
[41] C.J. Spedaliere, et al., The pseudouridine synthases: revisiting a mechanism that seemed settled, J. Am. Chem. Soc. 126 (40) (2004) 12758–12759.
[42] X. Gu, Y. Liu, D.V. Santi, The mechanism of pseudouridine synthase I as deduced from its interaction with 5-fluorouracil-tRNA, Proc. Natl. Acad. Sci. U.S.A. 96 (25) (1999) 14270–14275.
[43] J. Zhou, B. Liang, H. Li, Functional and structural consequences of uridine substitutions on H/ACA ribonucleoprotein particle pseudouridine synthase, Biochemistry 49 (29) (2010) 6276–6281.

[44] J. Duan, et al., Structural mechanism of substrate RNA recruitment in H/ACA RNA-guided pseudouridine synthase, Mol. Cell 34 (4) (2009) 427–439.
[45] B. Liang, et al., Structure of a functional ribonucleoprotein pseudouridine synthase bound to a substrate RNA, Nat. Struct. Mol. Biol. 16 (7) (2009) 740–746.
[46] C. Hoang, A.R. Ferré-D'Amaré, Cocrystal structure of a tRNA Ψ55 pseudouridine synthase: nucleotide flipping by an RNA-modifying enzyme, Cell 107 (7) (2001) 929–939.
[47] R. Rashid, et al., Crystal structure of a Cbf5-Nop10-Gar1 complex and implications in RNA-guided pseudouridylation and dyskeratosis congenita, Mol. Cell 21 (2) (2006) 249–260.
[48] T. Hamma, et al., The Cbf5-Nop10 complex is a molecular bracket that organizes box H/ACA RNPs, Nat. Struct. Mol. Biol. 12 (12) (2005) 1101–1107.
[49] X. Manival, et al., Crystal structure determination and site-directed mutagenesis of the Pyrococcus abyssi aCBF5–aNOP10 complex reveal crucial roles of the C-terminal domains of both proteins in H/ACA sRNP activity, Nucleic Acids Res. 34 (3) (2006) 826–839.
[50] K. Ye, H/ACA guide RNAs, proteins and complexes, Curr. Opin. Struct. Biol. 17 (3) (2007) 287–292.
[51] S.L. Reichow, et al., The structure and function of small nucleolar ribonucleoproteins, Nucleic Acids Res. 35 (5) (2007) 1452–1464.
[52] D.L. Baker, et al., RNA-guided RNA modification: functional organization of the archaeal H/ACA RNP, Genes Dev. 19 (10) (2005) 1238–1248.
[53] C. Wang, U.T. Meier, Architecture and assembly of mammalian H/ACA small nucleolar and telomerase ribonucleoproteins, EMBO J. 23 (8) (2004) 1857–1867.
[54] M.L. Bortolin, P. Ganot, T. Kiss, Elements essential for accumulation and function of small nucleolar RNAs directing site-specific pseudouridylation of ribosomal RNAs, EMBO J. 18 (2) (1999) 457–469.
[55] M. Xiao, et al., Functionality and substrate specificity of human box H/ACA guide RNAs, RNA 15 (1) (2009) 176–186.
[56] X. Ma, Pseudouridylation of yeast U2 snRNA is catalyzed by either an RNA-guided or RNA-independent mechanism, EMBO J. 24 (2005) 2403–2413.
[57] J. Ni, A.L. Tien, M.J. Fournier, Small nucleolar RNAs direct site-specific synthesis of pseudouridine in ribosomal RNA, Cell 89 (1997) 565–573.
[58] J. Ge, Y.-T. Yu, RNA pseudouridylation: new insights into an old modification, Trends Biochem. Sci. 38 (4) (2013) 210–218.
[59] J.R. Patton, Ribonucleoprotein particle assembly and modification of U2 small nuclear RNA containing 5-fluorouridine, Biochemistry 32 (34) (1993) 8939–8944.
[60] J.R. Patton, Multiple pseudouridine synthase activities for small nuclear RNAs, Biochem. J. 290 (Pt. 2) (1993) 595–600.
[61] V. Ségault, et al., In vitro reconstitution of mammalian U2 and U5 snRNPs active in splicing: Sm proteins are functionally interchangeable and are essential for the formation of functional U2 and U5 snRNPs, EMBO J. 14 (16) (1995) 4010–4021.
[62] G. DÖNmez, K. Hartmuth, R. LÜHrmann, Modified nucleotides at the 5′ end of human U2 snRNA are required for spliceosomal E-complex formation, RNA 10 (12) (2004) 1925–1933.
[63] Y.T. Yu, M.D. Shu, J.A. Steitz, Modifications of U2 snRNA are required for snRNP assembly and pre-mRNA splicing, EMBO J. 17 (19) (1998) 5783–5795.
[64] X. Zhao, Y.T. Yu, Pseudouridines in and near the branch site recognition region of U2 snRNA are required for snRNP biogenesis and pre-mRNA splicing in Xenopus oocytes, RNA 10 (2004) 681–690.

[65] C. Yang, D.S. McPheeters, Y.-T. Yu, ψ35 in the branch site recognition region of U2 small nuclear RNA Is important for Pre-mRNA splicing in Saccharomyces cerevisiae, J. Biol. Chem. 280 (8) (2005) 6655–6662.
[66] G. Wu, et al., U2 snRNA is inducibly pseudouridylated at novel sites by Pus7p and snR81 RNP, EMBO J. 30 (1) (2011) 79–89.
[67] J. Karijolich, Y.-T. Yu, Converting nonsense codons into sense codons by targeted pseudouridylation, Nature 474 (7351) (2011) 395–398.
[68] C.F. Lesser, C. Guthrie, Mutational analysis of pre-mRNA splicing in Saccharomyces cerevisiae using a sensitive new reporter gene, CUP1, Genetics 133 (1993) 851–863.
[69] A. Hüttenhofer, et al., RNomics: an experimental approach that identifies 201 candidates for novel, small, non-messenger RNAs in mouse, EMBO J. 20 (11) (2001) 2943–2953.
[70] P. Vitali, et al., Identification of 13 novel human modification guide RNAs, Nucleic Acids Res. 31 (22) (2003) 6543–6551.
[71] A.M. Kiss, et al., Human box H/ACA pseudouridylation guide RNA machinery, Mol. Cell. Biol. 24 (13) (2004) 5797–5807.
[72] T.M. Carlile, et al., Pseudouridine profiling reveals regulated mRNA pseudouridylation in yeast and human cells, Nature 515 (7525) (2014) 143–146.
[73] S. Schwartz, et al., Transcriptome-wide mapping reveals widespread dynamic-regulated pseudouridylation of ncRNA and mRNA, Cell 159 (1) (2014) 148–162.
[74] X. Li, et al., Chemical pulldown reveals dynamic pseudouridylation of the mammalian transcriptome, Nat. Chem. Biol. 11 (8) (2015) 592–597.
[75] A.F. Lovejoy, D.P. Riordan, P.O. Brown, Transcriptome-wide mapping of pseudouridines: pseudouridine synthases modify specific mRNAs in S. cerevisiae, PLoS One 9 (10) (2014). e110799.
[76] Y. Bykhovskaya, et al., Missense mutation in pseudouridine synthase 1 (PUS1) causes mitochondrial myopathy and sideroblastic anemia (MLASA), Am. J. Hum. Genet. 74 (6) (2004) 1303–1308.

CHAPTER SIX

Synthesis, Function, and Heterogeneity of snoRNA-Guided Posttranscriptional Nucleoside Modifications in Eukaryotic Ribosomal RNAs

Anthony K. Henras[1], Célia Plisson-Chastang, Odile Humbert, Yves Romeo, Yves Henry[1]

Laboratoire de Biologie Moléculaire Eucaryote, Centre de Biologie Intégrative, Université de Toulouse, CNRS, UPS, Toulouse, France

[1]Corresponding authors: e-mail address: henras@biotoul.fr; henry@biotoul.fr

Contents

Abstract

Ribosomal RNAs contain numerous 2′-O-methylated nucleosides and pseudouridines. Methylation of the 2′ oxygen of ribose moieties and isomerization of uridines into pseudouridines are catalyzed by C/D and H/ACA small nucleolar ribonucleoprotein particles, respectively. We review the composition, structure, and mode of action of archaeal and eukaryotic C/D and H/ACA particles. Most rRNA modifications cluster in functionally crucial regions of the rRNAs, suggesting they play important roles in translation. Some of these modifications promote global translation efficiency or modulate translation fidelity. Strikingly, recent quantitative nucleoside modification profiling methods have revealed that a subset of modification sites is not always fully modified. The finding of such

The Enzymes, Volume 41
ISSN 1874-6047
http://dx.doi.org/10.1016/bs.enz.2017.03.007

ribosome heterogeneity is in line with the concept of specialized ribosomes that could preferentially translate specific mRNAs. This emerging concept is supported by findings that some human diseases are caused by defects in the rRNA modification machinery correlated with a significant alteration of IRES-dependent translation.

1. INTRODUCTION

The ribosomes are universal molecular machines responsible for the synthesis of cellular proteins. In eukaryotic cells, translating ribosomes are composed of two ribonucleoprotein subunits: the small subunit (40S) contains the 18S ribosomal RNA (rRNA) associated with over 30 ribosomal proteins (RPs, 33 RPs in yeast and human), and the large subunit (60S) is composed of the 5S, 5.8S, and 25S/28S rRNA molecules combined with close to 50 RPs (45 RPs in yeast, 46 RPs in human) [1]. Ribosome synthesis is a complex multistep process, beginning with the coordinated synthesis of rRNAs, RPs, and numerous ribosome assembly and maturation factors (collectively referred to as RAFs, see below). These syntheses mobilize all three cellular RNA polymerases: the preribosomal RNA (pre-rRNA), a common precursor to the 18S, 5.8S, and 25S/28S RNAs, is generated by RNA polymerase I (Pol I); synthesis of the precursor to the fourth ribosomal RNA (5S) is performed by Pol III; transcription of the genes encoding RPs and RAFs relies on Pol II. In proliferating eukaryotic cells, the transcriptional activity of Pol I and that of Pol II and Pol III devoted to ribosome synthesis probably account collectively for more than 50% of the total transcriptional effort of the cells. During Pol I transcription, the nascent pre-rRNA cotranscriptionally assembles with a subset of RPs and RAFs to generate a large preribosomal particle precursor which will give rise to the mature ribosomal subunits after a complex maturation process. For recent reviews on ribosome synthesis in yeast and mammals, see Refs. [2–4].

One important series of events taking place during early stages of preribosome maturation is the synthesis of chemical modifications on specific nucleosides of the pre-rRNAs. Two abundant types of modifications are found in the rRNAs of eukaryotic cells: methylations of ribose 2′-hydroxyl moieties and conversions of uridine residues into pseudouridines. The enzymatic activities responsible for these modifications are the methyltransferase Nop1p/fibrillarin (yeast/human) and the pseudouridine synthase Cbf5p/dyskerin, which are embedded within numerous small nucleolar ribonucleoprotein particles (snoRNPs), also containing a small RNA guide called small nucleolar RNA (snoRNA). Consistent with the two most abundant types of chemical modifications, two populations of snoRNPs operate during ribosome biogenesis, which differ in their protein composition and in the nature of their RNA components. The snoRNPs containing box C/D snoRNAs synthesize ribose methylations and those containing box H/ACA snoRNAs isomerize uridine residues into pseudouridines. Each snoRNA contains a short antisense element allowing, via base-paring interactions with the pre-rRNA, the precise docking of the modifying enzyme at the site of modification within the preribosomal particles.

In this chapter, we will describe the composition, structural organization, and mode of action of C/D- and H/ACA-type snoRNPs. Since detailed structural information is mostly available for archaeal RNPs, these will also be reviewed in depth. We will then review the current literature on the roles of the modified nucleosides in ribosome function in yeast and vertebrates. We will finally present the concept of ribosome heterogeneity due to differential modifications and its potential functional implications.

2. INVENTORY OF rRNA NUCLEOSIDE MODIFICATIONS AND snoRNAs IN YEAST AND MAMMALS

Yeast rRNAs contain a total of 47 pseudouridine residues (Ψs), distributed within the sequences of 18S rRNA (14 Ψs including a hyper modified methyl^1-aminocarboxypropyl^3-pseudouridine at position 1191), 5S rRNA (1 Ψ), 5.8S rRNA (1 Ψ), and 25S rRNA (31 Ψs), [5] and references therein for a recent updated count). The 31 Ψs in the 25S rRNA include a novel pseudouridine residue at position 2345, recently identified through a high-throughput, mass spectrometry-based methodology [6] or by Ψ-seq [7]. This site has also been found to be methylated on the ribose moiety. All these modifications except one are guided by a total of 28 box H/ACA snoRNA guides, each directing one or two pseudouridylation(s) [8–10]. The exception is the pseudouridine residue Ψ50 in the 5S rRNA, synthesized by a snoRNP-independent mechanism involving the multisubstrate pseudouridine synthase Pus7 [11].

Yeast rRNAs contain a total of 55 nucleosides bearing a ribose 2′-O-methylation, distributed within the 18S rRNA (18) and 25S rRNA (37) [6,12]. Among these, a novel 2′-O-ribose methylation has been identified recently at position G562 in the 18S rRNA [13,14]. There is still an ambiguity concerning residue G1142 which is found methylated in some studies but not others [5,12,13]. All these modifications except one are synthesized by a total of 47 C/D-type snoRNPs, each directing the methylation of 1–3 positions [5,15,16]. The exception is residue G2922 in the 25S rRNA which is produced by the methyltransferase Spb1p [17,18]. Spb1p can also methylate an adjacent position, G2921, which is a peculiar site with a redundant methylation mechanism since it is also targeted by a snoRNP containing the snR52 box C/D snoRNA [18].

In mammalian cells, rRNAs are estimated to contain around a hundred pseudouridine residues and as many 2′-O-ribose methylations. With the recent advent of the RiboMeth-seq approaches, profiling studies revealed the presence of 106 2′-O-methylations in human rRNAs (HeLa cells), among which 103 could be assigned to a snoRNA antisense element ([19] and references therein for early studies). Concerning pseudouridylations, around 60 box H/ACA snoRNAs have been characterized so far in human or mouse that are responsible for the synthesis of over 80 of the estimated 97 rRNA pseudouridines [20–23].

More exhaustive information concerning yeast and human rRNA nucleoside modifications and the assigned snoRNAs can be found in two online databases [24,25].

3. STRUCTURE OF s(no)RNPs AND MODE OF CATALYSIS

3.1 C/D RNPs

2′-O-methylations within rRNAs are catalyzed by box C/D RNPs both in eukaryotes and in archaea [15,26–32]. The rRNA nucleoside to be methylated on the 2′ oxygen of its ribose is selected by the box C/D guide RNA component of the RNP [15]. This guide RNA features the conserved C (5′RUGAUGA3′, R being a purine) and D (5′CUGA3′) boxes positioned close to the 5′ and 3′ ends of the RNA, respectively, as well as internal C′ and D′ boxes that in eukaryotes deviate more from the consensus than the terminal C and D boxes [15,33–39]. The guide RNA can establish perfect, approximately 10–21-nucleoside-long base-pairing interaction(s) with the rRNA sequence(s) encompassing the target nucleoside(s) via the sequence immediately upstream from the D and/or D′ box. Such base-paring interactions direct the 2′-O-methylation of the rRNA nucleoside base-paired to the fifth guide nucleoside upstream from the D or D′ box [15,26,27] (Fig. 1A). Some exceptions to this rule have been found. The rRNA complementary sequence upstream from a D′ box in a few yeast box C/D snoRNAs can direct the methylation of two rRNA nucleosides, either adjacent or separated by one nucleoside [40]. Moreover, a subset of eukaryotic box C/D snoRNAs feature an

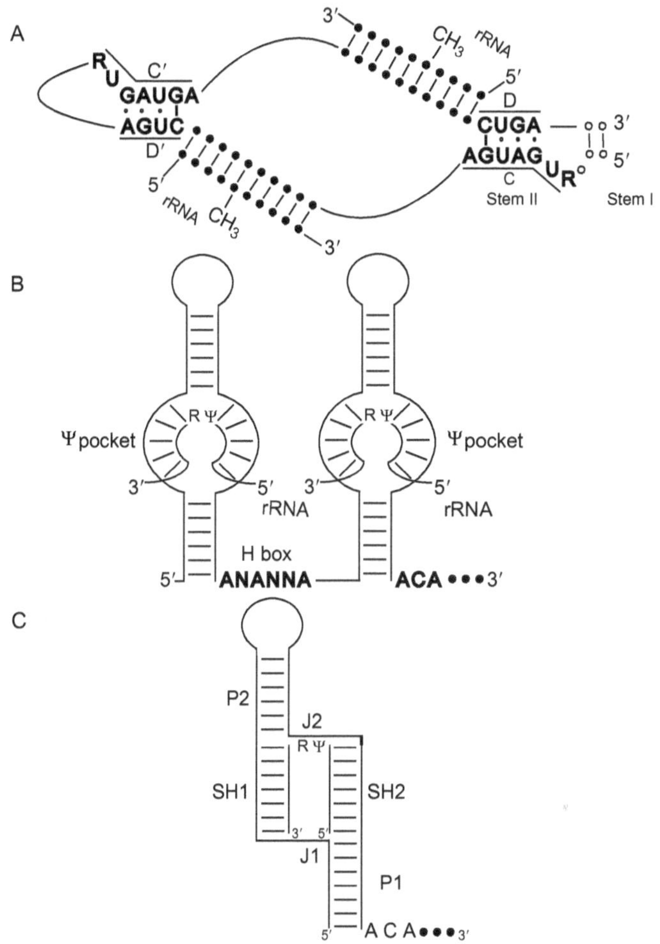

Fig. 1 rRNA nucleoside selection by box C/D and box H/ACA guide RNAs. (A) Schematics of a box C/D guide RNA interacting with two complementary rRNA sequences. The rRNA nucleosides selected for methylation following hybridization with the guide RNA antisense sequences are indicated ($-CH_3$). Conserved C, D, C′, and D′ boxes are highlighted in *bold characters*. (B) Schematics of a box H/ACA guide RNA bound to complementary rRNA regions via the antisense sequences of the pseudouridylation pockets. The rRNA uridine residues selected for pseudouridylation are indicated (ψ). The conserved H and ACA boxes are highlighted in *bold character*. (C) Schematics of a single hairpin ACA guide RNA bound to a substrate RNA showing coaxial stacking of the P2 stem and the substrate helix 1 (SH1) on the one hand and of the substrate helix 2 (SH2) and the P1 stem on the other. The *upper* (J2) and *lower* (J1) three-way junctions are indicated. *R*, purines.

additional conserved sequence, distinct from the D or D′ box-associated antisense sequences, which is complementary to a rRNA region located in the vicinity of the methylation site. These conserved additional complementary sequences stimulate methylation up to fivefold and likely function by stabilizing snoRNA/pre-rRNA interactions [35].

Eukaryotic box C/D snoRNPs contain a set of four common proteins, namely, Nop1p/fibrillarin (yeast/mammals), Nop56, Nop58, Snu13p/15.5K (yeast/mammals) [41–47]. Archaeal box C/D RNPs contain L7Ae instead of Snu13p/15.5K, aNop5 which is related to both Nop56 and Nop58, and aFibrillarin [48–51]. Nop1p/fibrillarin proteins contain a conserved *S*-adenosyl methionine (SAM)-binding motif and provide the catalytic activity of the box C/D RNPs [52–55].

3.1.1 Structure and Activity of Archaeal Box C/D RNPs

Active archaeal box C/D RNPs could be reconstituted using purified recombinant protein components and in vitro transcribed guide RNAs [49,56,57]. These studies indicate that L7Ae binds box C/D RNAs with high affinity and specificity. Further addition of aNop5 and aFibrillarin then yields RNPs able to catalyze site-specific ribose methylation using the rRNA complementary sequences upstream from both the D and D′ boxes. The set of three proteins (L7Ae/aNop5/aFibrillarin) can interact with each of the two halves of a guide RNA, constituted by the 5′/3′ terminal stem, boxes C and D and the rRNA antisense sequence upstream from the box D on the one hand and the boxes C′ and D′ as well as the two rRNA antisense sequences, on the other [56,58]. These data suggest that an archaeal box C/D RNP assembled on a full-length box C/D RNA contains at least two sets of core RNP proteins. The juxtaposition within the full RNP of two sets of these core proteins assembled around the box C/D and C′/D′ motifs seems required for optimal methylation guided from both antisense sequences [56,59]. The minimal module able to catalyze methylation, albeit with reduced efficiency, is constituted by the RNP assembled on the terminal box C/D motif and associated antisense sequence [56,57]. However, such a minimal RNP has lost the ability to select the correct substrate nucleoside [60], suggesting that the bipartite RNP structure is crucial for proper positioning of aFibrillarin.

Crystal structures of archaeal box C/D RNP core protein components [55,61–64] as well as those of halfmer RNPs (that is RNPs assembled on a fragment of the guide RNA consisting only of the terminal box C/D motif associated with a rRNA antisense sequence) [65,66] have been obtained.

The conclusions drawn from these studies are largely in agreement and can be summarized as follows. The canonical terminal box C/D RNA motif folds into a kink-turn structure (Figs. 1A and 2). L7Ae interacts with the K-turn by docking onto the protruding U and surrounding nucleosides. Often the internal C′ and D′ boxes of archaeal box C/D guide RNAs fold into a structure related to the K-turn, called the K-loop [68]. L7Ae also binds with high specificity and affinity to the internal K-loop [61,68], consistent with the notion that archaeal box C/D RNPs contain two L7Ae-binding sites. aNop5 and aFibrillarin form a tight complex that may exist before their integration into the assembling RNPs. aNop5 binds aFibrillarin via its N-terminal domain. Some aNop5 amino acids are found in close proximity to the *S*-adenosyl-L-methionine (Ado-Met) bound to aFibrillarin, suggesting that aNop5 N-terminal domain stabilizes binding of the methyl group donor. aNop5 also features a long antiparallel coiled–coiled domain,

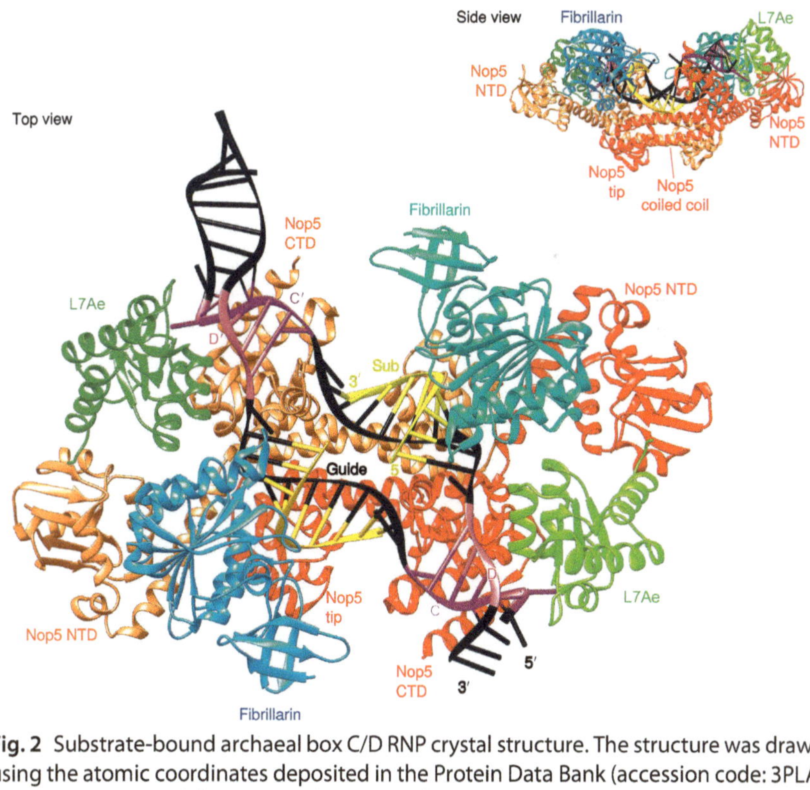

Fig. 2 Substrate-bound archaeal box C/D RNP crystal structure. The structure was drawn using the atomic coordinates deposited in the Protein Data Bank (accession code: 3PLA) [67]. *CTD*, C-terminal domain; *Guide*, RNA guide; *NTD*, N-terminal domain; *Sub*, RNA substrate. *Top* and *side views* are shown.

a tip domain, and a C-terminal domain. In all crystal structures solved, the coiled–coiled domain mediates the dimerization of aNop5. In the "halfmer" box C/D RNP, the highly conserved C-terminal domain of aNop5 contacts both L7Ae and the K-turn [66,67]. Hence, aFibrillarin and the N-terminal domain of aNop5 form the catalytic module, while L7Ae and the C-terminal domain of aNop5 form the RNA-binding module.

Models of full archaeal box C/D RNPs containing an entire bipartite box C/D RNA, free or bound to substrates RNAs, have been derived from crystallographic data [67], NMR spectroscopy and small angle neutron-scattering (SANS) [69], and cryo-electron microscopy [70]. In all these models, the associated coiled–coiled domains of two aNop5 molecules play a crucial structural role. Apart from this common feature, the three models are highly divergent.

The crystallized full RNP was obtained using a symmetrical bipartite RNA containing two box C/D motifs each associated with an antisense sequence hybridized to a substrate RNA [67]. The bipartite RNA itself was produced by hybridization of two linear strands. The crystallized RNP contains a single bipartite box C/D RNA and two copies of each of the core proteins (Fig. 2). The associated coiled–coiled domains of two aNop5 molecules (red and gold) form a flat platform that constitutes the backbone of the overall structure. Due to their association with the C-terminal domain of aNop5, the two K-turns with a bound L7Ae protein (green) each are placed at opposite ends of the coiled–coiled platform, while the two antisense sequences and associated substrate RNAs occupy a central position. Each antisense sequence/substrate duplex is bound by one aFibrillarin molecule (blue) that interacts with the minor groove in such a way that the correct target nucleoside is precisely placed in the active site of the enzyme. aFibrillarin is probably properly positioned relative the antisense sequence/substrate duplex by docking of the catalytic module against L7Ae on one side and the tip domain as well as part of the coiled–coiled domain of aNop5 on the other. In addition, the fixed spacing of the antisense sequence relative to the D box likely also plays a central role in correctly positioning the active site of aFibrillarin.

In contrast to the crystallized RNP structure featuring a single guide RNA, the RNPs studied by NMR and electron microscopy adopt a so-called di-sRNP structure, containing two bipartite box C/D RNAs and four copies of each of the core proteins [69–71]. Formation of the di-sRNP structure, at least in vitro, is obtained with a wide diversity of naturally occurring archaeal bipartite box C/D RNAs and seems dependent on the presence of the conserved loop connecting boxes C′ and D′ [58,71–73], which is absent from the two annealed RNAs used for assembling the crystallized RNP [67].

The substrate-bound RNP structure obtained by NMR/SANS [69,74] features a central platform formed by four aNop5 coiled–coiled domains (Fig. 3). Two bipartite box C/D molecules are positioned as a continuous long rod across one side of the platform. Two aFibrillarin molecules are located on one side of the platform, the other two on the other side. The two antisense sequence–substrate helices associated with the D′ boxes, close to the middle of the di-sRNP, are contacted by one aFibrillarin molecule

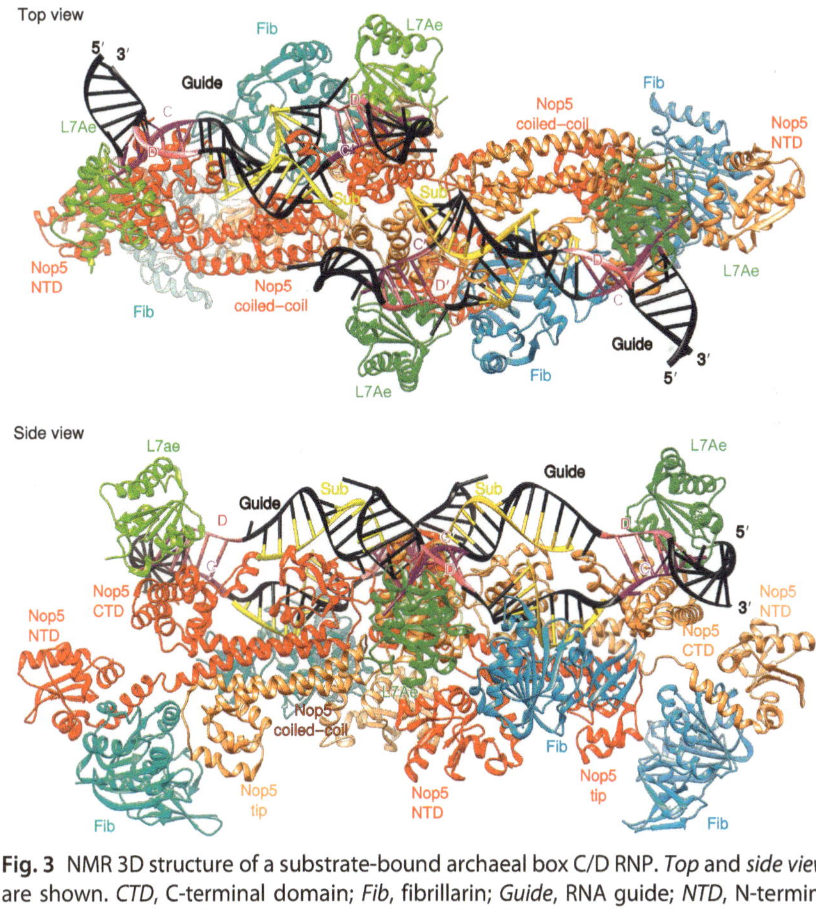

Fig. 3 NMR 3D structure of a substrate-bound archaeal box C/D RNP. *Top* and *side views* are shown. *CTD*, C-terminal domain; *Fib*, fibrillarin; *Guide*, RNA guide; *NTD*, N-terminal domain; *Sub*, RNA substrate. PDB accession code: 4BY9 [69].

each. The other two aFibrillarin molecules are not in contact with the remaining antisense sequence–substrate helices associated with the D boxes. Strikingly, each aFibrillarin molecule that contacts a D′ antisense sequence/substrate helix is bound to an aNop5 molecule interacting with the box C′/D′ motif of the other bipartite RNA. This structural model begs the question of how the substrate RNAs hybridized with the antisense sequences associated with the D boxes get modified.

The 3D reconstruction of a substrate-free archaeal box C/D RNP based on cryo-electron microscopy images [70] differs from the NMR-based model in the location and orientation of the box C/D RNAs (Fig. 4). While in the NMR model, the two RNAs lie longitudinally across only one face of the aNop5 coiled–coiled domain platform, in the cryo-EM model, the two RNAs run parallel to each other from one face of the platform to the other, perpendicular to aNop5 coiled–coiled domains. The two RNAs adopt the same orientation such that the base-paired terminal 5′ and 3′ ends of both RNAs are found on the same face of the platform. These terminal stem ends

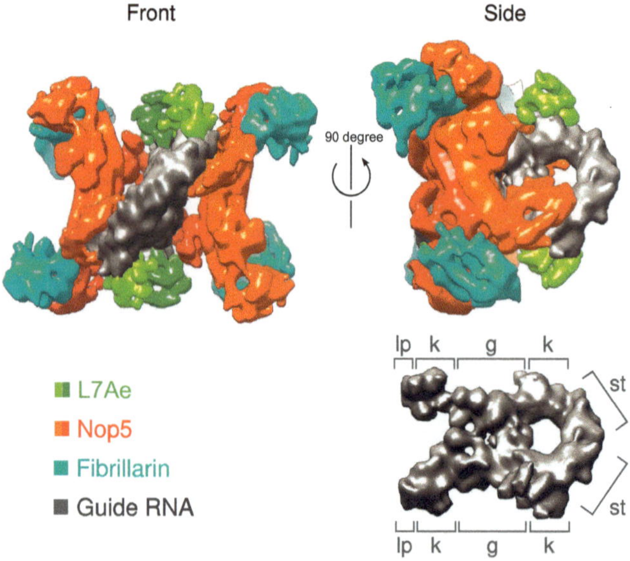

Fig. 4 Cryo-EM 3D structure of a substrate-free archaeal box C/D RNP. Surface views seen from the *front* (*left panel*) and the *side* (*right panel*) of the structure were drawn using the electron density map deposited in the Electron Microscopy Database (accession code: EMD-8146, [70]). *Lower right panel*: *side view* of the electron density attributed to the box C/D guide RNA. This density was segmented and extracted from the cryo-EM 3D structure using the Segger extension to the UCSF Chimera package [75,76]. *g*, guide antisense sequence; *k*, kink; *lp*, loop; *st*, terminal stems.

seem to interact, an interaction which appears important for formation and/or stability of the di-sRNP structure. So far, a cryo-EM-based structure of the substrate-bound RNP is not available.

3.1.2 Structure and Activity of Eukaryotic Box C/D snoRNPs

So far eukaryotic box C/D snoRNPs could not be reconstituted from purified recombinant proteins explaining why the biochemical roles of their individual components as well as structural studies have lagged behind those performed with archaeal systems. Nevertheless, it was demonstrated that purified box C/D snoRNPs from yeast can direct site-specific methylation in vitro [54]. The purified system appears to contain only the core box C/D snoRNP proteins and box C/D snoRNAs, suggesting that in eukaryotes as in archaea, a mature box C/D snoRNP is able to carry out faithful methylation on its own. Functional studies indicate that as in archaea, the eukaryotic terminal box C/D motif associated with an antisense sequence is the minimal module able to carry out methylation, while the module constituted by the internal box C′/D′ motif and associated antisense sequence must be linked to a terminal box C/D motif for activity [77]. Moreover, the relative spatial positioning of the two modules is important for C′/D′ motif activity, although the spatial constraints are not as strict as in archaea [78]. Contrary to the situation in archaea where L7Ae is able to directly bind both terminal C/D and internal C′/D′ motifs (see above), the human counterpart of L7Ae, the 15.5K protein, only interacts on its own with the C/D but not the C′/D′ motif [46,56,79]. Moreover, cross-linking studies in Xenopus oocytes suggest an asymmetric distribution of proteins in the box C/D snoRNPs, notably with interactions detected between box C and Nop58 and between box C′ and Nop56 [80]. The hypothesis of an exclusive interaction of Snu13p/15.5K protein with the terminal box C/D motif was challenged by immunoprecipitation results from yeast extracts showing that isolated box C/D and box C′/D′ motifs could both be precipitated via all four core box C/D snoRNP proteins [77]. This indicates that, at least in yeast, both C/D and C′/D′ RNP modules contain the full complement of core proteins. One way to reconcile the available data is to propose that Snu13p/15.5K is recruited to the box C′/D′ motif primarily via protein/protein interactions and that the association of Nop56 and Nop58 with both C/D and C′/D′ motifs is due to their dimerization via their coiled–coiled domain [77]. Thus, the box C/D module could be associated with Snu13p/15.5K, Nop58, and Nop1p/fibrillarin and the box C′/D′ module with Snu13p/15.5K, Nop56, and Nop1p/fibrillarin, the two modules being connected via both the snoRNA and the interacting coiled–coiled domains of Nop58 and Nop56. Assembly of the terminal box C/D RNP module is probably important, if not essential, for stable formation of the internal box C′/D′ RNP module [81].

Although structures of canonical eukaryotic box C/D snoRNPs are still lacking, a structural model of an atypical box C/D snoRNP, namely, the U3 box C/D snoRNP, has been proposed recently, based on cryo-EM images of purified 90S preribosomal particles from the eukaryotic thermophile *Chaetomium thermophilum* [82]. Unlike canonical box C/D snoRNPs, the U3 RNP does not direct 2′-O-methylation but functions as a chaperone promoting correct folding of 18S rRNA precursors within 90S particles. The U3 snoRNA contains two box motifs, an internal so-called box C/box B motif and a box C′/D motif, closely related to the box motifs found in canonical methylation guide snoRNAs. In the 90S particle, the U3 snoRNP is present as a monomer. It contains two molecules of Nop1p and Snu13p and one molecule of Nop56p and Nop58p. The box C/box B motif is bound by Nop56p and one molecule of Snu13p and Nop1p, while the box C′/D motif is bound by Nop58p and the other molecule of Snu13p and Nop1p. This structure of the U3 snoRNP lends support to the hypothesis that canonical eukaryotic methylation guide box C/D snoRNPs adopt a similar monomeric symmetrical structural organization.

3.2 H/ACA RNPs

3.2.1 Overall Organization of H/ACA RNPs

H/ACA RNPs catalyze the site-specific isomerization of uridines into pseudouridines [20]. They contain a small guide RNA (typically 60–150 nucleosides in length, sometimes more in yeast) [48,50,83–85], the canonical version of which possesses the conserved H (consensus 5′ANANNA3′) and ACA boxes, hence their name [33,86] (Fig. 1B). These canonical box H/ACA small RNAs fold into a conserved structure featuring two stem–loop structures containing an internal bulge separated by a single-stranded hinge region containing the H box and followed by a tail containing the ACA box positioned three nucleosides upstream from the RNA 3′ end (Fig. 1B) [33,86]. Many exceptions to this canonical structure exist. Box H/ACA-like RNAs featuring a single hairpin have been found in some trypanosomes [87] or in the protist *Euglena gracilis* [88], while in some archaea, RNAs with one, two, or three hairpins have been identified [31,89–91]. The sequence of the 5′ACA3′ motif can vary at the second position and hence is sometimes termed the "ANA" terminal motif [48]. Moreover, this motif is even absent in some archaeal pseudouridylation guide RNAs [91]. Eukaryotic H/ACA RNPs all contain a conserved set of four proteins, termed Cbf5p, Nhp2p, Gar1p, and Nop10p in the yeast *Saccharomyces cerevisiae* [33,92–95]. The orthologues of yeast Cbfp5p in *Drosophila*, rat, and human are called Nop60B [96,97], NAP57 [98,99], and dyskerin [100], respectively, while the remaining three H/ACA RNP proteins in higher eukaryotes were attributed the yeast names [101,102]. Archaeal H/ACA RNPs also contain the orthologues of Cbf5p, Gar1p, and Nop10p as well as the ribosomal protein L7Ae instead of Nhp2p [90].

The RNA component of H/ACA RNPs selects the uridine to be converted into pseudouridine by establishing in most cases (but not all, [91]) two short base-pairing interactions on each side of the substrate uridine. The H/ACA RNA sequences that base-pair with the substrate RNA are the 5′ and 3′ sequences of the internal bulge regions found in the middle of the hairpin structures, regions hence termed "pseudouridylation pockets" [20]. The targeted uridine and the 3′ adjacent nucleoside are exposed at the centre of the upper three-way junction generated by the guide RNA/substrate RNA interactions (Fig. 1B and C). The target uridine is not engaged in Watson–Crick base-pairing interactions, a prerequisite for the uridine to pseudouridine isomerization reaction. The actual catalysis is carried out by Cbf5p/NAP57/Dyskerin, who belong to the TruB family of pseudouridine synthases [103]. In addition to the catalytic domain, enzymes of this family contain a C-terminal PUA (pseudouridine and archaeosine tRNA guanine transglycosylase) domain implicated in substrate RNA binding.

3.2.2 Structure and Activity of Archaeal H/ACA RNPs

Numerous biochemical and structural studies have provided detailed information on the activity and structure of archaeal H/ACA RNPs. Fully active H/ACA RNPs have been reconstituted in vitro from isolated purified archaeal protein components and an in vitro transcribed single hairpin pseudouridylation guide RNA containing an ACA triplet [104,105]. These studies demonstrated that archaeal Cbf5 (aCbf5) binds on its own with high specificity to the guide RNA and that this binding is dependent on the integrity of the ACA box. L7Ae also binds independently directly and specifically to the guide RNA [90,104–106]. In this case, binding is dependent on the K-loop or K-turn structure present in the helical stem located above the pseudouridylation pocket. Archaeal Gar1 (aGar1) and archaeal Nop10 (aNop10) cannot bind on their own to the guide RNA but both interact independently with aCbf5. Crystal structures of the aCbf5/aNop10 complex [107,108], the aCbf5/aNop10/aGar1 complex [109], and of the full substrate-free RNP [110] (Fig. 5) have confirmed the conclusions drawn from the biochemical studies and provided additional structural information. aCbf5 and the guide RNA form the core of the structure. The PUA domain of aCbf5 contacts the ACA box and lower (P1) stem of the guide RNA. The two conserved adenines of the ACA box are bound by aCbf5 PUA domain in a sequence-specific manner. The catalytic domain of aCbf5 can be further subdivided into subdomains D1 and D2, separated by the active site cleft. aCbf5 subdomain D1 contacts the pseudouridylation pocket and the lower part of the guide RNA stem above the pseudouridylation pocket (stem P2). The small aNop10 protein adopts an extended structure consisting of an N-terminal zinc-binding domain that interacts with aCbf5 D2, a linker region and a C-terminal alpha-helix that both bind aCbf5 D1. Binding of aNop10 to aCbf5 stabilizes aNop10, since the central and C-terminal

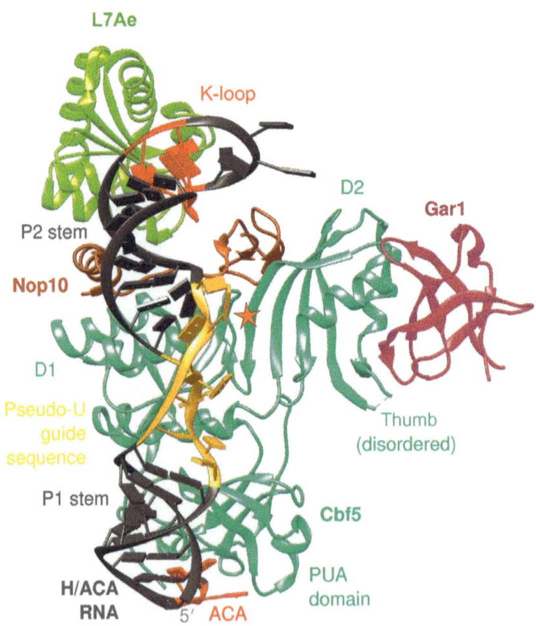

Fig. 5 Substrate-free archaeal box H/ACA RNP crystal structure. The structure was drawn using the atomic coordinates deposited in the Protein Data Bank (accession code: 2HVY) [110]. The tip of the thumb of Cbf5, not resolved in the structure, is not shown and replaced by a *dotted line*. The *star* denotes the active site.

regions of both aNop10 and yeast Nop10p are partially unstructured in the free proteins [111,112]. Conversely, aNop10 binding to aCbf5 seems to buttress its active site and to extend the active site cleft. In the RNP structure, aNop10 also interacts with the P2 stem and the L7Ae protein, while aNop10 on its own is unable to bind either the guide RNA or L7Ae. As predicted from biochemical studies, L7Ae is bound in the RNP to the K-loop motif at the top of the P2 stem. Thus, sequential interactions between L7Ae, aNop10, and aCbf5 provide a protein surface that binds one face of the guide RNA, at the level of the P2 stem (L7Ae, aNop10, and aCbf5 D1), the P1 stem and the ACA box (aCbf5 PUA domain). This mode of two-end binding induces a bend in the structure of the guide RNA and places the pseudouridylation pocket in loose contact with aCbf5 D1. The six-stranded beta barrel aGar1 protein is the only RNP member not in contact with the guide RNA. Instead, it interacts with Cbf5 D2 domain.

NMR structures of substrate RNA bound to a guide RNA [113,114] and crystal structures of partial [115–117] or full arachaeal RNPs [115] containing a bound substrate RNA have provided insights into the mode of substrate binding and the potential functions of the aNop10, L7Ae, and aGar1 proteins during the pseudouridylation reaction. Watson–Crick base-pairing of the 3′ and 5′ arms of the substrate RNA with complementary sequences in the 5′ and 3′ halves of the pseudouridylation pocket creates helices SH1 (for "substrate helix" 1) and SH2, respectively (Fig. 1C). SH1 and SH2, respectively, stack on and are coaxial with the upper and lower stems of the guide RNA, P2 and P1. Two three-way junctions, the upper J2 and the lower J1, are created. The substrate RNA, which interacts only with the guide RNA and aCbf5, adopts a V-shaped conformation, the target uridine being placed at the tip of this V and inserted inside the catalytic cleft of aCbf5 (Fig. 6). In active RNPs, the target residue makes specific contacts with the catalytic aspartate of the enzyme. Apart from the target nucleoside, aCbf5 does not recognize any other specific sequence feature of the substrate, consistent with the fact that only the ability to base-pair on each side of the target U determines substrate selection. The bend in the main axis of the RNA structure induced by protein binding is required for the entry of the substrate RNA inside the catalytic cleft.

In a partial RNP containing a bound substrate RNA but lacking L7Ae [116], the target uridine is found 11 Å away from the catalytic aspartate residue of aCbf5. Hence, it was proposed that L7Ae binding is required for correct placement of the substrate RNA in the active site of aCbf5 [116,118]. This proposal is fully consistent with the finding that partial

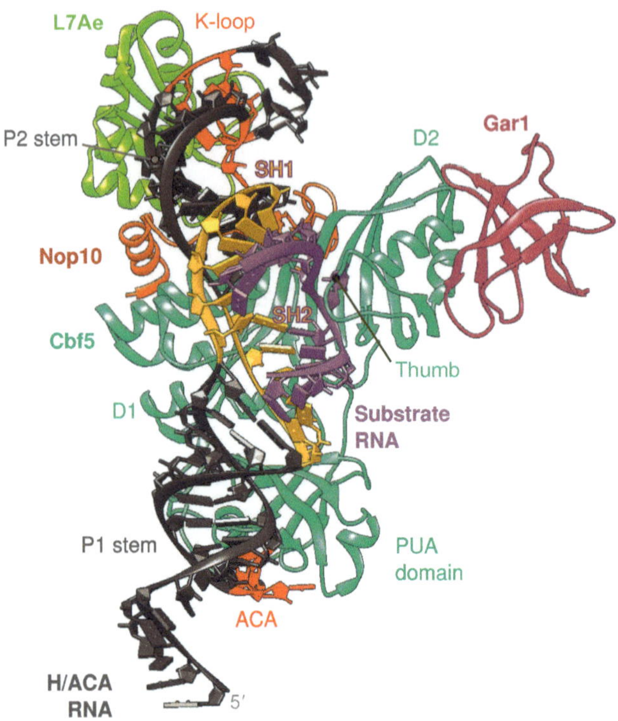

Fig. 6 Crystal structure of an archaeal substrate-bound box H/ACA RNP. The structure was drawn using the atomic coordinates deposited in the Protein Data Bank (accession code: 3HAY) [115].

RNPs lacking L7Ae are significantly less active in pseudouridylation assays than complete RNPs [105]. Comparisons between the structures of the substrate-free RNP [110], active RNPs with and without aGar1 bound to diverse substrate RNAs [115,119] suggest that aGar1 and the so-called thumb loop of aCbf5 are involved in controlling substrate placement and turnover. In the substrate-bound active RNP (Fig. 6), the thumb loop of aCbf5, a loop protruding from aCbf5 D2, adopts a closed conformation, where it interacts with the lower part of the V-shaped substrate RNA and locks it inside the catalytic cleft. In the substrate-free RNP (Fig. 5), the thumb loop is displaced toward aGar1, some residues at the N-terminal root of the thumb establishing interactions with a hydrophobic patch of the protein. In such an open state, the thumb loop is predicted not to be able to interact with a bound substrate RNA thus favoring substrate release. The proposal of a role of aGar1 in substrate release is supported by the finding that the partial RNP lacking aGar1, although still active under single turnover conditions, becomes extremely defective under multiple turnover conditions. Point mutations in aCbf5 thumb loop or aGar1 predicted to inhibit thumb loop/aGar1 interactions inhibit the activity of the full RNP under multiple, but not single turnover conditions, consistent with the proposal that these interactions are important for substrate release [115]. However, thumb loop mutations designed to specifically inhibit the closed state have the same consequences on activity. Hence, the phenotypes induced by these aCbf5 and aGar1 mutations should be interpreted with caution. The finding that partial RNPs lacking aGar1 are substantially less active than the full RNPs even under single turnover conditions suggests that the role of aGar1 is not restricted to substrate release. Indeed, kinetic measurements indicate that aGar1, while not crucial for substrate loading, is important for the catalytic step [120].

Altogether, structural data and kinetic measurements suggest a four-step mechanism for substrate loading, modification, and release, where the conformation of the thumb loop of aCbf5 plays a crucial role [115–118,120,121]. According to this mechanism, initial interactions of the substrate RNA with the H/ACA RNP occur primarily via base-pairing interactions with the complementary sequences of the pseudouridylation pocket. The substrate then interacts with aCbf5 active site cleft and the root of the thumb loop, followed by closing of the tip of the thumb. Once the isomerization reaction is completed, the pseudouridine product induces the opening of the thumb and the open state is stabilized by docking of the root of the thumb on a hydrophobic patch of aGar1, thus promoting release of the modified substrate.

3.2.3 Structure and Activity of Eukaryotic H/ACA RNPs

Our understanding of the structure and mode of action of eukaryotic H/ACA RNPs is less advanced than that of archaeal RNPs, chiefly because it is only recently that reconstitution of active eukaryotic H/ACA RNPs from purified recombinant components could be achieved [122]. The available data indicate that, as in archaea, yeast and mammalian Nop10p/NOP10 and Gar1p/GAR1 can interact with Cbf5p/NAP57 [122–124]. Contrary to the situation in archaea, yeast Nhp2p and mammalian NHP2, the counterparts of archaeal L7Ae, can interact with a Nop10p/Cbf5p or NOP10/NAP57 complex in the absence of the H/ACA guide RNA [122,124]. One additional major difference between eukaryotic Nhp2p/NHP2 and

L7Ae is that the former proteins interact nonspecifically on their own with RNA, while L7Ae specifically binds K-turn or K-loop motifs within the upper stem of archaeal H/ACA guide RNAs, motifs absent from eukaryotic H/ACA RNAs [90,124,125]. Nevertheless, cross-linking studies suggest that yeast Nhp2p is bound to Nop10p and occupies approximately the same position in yeast H/ACA RNPs as L7Ae does in archaeal RNPs, presumably contacting the upper stem of the guide RNA [122]. So far, a complete crystal structure of a eukaryotic H/ACA RNP has not been reported. However, the NMR structure of yeast Nhp2p [126] and the crystal structure of the yeast Cbf5p/Gar1p/Nop10p complex have been solved [122]. The structure of this complex is similar to that of the equivalent archaeal complex, with some important differences. Contrary to archaeal aNop10, yeast Nop10p lacks zinc-binding cysteine residues and does not bind zinc. Yeast Gar1p contains N- and C-terminal glycine–arginine-rich domains, termed GAR domains, as well as N- and C-terminal extensions (NTE and CTE) to the core domain that are all absent from aGar1. Importantly, the CTE of yeast Gar1p contacts the thumb loop of Cbf5p and induces an open conformation of the thumb that is different from that found in archaeal H/ACA RNPs and is in fact more similar to the archaeal closed conformation. Hence, during the catalytic cycle, the thumb of yeast Cbf5p may undergo more restrained conformational transitions than the archaeal thumb.

In vitro pseudouridylation assays using reconstituted yeast H/ACA RNPs lacking Nhp2p or Gar1p indicate that the relative importance of these two proteins for catalysis is inverted in yeast and archaea. While L7Ae is crucial for archaeal H/ACA RNP activity, yeast H/ACA RNPs devoid of Nhp2p are still highly active, albeit less so than the full RNPs [122]. On the contrary, yeast H/ACA RNPs lacking Gar1p are strongly defective even under single turnover conditions [122], whereas archaeal RNPs missing aGar1 retain substantial activity under such conditions [104,105]. In spite of this difference, both aGar1 and yeast Gar1p are important for substrate turnover [122]. Moreover, as in archaea, the interaction between yeast Gar1p and the thumb loop of Cbf5p is important for substrate turnover since substitutions of key amino acids involved in the Gar1p/thumb loop interaction impair enzyme activity under multiple turnover conditions [122].

Contrary to the situation in archaea where single stem–loop guide RNAs are common, the vast majority of eukaryotic H/ACA RNAs are composed of two stem–loops. Moreover, altering key elements (conserved box or stems) of only one-half of a eukaryotic bipartite guide RNA abolishes pseudouridylation events directed by both 5′ and 3′ pseudouridylation pockets in yeast cells, suggesting that both parts of eukaryotic H/ACA RNPs are highly interdependent in vivo [127]. However, active RNPs can be reconstituted in vitro with the 5′ or 3′ half of a yeast H/ACA snoRNA and yeast H/ACA snoRNP proteins, indicating that both in archaea and yeast, the particle assembled on a single hairpin guide RNA constitutes a fully functional entity [122].

4. ROLE OF rRNA NUCLEOSIDE MODIFICATIONS IN RIBOSOME FUNCTION

4.1 Chemical Properties of the Modified Nucleosides

Most pseudouridines in rRNAs result from the posttranscriptional isomerization of the base of uridines by the pseudouridine synthase Cbf5p/dyskerin of H/ACA RNPs. The isomerization reaction consists in the breakage of the uridine C1′–N1 glycosidic bond followed by rotation of the base around the C6–N3 axis and reformation of a C1′–C5 bond (Fig. 7). This rearrangement does not change the base-pairing properties of pseudouridines toward adenosines, but it increases their hydrogen bonding capacity since, in addition to N3—H, a second N1 imino proton becomes available to establish hydrogen bonding interactions. The presence of pseudouridines within RNA molecules has been shown to induce a local stabilization of nucleoside stacking, independently of the structural context, but only within RNA molecules of substantial length [128,129]. This effect is correlated with a strong stabilization of the ribose C3′-endo conformation compared to uridines [129].

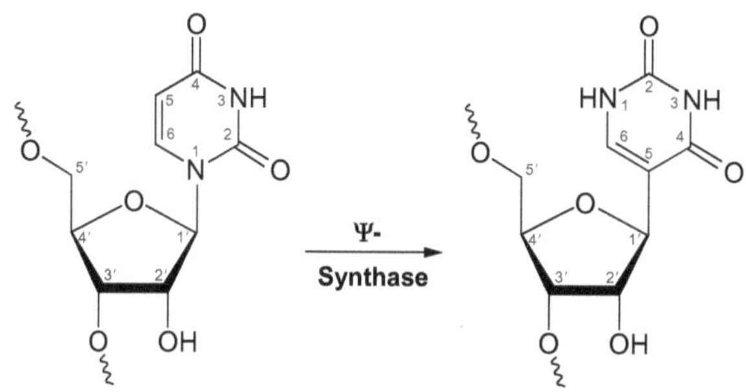

Fig. 7 Conversion of uridine into pseudouridine. Isomerization of uridines into pseudouridines requires breakage of the uridine C1′–N1 glycosidic bond followed by rotation of the base around the C6–N3 axis and reformation of a C1′–C5 bond. This rearrangement creates an additional hydrogen-bond donor group since a second N1—H imino proton becomes available in addition to the preexisting N3—H.

Importantly, NMR studies showed that the imino N1—H proton of pseudouridines in solution is in slow exchange with solvent, whatever the RNA environment [130,131]. This is due to the formation of a water-mediated hydrogen bond between the imino N1 proton and acceptor oxygens of the phosphate backbone which restricts the conformations of the pseudouridine and several additional 5′ nucleosides.

The posttranscriptional addition of methyl groups to ribose 2′-hydroxyl groups also induces a local stabilization of the RNA structure. In this case also, the ribose is stabilized in the C3′-endo conformation to limit steric hindrance between the methyl group, the base, and the 3′ phosphates [132]. This confers conformational rigidity to the ribose 2′-O-methylated nucleoside. In addition, methylation of the 2′-hydroxyl group abrogates its ability to engage in H-bonding interactions and increases the hydrophobicity of the nucleoside, thereby changing its interaction properties.

4.2 Evolutionary Conservation of the Modified Nucleosides in rRNAs

Pseudouridines and 2′-O-ribose methylations are synthesized for the very most part on highly conserved rRNA nucleosides, and some of these modifications are themselves universally conserved. In addition, the vast majority of nucleoside modifications are clustered in functionally important regions of the ribosome such as the decoding center (DC), the peptidyl transferase center (PTC), the peptide-exit tunnel, and several intersubunit bridges [133]. The DC is the domain of the small ribosomal subunit surrounding the sites of codon–anticodon interactions between mRNA and the tRNAs in the A-, P-, and E-site of the ribosome. This domain is composed essentially by RNA elements of the 18S rRNA and includes about 10 modified nucleosides in yeast. The PTC is the active site of the ribosome orchestrating formation of the peptide bond between the nascent polypeptide chain carried by the P-site tRNA and the incoming aminoacylated tRNA in the A-site. The PTC involves domain V of the 25S rRNA which surrounds the acceptor stems of the A- and P-site tRNAs and contains about 10 modified nucleosides. The peptide-exit tunnel comprises several domains of the 25S rRNA (I–V) and includes more than 30 modified nucleosides. Two particular intersubunit bridges contain modified nucleosides: one involving helix 69 of the 25S rRNA which contacts helix 44 of the 18S rRNA and one involving the A-site finger (helix 38 of the 25S rRNA).

4.3 Functional Role of rRNA Nucleoside Modifications in Translation

4.3.1 Impact of rRNA Nucleoside Modification Loss on Yeast Cell Growth

The evolutionary conservation of modified nucleosides and their clustering into functionally important regions of the ribosome predict an important role in ribosome function. The vast majority of the nucleoside modifications of both ribosomal subunits are localized at the interface between subunits, an area of the ribosome largely devoid of ribosomal proteins. This property indicates that these modifications do not influence RNA–ribosomal protein interactions within translating ribosomes but rather RNA–RNA interactions taking place between rRNA elements and the mRNA and tRNAs. Most of the current knowledge on the role of rRNA nucleoside modifications on the synthesis and function of eukaryotic ribosomes emanates from studies performed in yeast, which proved to be a tractable model organism to study the phenotypes resulting either from the loss of individual modifications (through deletion of the corresponding snoRNA gene) or from global defects in the synthesis of pseudouridines or 2′-O-ribose methylations generated by mutations in Cbf5p or Nop1p, respectively. Global pseudouridylation loss was obtained by introducing the L94A point mutation in a catalytic motif of Cbf5p [136]. This mutation almost completely abolishes rRNA pseudouridylation at 37°C and strongly inhibits growth. Importantly, this mutation does not affect the stability of H/ACA snoRNPs and does not inhibit synthesis of the 18S and 25S rRNAs. Therefore, ribosomes seem to be efficiently produced in cells expressing Cbf5p L94A, and the lethality associated with this mutation is therefore most likely due to pseudouridine loss. On this basis, it was concluded that global rRNA pseudouridylation is essential for ribosome function and growth. Similar conclusions seem to apply to global defects in the synthesis of rRNA 2′-O-ribose methylations [52]. A point mutation in the methyltransferase Nop1p (nop1–3) induces a temperature-sensitive growth defect. In these cells, incorporation of methyl groups in rRNAs is fully abolished at nonpermissive temperature, but pre-rRNA processing is only slightly affected, and the 18S and 25S rRNAs are efficiently produced. Therefore, ribose methylation is not required for ribosome production, but heavily hypomethylated ribosomes are not functional in translation and cannot support growth.

These two studies strongly suggest that global defects in the synthesis of ribose methylations or pseudouridines do not prevent ribosome production but yield nonfunctional ribosomes that do not support growth. In contrast, loss of single or limited numbers of modifications clustering into functionally important regions of the ribosome is generally not sufficient to induce detectable growth defects under laboratory conditions. Indeed, early studies showed that combined deletion of up to seven box C/D or H/ACA snoRNA genes does not reduce growth rate, indicating that loss of a limited number of modifications does not perturb ribosome function to a level sufficient to yield translation defects that interfere with growth [137–140]. More recent studies (see below) confirmed that individual deletion of numerous snoRNA genes does not impact cell growth, one notable exception being box H/ACA snoRNA snR35, guiding the pseudouridylation of the hypermodified uridine 1191 ($m^1acp^3\Psi1191$) in the DC [141]. Synthesis of this complex nucleoside involves conversion of uridine 1191 into pseudouridine by the snR35-containing H/ACA snoRNP, methylation by Emg1p of the base at the N1 position released following pseudouridylation of the nucleoside [142,143], and further addition of an aminocarboxypropyl (acp) group at the N3 position by Tsr3p [144]. Loss of snR35, which inhibits both the pseudouridylation of U1191 and its methylation by Emg1p, induces a modest (10%) reduction in the growth rate at 30°C [141].

Besides *SNR35*, the apparently neutral phenotypes resulting from snoRNA gene deletions are intriguing since the modifications have been individually selected during evolution and some of them are universally conserved, indicating that they must confer a selective advantage on their own. A reasonable explanation for this paradox is that the growth conditions tested in the laboratory are far different and less complex than those in the natural environment, where cells are constantly subjected to stresses such as frequent changes in temperature, desiccation, exposure to UV lights, and competition with other microorganisms. Lack of individual modifications probably induces very subtle defects in ribosome performance that are sufficient to affect growth in the natural environment, resulting in a counter-selection of the mutant cells during evolution. In agreement with this hypothesis, several studies indeed reported that individual loss of some nucleoside modifications induces subtle but detectable defects in global translation and ribosome accuracy (see below). In addition, an interesting example of growth defect under conditions of competition has been provided by the loss of a single snoRNA, snR191 [145]. This snoRNA is responsible for the synthesis of two highly conserved pseudouridines (Ψ2258 and Ψ2260) located in the loop of helix 69 of the 25S rRNA which forms an intersubunit bridge with H44 of the 18S rRNA in the 80S ribosome and directly contacts the acceptor site tRNA. Deletion of *SNR191* does not induce growth defects at the different temperatures tested, but the mutant strain is clearly disadvantaged in growth competition assays when cocultured with a wild-type strain [145].

Altogether, these data strongly suggest that each single rRNA modification contributes very modestly to different aspects of ribosome function depending on its localization and likely participates in the fine-tuning of the translation mechanisms. Given these subtle individual roles, understanding the precise function of each single modification in translation proved very challenging. Several groups nevertheless undertook this difficult task using yeast cells as a model system. The approaches consisted in preventing synthesis of single or multiple nucleoside modifications clustering in functionally important regions of the ribosome by deleting the corresponding snoRNA genes, and assessing the consequences on several testable features of ribosome structure and function. Changes in ribosome structure were addressed indirectly by analyzing the sensitivity or resistance to antibiotics, and more directly by measuring the accessibility of rRNA nucleosides to chemicals such as DMS. Global translation levels were assessed by measuring the rate of incorporation of ^{35}S methionine. More subtle aspects of ribosome function such as the accuracy of translation were studied using reporter systems allowing to measure the fidelity of elongation (+1 and −1 frameshifting) or termination (stop codon read-through). Defects in subunit joining and polysome formation could be addressed by assessing the ribosomal profiles using sedimentations on density gradients. These studies allowed a better characterization of the role of modified nucleosides in translation, and the next sections describe the major findings.

4.3.2 Role of the Nucleoside Modifications in the Decoding Center of the Yeast Ribosome

As mentioned earlier, the decoding center of the ribosome is the domain of the small subunit establishing contacts with mRNAs and tRNAs, and ensuring proper interactions between mRNA codons and tRNA anticodon loops in the A- and P-sites of the ribosome. This domain is formed by regions of the 18S rRNA containing eight modified nucleosides (Fig. 8). Four of these surround the A-site: two are close to the A-site codon–anticodon interaction and two others are located above A-site tRNA (refered to as "aA" below) (Fig. 8). The four remaining DC modifications are in the vicinity of the P-site (2) and E-site (2) (Fig. 8). It can be anticipated from their localization that these modifications participate in the correct selection of tRNAs and their translocation during translation.

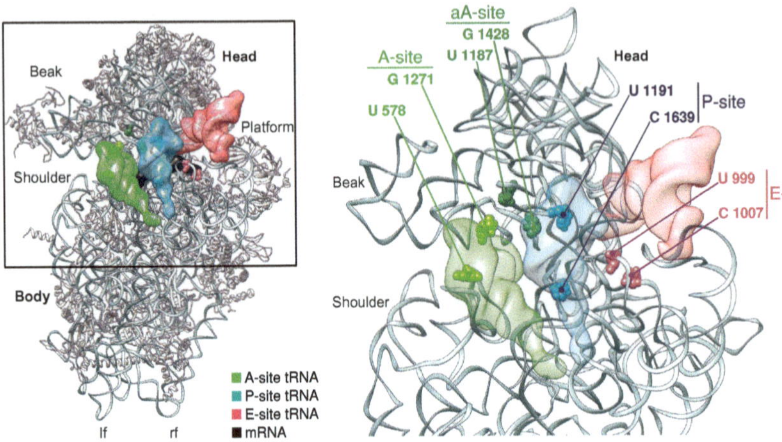

Fig. 8 Position of selected nucleoside modifications in the decoding center (DC) of the small ribosomal subunit. *Left*: ribbon representation of the three-dimensional structure of the small ribosomal subunit (*white*, ribosomal proteins; *pale blue*, 18S rRNA) extracted from the yeast 80S ribosome crystal structure (PDB accession code 4V88, [134]), viewed from the 60S subunit interface. Electron densities corresponding to tRNAs occupying the A- (*green*), P- (*blue*), and E- (*pink*) sites, as well as mRNA (*black*) have been extracted and modeled from the crystal structure of the *Thermus thermophilus* 70S ribosome (PDB accession code 1VY4 [135]). The position of selected nucleosides undergoing posttranscriptional modifications in the DC region is highlighted. *lf*, left foot; *rf*, right foot. *Right*: zoom in the DC region framed on the *left panel*. Residues Um578 and Gm1271 are close to the A-site tRNA (*lime green*); residues Ψ1187 and Gm1428 (*green*) are above A-site tRNA; residues Ψ1191 and Cm1639 (*blue*) are proximal to P-site tRNA; residues Ψ999 and Cm1007 (*pink*) are in the vicinity of E-site tRNA.

Deletion of the gene encoding snR35, guiding pseudouridylation of the hypermodified pseudouridine residue $m^1acp^3\Psi1191$ in the P-site region (helix 31 of 18S rRNA), slightly affects growth as mentioned earlier and induces a detectable translation defect as attested by a 7% reduction in the incorporation rate of ^{35}S methionine in vivo [141]. In addition, the strain lacking Ψ1191 displays defects in the accuracy of translation elongation as judged by the increased +1 frameshifting [146]. Surprisingly, and somehow contradictorily with earlier data indicating that global loss of pseudouridine synthesis does not interfere with production of the ribosomal subunits [136], cells lacking snR35 alone display a significant defect in the production of the 40S subunit. This defect is correlated with a delay in the maturation of the 20S precursor to the mature 18S rRNA [141]. Whether production of the 40S subunit requires the synthesis of Ψ1191 or the base-pairing of the snR35 antisense element with the pre-rRNA remains unresolved. It is also unclear whether the translation and growth defects observed upon loss of Ψ1191 result from reduced amounts of 40S subunits, production of partially defective 40S subunits, or a combination of both.

Individual loss of several other nucleoside modifications in the DC does not affect cell growth under laboratory conditions nor 40S subunit production but does affect some aspects of ribosome function. One example is the methylation of C1639, located like Ψ1191 in the P-site region of the DC (helix 44 of 18S rRNA), guided by snR70. Deletion of *SNR70* alone reduces by 5% the rate of incorporation of ^{35}S methionine in vivo [141] and slightly impairs the accuracy of translation elongation (increased +1 and −1 frameshift activity) and of translation termination (increase in stop codon read-through) [146]. In addition, individual loss of either Ψ1187 (guided by snR36) or Gm1428 (guided by snR56), both located above the A-site tRNA (aA) increases specifically −1 frameshifting [146]. Interestingly, lack of the two snoRNAs guiding modifications in the E-site substantially decreases +1 frameshift activity indicating that some modifications may reduce translation accuracy [146].

Although individual loss of all the DC modifications except $m^1acp^3Ψ1191$ does not affect growth, moderate to strong growth defects were observed upon the combined depletion of 2 or 3 modifications in the aA- and P-sites of the DC (including or not $m^1acp^3Ψ1191$), indicating cumulative effects [141,146]. These growth defects are correlated with more drastic impairments in ribosome accuracy [146]. Interestingly in several instances, the defects in ribosome function caused by the absence of individual or limited number of modifications are rescued or alleviated by simultaneous removal of other nucleoside modifications in the DC. These observations indicate that some modifications can influence translation positively or negatively and that the phenotypes resulting from loss of nucleoside modifications depend on the specific pattern of the remaining modifications, illustrating the complexity of the role of these modifications [141,146]. Several mutant strains lacking modifications in the aA- and P-sites are more sensitive to neomycin, an antibiotic that interacts with the A-site of the DC [141] and on the contrary, lack of $m^1acp^3Ψ1191$ confers increased resistance to the tetracycline analogue doxycycline which targets the P-site region [146]. This probably reflects structural changes in the environment of the DC. Another important issue to consider is that, among the different strains affected in translation, some display defects in the production of the 40S subunit (those with deletion of *SNR35*, for example) and others do not. The translation defects observed can therefore result either from reduced levels of 40S subunits, from deficient 40S subunits accumulating at wild-type levels, or from a combination of both.

In summary, some of the modifications in the DC, especially those located above the A-site and close to the P-site, probably participate in the proper organization of the ribosome structure around these sites to maintain optimal ribosome accuracy.

4.3.3 Role of Nucleoside Modifications in the PTC of the Yeast Ribosome

The PTC is the region of the large ribosomal subunit catalyzing formation of the peptide bond between the nascent peptide chain and the next amino acid carried by the A-site tRNA. Six pseudouridines and six 2′-O-ribose methylated nucleosides are present in this region of the 25S rRNA (domain V, Fig. 9), although not close enough to the site of peptide bond formation to take part directly in the chemistry of the reaction. Given their position, it can be anticipated that these modifications participate in the proper incorporation of tRNAs in the A- and P-sites, correct positioning of the acceptor stem of tRNAs, tRNA translocation, and release (E-site). The six pseudouridines are synthesized by five snoRNAs: snR10 (Ψ2923), snR34 (guiding both Ψ2826 and Ψ2880), snR37 (Ψ2944), snR42 (Ψ2975), and snR46 (Ψ2865). Importantly, snR10 fulfills a dual function in ribosome synthesis: it directs the synthesis of Ψ2923 via its 3′ pseudouridylation pocket and also functions as a chaperone via a 7 nucleoside element located in the internal bulge of the 5′ hairpin partially complementary to the pre-rRNA [147]. Loss of snR10 impairs production of the 18S rRNA and therefore affects growth

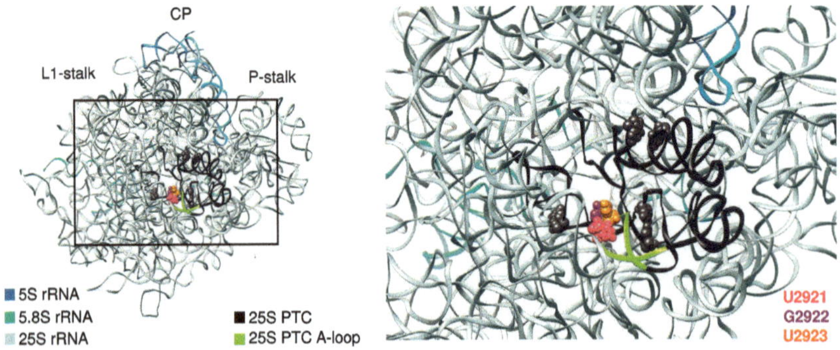

Fig. 9 Position of selected nucleoside modifications in the PTC region of the large ribosomal subunit. *Left*: ribbon representation of the whole three-dimensional structure of the 25S (*gray*), 5.8S (*light blue*), and 5S (*marine blue*) rRNA extracted from the yeast 80S ribosome crystal structure (PDB accession code 4V88, [134]) and seen from the 40S subunit interface. The PTC region is colored in *black* with the crucial A-loop in *green*. Position of selected nucleosides undergoing posttranscriptional modifications is highlighted. *Right*: zoom in the PTC region framed on the *left panel*. Residues Um2921, Gm2922, and Ψ2923 are highlighted in *pink*, *purple*, and *orange*, respectively. The five other pseudouridines of the PTC region are indicated by *gray spheres*.

[148,149]. Interestingly, genetic separation of the processing and modification functions of snR10 showed that the absence of Ψ2923 in the crucial A-loop of the PTC induces translation defects [137]. A yeast strain was engineered to express a mutant version of snR10 bearing a single nucleoside deletion in the antisense elements of the 3′ pseudouridylation pocket preventing synthesis of Ψ2923. This strain does not display any growth defect, and pre-rRNA processing and production of the mature 40S and 60S subunits are comparable to those in wild-type cells. However, this strain shows an aberrant polysome profile on sucrose gradients, characterized by the presence of halfmer polysomes suggested to reflect defective recruitment of 60S subunits on 43S preinitiation complexes. This altered profile is correlated with reduced levels of incorporation of ^{35}S methionine in cellular proteins. Therefore, loss of a single pseudouridine residue in the A-loop of the 25S rRNA seems to inhibit translation by affecting subunit joining and formation of 80S ribosomes [137]. A yeast strain lacking all five snoRNAs targeting the six pseudouridines of the PTC grows more slowly than the strain lacking snR10 only, indicating that besides Ψ2923, cumulative loss of the other pseudouridines affects translation. Indeed, the rate of ^{35}S methionine incorporation is reduced in cells lacking all PTC pseudouridines except Ψ2923. This translation defect is not due to a shortage of 60S ribosomal subunits, but DMS structure probing experiments suggested that the structure of 60S subunits is modified as attested by the increased reactivity of specific nucleosides of domain V of the 25S rRNA. Polysome profiles showed that in comparison with cells lacking snR10 only, which display reduced 80S and polysome peaks due to the shortage of 40S subunits, lack of all six pseudouridines results in a strong increase in the 80S peak, suggestive of an accumulation of preinitiation complexes [137].

In the A-loop of the PTC, Ψ2923 (guided by snR10) is preceded by two 2′-O-ribose methylated nucleosides on the 25S rRNA, Um2921 modified by the box C/D snoRNP containing the snR52 snoRNA, and Gm2922 modified by the site-specific methyltransferase Spb1p. Interestingly, lack of snR52 does not prevent synthesis of Um2921, which can still be catalyzed by Spb1p in these conditions [18]. The absence of snR52 does not affect growth but results in substantial defects in pre-rRNA processing in the pathway leading to production of the 40S subunit, possibly due to the fact that snR52 also directs the methylation of Am420 in the 18S rRNA in addition to Um2921. Expression of a catalytic-dead mutant of Spb1p (D52A), which abrogates the synthesis of Gm2922 but not Um2921, induces substantial growth defects at different temperatures and strongly increases sensitivity to the ribosome-based antibiotic paromomycin [17,18]. This evidence may suggest that methylation of G2922 is important for translation. However, the growth defects are correlated with defects in the production of the 60S subunits, suggesting either that Gm2922 is required for proper production of the 60S subunits or that the catalytic activity of Spb1p fulfills other functions important for this production.

In summary, individual loss of Ψ2923 or combined loss of this and other pseudouridines clustering at the PTC region seem to induce defects in subunit joining during translation initiation or entry into elongation.

4.3.4 Role of Nucleoside Modifications in Intersubunit Bridges of the Yeast Ribosome

Several RNA bridges are established between the small and large subunits in translating ribosomes. They probably mediate communications between the subunits, and some of them are close to tRNAs in the A- and P-site. Among these, bridge B2a is formed between helix 69, a short stem–loop structure of domain IV of the 25S rRNA positioned in the vicinity of the A- and P-site tRNAs, and establishing contacts with helix 44 of the small subunit (Fig. 10). Helix 69 contains five nucleoside modifications in yeast, four pseudouridines (Ψ2258, Ψ2260, Ψ2264, Ψ2266) and one ribose methylation (Am2256), directed by four snoRNAs: snR191 (Ψ2258, Ψ2260), snR3 (Ψ2264), snR84 (Ψ2266), and snR63 (Am2256). Growth defects become detectable upon the combined loss of three or more of the snoRNAs guiding these modifications [150]. These mutant strains display increased sensitivity to different ribosome-based antibiotics such as neomycin or sparsomycin, targeting the DC and PTC, respectively, suggesting that ribosome structure is affected. Chemical probing of ribosome structure confirmed that residue A2259, flanked by two pseudouridines (Ψ2258, Ψ2260), displays increased reactivity in the absence of these modifications, suggesting that the pseudouridines are responsible for a local stabilization. The rate of incorporation of ^{35}S methionine is reduced in strains lacking nucleoside modifications in helix 69, and the extent of the reduction correlates directly with the severity of the growth defects, suggesting that translation capacity is the primary determinant of growth rate. More precise analyses indicate that combined loss of three or five nucleoside modifications in helix 69 strongly increases stop codon read-through but has only a minor impact on +1 frameshifting [150]. However, the effects seem to depend on the specific combination of remaining modifications, and conflicting results have been reported. Indeed, individual loss of any of these snoRNAs has been reported by others to induce a decrease in stop codon read-through [146]. In addition, −1 and +1 frameshifting are also decreased in the mutant strain lacking Ψ2266

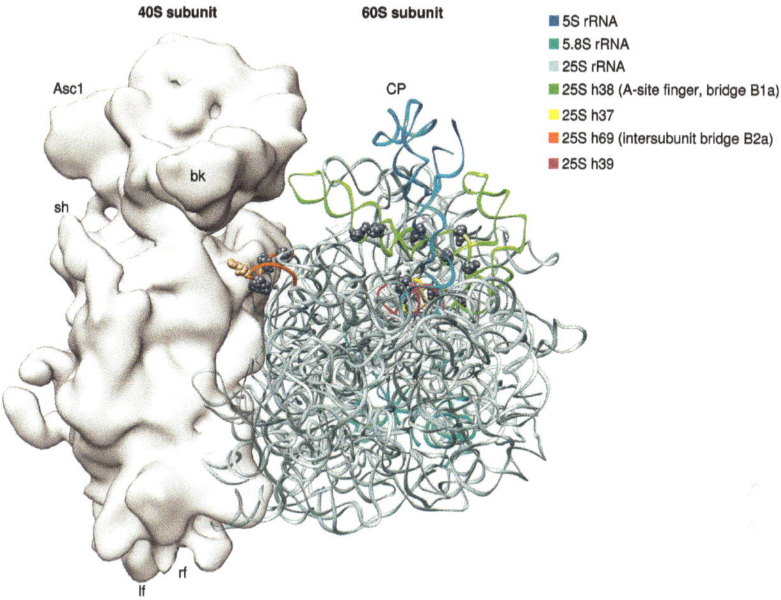

Fig. 10 Position of selected nucleoside modifications in the vicinity of ribosome intersubunit bridges B1a and B2a. Surface view of the 40S ribosomal subunit associated with a ribbon representation of the 25S (*gray*), 5.8S (*light blue*), and 5S (*marine blue*) rRNA extracted from the yeast 80S ribosome crystal structure (PDB accession number 4V88, [134]) seen from the beak (bk) of the small ribosomal subunit. *lf*, left foot; *rf*, right foot; *sh*, shoulder. Helix 38 of the A-site finger involved in intersubunit bridge B1a is colored in *green*, and the adjacent helices 37 and 39 are highlighted in *yellow* and *pink*, respectively. The 10 pseudouridines in this environment are labeled in *gray*. Helix 69 involved in intersubunit bridge B2a is colored in *red* with pseudouridines and methylations shown in *gray* and *gold*, respectively.

[146]. This increased ribosome fidelity is correlated with increased sensitivity to aminoglycoside antibiotics such as kanamycin or G418 [146]. Importantly, loss of modifications in helix 69 decreases the steady-state levels of both the 18S and 25S rRNAs. Pre-rRNA processing analyses revealed only minor delays in the maturation of the 18S and 25S rRNA precursors, and the most important contribution to the decreased steady-state levels of rRNAs is an increased instability, probably caused by partial degradation of functionally impaired ribosomes [150].

In summary, loss of modifications in helix 69 of the 25S rRNA involved in the B2a intersubunit bridge alters ribosome structure and affects global translation, resulting in an increased turnover of deficient ribosomes. The precise role of these modifications on translation accuracy still remains controversial.

Another intersubunit bridge (B1a) involves the A-site finger (ASF) corresponding to the distal portion of helix 38 in domain II of the 25S rRNA. This ribosome region includes three consecutive helices (H37, H38, and H39). The distal portion of H38 contacts tRNA in the A-site (hence the name "A-site finger") and further protrudes in the intersubunit space where it establishes the B1a intersubunit bridge (Fig. 10). Although the distal portion of H38 does not contain modified nucleosides, 10 pseudouridine residues are present in the base and mid portions of H38 and in adjacent helices H37 and H39. On the basis of this localization, it can be anticipated that these modifications influence tRNA positioning and movement during translocation as well as intersubunit communications. As in several other instances, growth defects are observed upon loss of at least three modifications [151], and in some instances, the lack of additional modifications does not aggravate but rescues the growth phenotype. The extent of the translation defects therefore results from the specific pattern of remaining modifications. Loss of at least three pseudouridines in the vicinity of the ASF is associated with increased sensitivity to the antibiotics paromomycin or sparsomycin [151]. Since these antibiotics do not target the A-site finger region (paromomycin targets the DC and sparsomycin the PTC), these results indicate that the absence of pseudouridines probably induces local structural alterations that impinge on the other functional sites of the ribosome. Translational activity is reduced in cells lacking pseudouridines in the ASF and polysome profiles on density gradients revealed a minor shortage of 60S subunits (15% reduction), and the presence of halfmer polysomes proposed to reflect delays in subunit joining during formation of 80S ribosomes [151]. Metabolic labeling experiments did not reveal any defect in the production of the 60S subunit suggesting that 60S subunit shortage rather results from faster turnover by quality control mechanisms detecting functionally impaired ribosomal subunits. Since pseudouridines have been shown to stabilize helical RNA structures, the authors propose that they may rigidify the ASF and thereby optimize its interactions with tRNA and other interacting elements during ribosome translocation [151]. An independent study showed that loss of six pseudouridines in helices H37, H38, and H39 did not induce drastic defects in ribosome accuracy except for increased read-through of the UGA nonsense codon, the least efficient stop codon in eukaryotes [146]. The authors propose that by rigidifying the helical structures around the ASF (H37, H38, and H39), pseudouridine residues may favor selection of the eRF1 translation termination factor at the expense of near cognate tRNAs and thereby prevent UGA stop codon read-through.

4.3.5 Role of rRNA Nucleoside Modifications in Vertebrates

Although most of the current knowledge on the function of rRNA nucleoside modifications originates from yeast studies, the consequences of global defects in rRNA nucleoside modifications or individual snoRNA loss have also been addressed to some extent in vertebrates.

Early studies reported that methionine deprivation or treatment with ethionine, a toxic analog of methionine, strongly affects the production of rRNAs in HeLa cells [152–154]. However, as an essential amino acid, methionine starvation surely induces pleiotropic defects, limiting the interpretations of these data. Treatment of HeLa cells with cycloleucine, an inhibitor of SAM synthesis, also induces substantial processing defects and inhibits the production of mature rRNAs [155]. These data suggest that in contrast to the yeast situation, global methylation of the pre-rRNAs is required for pre-rRNA processing in HeLa cells. The discrepancy between the studies in yeast or human cells may reflect biological differences or stem from the different experimental designs. In the yeast study, a point mutation has been introduced in Nop1p, whereas in human cells, a depletion of SAM has been induced, which inhibits all cellular methyltransferase activities in addition to fibrillarin and probably induces pleiotropic defects. Full depletion of fibrillarin in mouse has been reported to be inviable since homozygous mutant embryos arrest development at premorula stages [156]. Interestingly, although a proportion of heterozygous mutant embryos die at early developmental stages, live-born mice do not display obvious phenotypes suggesting that fibrillarin haploinsufficiency can be tolerated. In MEFs derived from these heterozygous viable mutants, fibrillarin protein levels are substantially decreased, but pre-rRNA processing is not affected, and production of mature rRNAs is as efficient as in wild-type cells [156]. Although levels of the U3 snoRNA are not affected in these MEFs, accumulation of the intron-encoded U76 box C/D snoRNA guiding a methylation in the 28S rRNA is reduced to some extent, and it is likely that the accumulation of other methylation guides is also affected. These data suggest that partial depletion of fibrillarin, which may affect the accumulation levels of several box C/D guide snoRNAs does not impair pre-rRNA processing in MEFs.

Concerning the role of rRNA pseudouridylation in vertebrates, studies aiming at understanding the etiology of the human disease Dyskeratosis Congenita (DC) led to important discoveries on the role of rRNA pseudouridylation in ribosome function in mammalian cells. X-linked DC is a genetic disease caused by point mutations in *DKC1*, a gene encoding dyskerin, the pseudouridine synthase of H/ACA snoRNPs. Dyskerin is also a component of the telomerase holoenzyme, which contains a H/ACA-like

RNP domain in vertebrates in addition to the catalytic domain ensuring telomere elongation. It was suggested that the onset of DC is merely caused by defects in ribosome synthesis or function and that telomere maintenance defects contribute to the severity of the disease [157]. Mouse embryonic stem cells bearing one of the most frequent *DKC1* mutations found in X-DC patients display reduced accumulation levels of dyskerin correlated with reduced steady-state levels of telomerase RNA and several box H/ACA snoRNAs [158]. As expected, these phenotypes are associated with defects in telomerase activity, resulting in shorter telomere lengths, and with a global reduction of rRNA pseudouridylation. Interestingly, metabolic labeling experiments showed that the maturation of the pre-rRNAs is only modestly affected and production of the mature 18S and 28S rRNAs occur with kinetics comparable to the wild-type control. These data indicate that (i) the accumulation levels of the few snoRNAs required for pre-rRNA processing are not substantially decreased in these cells and (ii) that a global reduction in pre-rRNA pseudouridylation does not severely impair pre-rRNA processing, consistent with the observations in yeast [136].

To further elucidate the molecular origins of DC, a mouse model bearing a hypomorphic mutation in *DKC1* ($Dkc1^{m}$) that reduces expression of dyskerin was generated. These mice recapitulate some of the clinical symptoms of X-DC patients. They display a 10%–40% reduction in pseudouridylation of both the 18S and 28S rRNAs, and in this case, pre-rRNA processing is affected compared to control cells. It remains unclear whether the processing anomalies result from the global pseudouridylation defect or from a scarcity of the subset of box H/ACA snoRNPs required for pre-rRNA processing. In lymphocytes derived from $Dkc1^{m}$ mice, global translation is not affected [159]. Interestingly, however, more subtle and mRNA-specific translation defects could be detected. Internal ribosome entry site (IRES)-dependent translation of p53 mRNA was substantially impaired during oncogene-induced senescence in these cells, and a similar conclusion was drawn from studies performed using lymphoblasts derived from X-DC patients [160]. A subset of other mRNAs was less translated in $Dkc1^{m}$ cells, in particular those encoding the antiapoptotic factors XIAP and Bcl-xL or the tumor suppressor p27. These mRNAs have in common the presence of an IRES in their 5′ region, as in the p53-encoding mRNA. These observations and additional experiments, not only in $Dkc1^{m}$ mice but also in cells derived from patients affected with X-DC, led to the conclusion that ribosomes with low pseudouridine content are deficient for

IRES-dependent translation [159]. This was confirmed in other cell types such as yeast cells expressing a mutant version of Cbf5p (D95A), which display a strong defect in rRNA pseudouridylation, or human cell lines treated with DKC1 siRNAs [161]. More detailed analyses showed that the mature 40S subunits purified from yeast cells expressing Cbf5p-D95A or from mouse $Dkc1^{m}$ cells display a reduced affinity for IRES structures. In cytoplasmic extracts prepared from mouse $Dkc1^{m}$ cells, formation of the 48S preinitiation complex onto an IRES reporter was reduced by 50% compared to controls. The authors propose that hypopseudouridylated ribosomes are deficient in the interaction with IRES structures during cap-independent translation initiation in yeast and mammals [161]. Other ribosome dysfunctions could be detected such as decreased translation fidelity. These data support the view that rRNA pseudouridines are required for optimal ribosome-IRES interactions in yeast and human cells. The picture is probably more complex, since a more recent study reported that dyskerin depletion increases, on the contrary, translation of the IRES-containing VEGF mRNA, indicating that partial rRNA pseudouridylation loss can influence either negatively or positively translation initiation at IRES elements [162]. Another set of experiments, using a reconstituted in vitro translation system with ribosomes purified from dyskerin-depleted cells, confirmed the positive or negative effects of pseudouridine loss on IRES-dependent translation as well as the decreased translation fidelity [163]. Furthermore, these ribosome dysfunctions could not be attributed to a different composition in ribosomal proteins, as assessed by quantitative mass spectrometry analyses [163]. Accordingly, and consistent with the fact that global, cap-dependent translation is not affected, this study also revealed that hypopseudouridylated ribosomes are not defective in translation elongation.

Altogether, these data support the notion that partial loss of rRNA pseudouridines induces intrinsic ribosome dysfunctions affecting IRES-dependent translation initiation and the overall translational fidelity. Interestingly, X-DC patients bearing different point mutations in *DKC1* display heterogeneous box H/ACA snoRNA accumulation profiles that vary depending on the patients and even depending on the nature of the tissue for a given patient [164]. These global variations in snoRNA accumulation profiles are precisely correlated with concordant variations in the pseudouridylation levels of the corresponding positions in rRNAs, as assessed by mass spectrometry [164]. Different *DKC1* mutations therefore result in the production of heterogeneous pools of ribosomes differentially pseudouridylated, depending both on the patients and on the tissues within a given patient. This heterogeneity may account for the multiple clinical features and tissue specificity of DC symptoms. A specific subset of H/ACA snoRNAs is nevertheless systematically downregulated in all X-DC patients, targeting modifications in two functional regions of the ribosome, expansion segment ES6 of the 18S rRNA and domain II of the 28S rRNA [164]. Loss of these pseudouridines may have a stronger contribution to ribosome dysfunctions.

Studies performed in the context of the vertebrate model zebrafish led to the conclusions that individual rRNA nucleoside modifications in zebrafish are crucial for development [165]. In zebrafish, the u22hg and gas5 genes are two "noncoding" genes hosting in their introns several snoRNA sequences, including U26 and U78, which direct 2′-O-ribose methylations of 28S rRNA nucleosides and U44, directing methylation of an 18S rRNA position. Injection of antisense morpholinos designed to inhibit expression of U26, U44, and U78 snoRNAs induces development delays and leads to embryonic lethality. These data may suggest that a single nucleoside modification in the rRNAs is essential for the developmental program in zebrafish. In contrast, deletion of four methylation guide snoRNAs (U32a, U33, U34, and U35a [166]) from intronic sequences of *Rpl13a* gene has no observable consequence on mouse development since live-born mice, obtained at normal Mendelian ratios, are indistinguishable from wild-type mice [167]. However, these mice display altered glucose homeostasis correlated with enhanced insulin secretion, but these phenotypes have been proposed to reflect loss of noncanonical, cytosolic functions of snoRNAs [167], and references therein.

5. EMERGENCE OF THE CONCEPT OF HETEROGENEOUS/ SPECIALIZED RIBOSOMES

In the past few years, high-throughput studies aiming to identify all nucleoside modifications in rRNAs have been reported. Pseudouridylation [168] or methylation [19] sites were mapped independently or together [6,169]. These profiling studies were performed in yeast and humans and have been made possible by progress in transcriptome-wide sequencing tools and mass spectrometry-based methods. The data collected mostly confirmed the presence of previously described sites and allowed the identification of very few new sites. A previously unidentified pseudouridine was detected at position 2345 in *S. cerevisiae* 25S rRNA [6] and two novel

2′-O-methylation sites were mapped in rRNAs of HeLa cells [19]. Some profiling methods allowed the evaluation of the extent of modification at every site on the rRNAs. The overall picture that emerges from the profiling studies is that the modified sites are largely conserved. Quantitative mapping of pseudouridines and/or 2′-O-methylated nucleosides showed that the extent of modification at a given position on rRNAs is very similar throughout eukaryotes [6,19,168]. This substantial conservation suggests that both the presence and relative abundance of modified sites are under strong selection.

Interestingly, a subset of rRNA positions is not always fully modified [6,19]. Indeed, out of the 112 sites of posttranscriptional modifications mapped in *S. cerevisiae* rRNAs, 18 were modified to a lesser extent (5%–85%) [6]. Likewise, the set of 2′-O-methylation sites in human rRNAs is divided in 2/3 of fully methylated sites, forming the core of constitutively methylated positions, and 1/3 of partially methylated sites [19]. This observation suggests a coordinated regulation of the modification levels of specific rRNA nucleosides, and beyond, leads to the emergence of the concept of ribosome heterogeneity. Ribosome heterogeneity is observed both under normal physiological conditions or following environmental perturbations. For example, a study carried out using actively growing *S. cerevisiae* cells showed that 1/3 of nucleoside A100 of 18S rRNA were not 2′-O-ribose methylated [170]. Interestingly 40S ribosome subunits lacking this methylation can participate in the translation process since they were detected in polysomes. It was also shown that nucleoside modification levels in *Schizosaccharomyces pombe* rRNAs are altered following changes in growth temperature [169].

The concept of ribosome heterogeneity also applies to some human pathologies associated with alterations in rRNA nucleoside modification mechanisms. A growing number of studies has recently suggested that the machineries required for rRNA modifications are altered in several human diseases, especially in cancer. For instance, deregulated expression of specific subsets of H/ACA snoRNAs has been reported in a number of hematological disorders, including leukemia, lymphoma, myeloma, and X-DC, as well as in nonsmall cell lung, colorectal, and prostate cancer (reviewed in Ref. [171]). In addition, deregulated expression of C/D snoRNAs has been reported in human cancers [172–174]. Such changes in C/D or H/ACA snoRNA levels would directly affect the pattern of rRNA methylations and pseudouridylations. For example, decreased expression of snoRNA15 was associated with a reduced level of pseudouridylation at nucleoside U1367 in cells from X-DC [164].

Functional alterations of fibrillarin or dyskerin were both associated with human diseases. Elevated expression levels of fibrillarin were (i) observed in primary and metastatic prostate cancer and squamous cell cervical carcinoma, (ii) reported to promote anchorage-independent cell proliferation and chemoresistance of breast cancer cells [175], and (iii) associated with a poor survival in primary breast tumors [176]. Fibrillarin-dependent rRNA modifications were shown to be under the control of p53, since depletion of p53 in human mammary epithelial cells resulted in both increased fibrillarin expression and modulation of rRNA 2′-O-methylation pattern [176]. In addition, p53-depleted cells displayed defects in translational fidelity and increased translation of IRES-containing mRNAs, such as insulin-like growth factor 1 receptor (IGF1R) and c-MYC. Collectively, these recent data led to a model in which alterations of rRNA methylation pattern through p53-dependent changes in fibrillarin expression in cancer cells would drive a specific translation program of mRNAs encoding oncogenic proteins that favor tumor development. Moreover, cells displaying decreased dyskerin expression exhibit both reduced rRNA pseudouridylation [177] and decreased IRES-dependent translation of key tumor suppressor genes, such as p53 [163], highlighting a direct cause–effect link between rRNA pseudouridylation and alterations in translation of specific mRNAs.

Although these recent reports highlight an emerging role for selective modifications of rRNA in human diseases, the exact molecular mechanisms leading to translational program rewiring are not well documented. These data reinforce the emerging notion of specialized ribosomes exhibiting differences in ribosomal protein composition or in rRNA methylation and pseudouridylation [178]. Specialized ribosomes were proposed to drive translation of specific subsets of mRNAs involved in cellular and physiopathological processes, thereby introducing an additional level of regulation of gene expression. Thus, alterations in rRNA modification pattern in human diseases would participate to ribosome heterogeneity and actively contribute to cellular transformation and tumorigenesis by perturbing translational programs. Although very exciting, the existence of such specialized ribosomes has not been clearly proven, and the molecular mechanism by which such ribosomes specifically recognize their mRNA targets remains elusive.

REFERENCES

[1] N. Ban, et al., A new system for naming ribosomal proteins, Curr. Opin. Struct. Biol. 24 (2014) 165–169.

[2] A. Fernandez-Pevida, D. Kressler, J. de la Cruz, Processing of preribosomal RNA in *Saccharomyces cerevisiae*, Wiley Interdiscip. Rev. RNA 6 (2) (2015) 191–209.

[3] A.K. Henras, et al., An overview of pre-ribosomal RNA processing in eukaryotes, Wiley Interdiscip. Rev. RNA 6 (2) (2015) 225–242.

[4] J.L. Woolford Jr., S.J. Baserga, Ribosome biogenesis in the yeast *Saccharomyces cerevisiae*, Genetics 195 (3) (2013) 643–681.

[5] S. Sharma, D.L. Lafontaine, 'View from a bridge': a new perspective on eukaryotic rRNA base modification, Trends Biochem. Sci. 40 (10) (2015) 560–575.

[6] M. Taoka, et al., The complete chemical structure of *Saccharomyces cerevisiae* rRNA: partial pseudouridylation of U2345 in 25S rRNA by snoRNA snR9, Nucleic Acids Res. 44 (18) (2016) 8951–8961.

[7] A.F. Lovejoy, D.P. Riordan, P.O. Brown, Transcriptome-wide mapping of pseudouridines: pseudouridine synthases modify specific mRNAs in *S. cerevisiae*, PLoS One 9 (10) (2014) e110799.

[8] D.A. Samarsky, M.J. Fournier, A comprehensive database for the small nucleolar RNAs from *Saccharomyces cerevisiae*, Nucleic Acids Res. 27 (1) (1999) 161–164.

[9] P. Schattner, et al., Genome-wide searching for pseudouridylation guide snoRNAs: analysis of the *Saccharomyces cerevisiae* genome, Nucleic Acids Res. 32 (14) (2004) 4281–4296.

[10] C. Torchet, et al., The complete set of H/ACA snoRNAs that guide rRNA pseudouridylations in *Saccharomyces cerevisiae*, RNA 11 (6) (2005) 928–938.

[11] W.A. Decatur, M.N. Schnare, Different mechanisms for pseudouridine formation in yeast 5S and 5.8S rRNAs, Mol. Cell. Biol. 28 (10) (2008) 3089–3100.

[12] V. Marchand, et al., Illumina-based RiboMethSeq approach for mapping of 2′-O-Me residues in RNA, Nucleic Acids Res. 44 (16) (2016) e135.

[13] U. Birkedal, et al., Profiling of ribose methylations in RNA by high-throughput sequencing, Angew. Chem. Int. Ed. Engl. 54 (2) (2015) 451–455.

[14] J. Yang, et al., Identification of a new ribose methylation in the 18S rRNA of *S. cerevisiae*, Nucleic Acids Res. 43 (4) (2015) 2342–2352.

[15] Z. Kiss-Laszlo, et al., Site-specific ribose methylation of preribosomal RNA: a novel function for small nucleolar RNAs, Cell 85 (7) (1996) 1077–1088.

[16] T.M. Lowe, S.R. Eddy, A computational screen for methylation guide snoRNAs in yeast, Science 283 (5405) (1999) 1168–1171.

[17] C. Bonnerot, L. Pintard, G. Lutfalla, Functional redundancy of Spb1p and a snR52-dependent mechanism for the 2′-O-ribose methylation of a conserved rRNA position in yeast, Mol. Cell 12 (5) (2003) 1309–1315.

[18] B. Lapeyre, S.K. Purushothaman, Spb1p-directed formation of Gm2922 in the ribosome catalytic center occurs at a late processing stage, Mol. Cell 16 (4) (2004) 663–669.

[19] N. Krogh, et al., Profiling of 2′-O-Me in human rRNA reveals a subset of fractionally modified positions and provides evidence for ribosome heterogeneity, Nucleic Acids Res. 44 (16) (2016) 7884–7895.

[20] P. Ganot, M.L. Bortolin, T. Kiss, Site-specific pseudouridine formation in preribosomal RNA is guided by small nucleolar RNAs, Cell 89 (5) (1997) 799–809.

[21] A. Huttenhofer, et al., RNomics: an experimental approach that identifies 201 candidates for novel, small, non-messenger RNAs in mouse, EMBO J. 20 (11) (2001) 2943–2953.

[22] A.M. Kiss, et al., Human box H/ACA pseudouridylation guide RNA machinery, Mol. Cell. Biol. 24 (13) (2004) 5797–5807.

[23] P. Schattner, S. Barberan-Soler, T.M. Lowe, A computational screen for mammalian pseudouridylation guide H/ACA RNAs, RNA 12 (1) (2006) 15–25.
[24] L. Lestrade, M.J. Weber, snoRNA-LBME-db, a comprehensive database of human H/ACA and C/D box snoRNAs, Nucleic Acids Res. 34 (Database issue) (2006) D158–D162.
[25] D. Piekna-Przybylska, W.A. Decatur, M.J. Fournier, The 3D rRNA modification maps database: with interactive tools for ribosome analysis, Nucleic Acids Res. 36 (Database issue) (2008) D178–D183.
[26] J. Cavaille, M. Nicoloso, J.P. Bachellerie, Targeted ribose methylation of RNA in vivo directed by tailored antisense RNA guides, Nature 383 (6602) (1996) 732–735.
[27] K.T. Tycowski, et al., A small nucleolar RNA requirement for site-specific ribose methylation of rRNA in Xenopus, Proc. Natl. Acad. Sci. U.S.A. 93 (25) (1996) 14480–14485.
[28] C. Gaspin, et al., Archaeal homologs of eukaryotic methylation guide small nucleolar RNAs: lessons from the Pyrococcus genomes, J. Mol. Biol. 297 (4) (2000) 895–906.
[29] A.D. Omer, et al., Homologs of small nucleolar RNAs in Archaea, Science 288 (5465) (2000) 517–522.
[30] A.D. Omer, et al., RNA-modifying machines in archaea, Mol. Microbiol. 48 (3) (2003) 617–629.
[31] T.H. Tang, et al., Identification of 86 candidates for small non-messenger RNAs from the archaeon Archaeoglobus fulgidus, Proc. Natl. Acad. Sci. U.S.A. 99 (11) (2002) 7536–7541.
[32] P.P. Dennis, A. Omer, T. Lowe, A guided tour: small RNA function in Archaea, Mol. Microbiol. 40 (3) (2001) 509–519.
[33] A.G. Balakin, L. Smith, M.J. Fournier, The RNA world of the nucleolus: two major families of small RNAs defined by different box elements with related functions, Cell 86 (5) (1996) 823–834.
[34] Z. Kiss-Laszlo, Y. Henry, T. Kiss, Sequence and structural elements of methylation guide snoRNAs essential for site-specific ribose methylation of pre-rRNA, EMBO J. 17 (3) (1998) 797–807.
[35] R.W. van Nues, et al., Box C/D snoRNP catalysed methylation is aided by additional pre-rRNA base-pairing, EMBO J. 30 (12) (2011) 2420–2430.
[36] E. Caffarelli, et al., Processing of the intron-encoded U16 and U18 snoRNAs: the conserved C and D boxes control both the processing reaction and the stability of the mature snoRNA, EMBO J. 15 (5) (1996) 1121–1131.
[37] J. Cavaille, J.P. Bachellerie, Processing of fibrillarin-associated snoRNAs from pre-mRNA introns: an exonucleolytic process exclusively directed by the common stem-box terminal structure, Biochimie 78 (6) (1996) 443–456.
[38] K. Tyc, J.A. Steitz, U3, U8 and U13 comprise a new class of mammalian snRNPs localized in the cell nucleolus, EMBO J. 8 (10) (1989) 3113–3119.
[39] N.J. Watkins, et al., Elements essential for processing intronic U14 snoRNA are located at the termini of the mature snoRNA sequence and include conserved nucleotide boxes C and D, RNA 2 (2) (1996) 118–133.
[40] R.W. van Nues, N.J. Watkins, Unusual C'/D' motifs enable box C/D snoRNPs to modify multiple sites in the same rRNA target region, Nucleic Acids Res. 45 (4) (2017) 2016–2028.
[41] T. Gautier, et al., Nucleolar KKE/D repeat proteins Nop56p and Nop58p interact with Nop1p and are required for ribosome biogenesis, Mol. Cell. Biol. 17 (12) (1997) 7088–7098.
[42] D.L. Lafontaine, D. Tollervey, Nop58p is a common component of the box C+D snoRNPs that is required for snoRNA stability, RNA 5 (3) (1999) 455–467.
[43] S.K. Lyman, L. Gerace, S.J. Baserga, Human Nop5/Nop58 is a component common to the box C/D small nucleolar ribonucleoproteins, RNA 5 (12) (1999) 1597–1604.

[44] D.R. Newman, et al., Box C/D snoRNA-associated proteins: two pairs of evolutionarily ancient proteins and possible links to replication and transcription, RNA 6 (6) (2000) 861–879.

[45] T. Schimmang, et al., A yeast nucleolar protein related to mammalian fibrillarin is associated with small nucleolar RNA and is essential for viability, EMBO J. 8 (13) (1989) 4015–4024.

[46] N.J. Watkins, et al., A common core RNP structure shared between the small nucleoar box C/D RNPs and the spliceosomal U4 snRNP, Cell 103 (3) (2000) 457–466.

[47] P. Wu, et al., Nop5p is a small nucleolar ribonucleoprotein component required for pre-18 S rRNA processing in yeast, J. Biol. Chem. 273 (26) (1998) 16453–16463.

[48] L. Lui, T. Lowe, Small nucleolar RNAs and RNA-guided post-transcriptional modification, Essays Biochem. 54 (2013) 53–77.

[49] A.D. Omer, et al., In vitro reconstitution and activity of a C/D box methylation guide ribonucleoprotein complex, Proc. Natl. Acad. Sci. U.S.A. 99 (8) (2002) 5289–5294.

[50] N.J. Watkins, M.T. Bohnsack, The box C/D and H/ACA snoRNPs: key players in the modification, processing and the dynamic folding of ribosomal RNA, Wiley Interdiscip. Rev. RNA 3 (3) (2012) 397–414.

[51] J.F. Kuhn, E.J. Tran, E.S. Maxwell, Archaeal ribosomal protein L7 is a functional homolog of the eukaryotic 15.5kD/Snu13p snoRNP core protein, Nucleic Acids Res. 30 (4) (2002) 931–941.

[52] D. Tollervey, et al., Temperature-sensitive mutations demonstrate roles for yeast fibrillarin in pre-rRNA processing, pre-rRNA methylation, and ribosome assembly, Cell 72 (3) (1993) 443–457.

[53] H. Wang, et al., Crystal structure of a fibrillarin homologue from Methanococcus jannaschii, a hyperthermophile, at 1.6 A resolution, EMBO J. 19 (3) (2000) 317–323.

[54] S. Galardi, et al., Purified box C/D snoRNPs are able to reproduce site-specific 2′-O-methylation of target RNA in vitro, Mol. Cell. Biol. 22 (19) (2002) 6663–6668.

[55] M. Aittaleb, et al., Structure and function of archaeal box C/D sRNP core proteins, Nat. Struct. Biol. 10 (4) (2003) 256–263.

[56] E.J. Tran, X. Zhang, E.S. Maxwell, Efficient RNA 2′-O-methylation requires juxtaposed and symmetrically assembled archaeal box C/D and C'/D' RNPs, EMBO J. 22 (15) (2003) 3930–3940.

[57] R. Rashid, et al., Functional requirement for symmetric assembly of archaeal box C/D small ribonucleoprotein particles, J. Mol. Biol. 333 (2) (2003) 295–306.

[58] F. Bleichert, S.J. Baserga, Dissecting the role of conserved box C/D sRNA sequences in di-sRNP assembly and function, Nucleic Acids Res. 38 (22) (2010) 8295–8305.

[59] A.D. Omer, et al., Probing the structure and function of an archaeal C/D-box methylation guide sRNA, RNA 12 (9) (2006) 1708–1720.

[60] J.W. Hardin, R.T. Batey, The bipartite architecture of the sRNA in an archaeal box C/D complex is a primary determinant of specificity, Nucleic Acids Res. 34 (18) (2006) 5039–5051.

[61] C. Charron, et al., The archaeal sRNA binding protein L7Ae has a 3D structure very similar to that of its eukaryal counterpart while having a broader RNA-binding specificity, J. Mol. Biol. 342 (3) (2004) 757–773.

[62] T. Moore, et al., Molecular basis of box C/D RNA-protein interactions; cocrystal structure of archaeal L7Ae and a box C/D RNA, Structure 12 (5) (2004) 807–818.

[63] S. Oruganti, et al., Alternative conformations of the archaeal Nop56/58-fibrillarin complex imply flexibility in box C/D RNPs, J. Mol. Biol. 371 (5) (2007) 1141–1150.

[64] J. Suryadi, et al., The crystal structure of the Methanocaldococcus jannaschii multifunctional L7Ae RNA-binding protein reveals an induced-fit interaction with the box C/D RNAs, Biochemistry 44 (28) (2005) 9657–9672.

[65] S. Xue, et al., Structural basis for substrate placement by an archaeal box C/D ribonucleoprotein particle, Mol. Cell 39 (6) (2010) 939–949.
[66] K. Ye, et al., Structural organization of box C/D RNA-guided RNA methyltransferase, Proc. Natl. Acad. Sci. U.S.A. 106 (33) (2009) 13808–13813.
[67] J. Lin, et al., Structural basis for site-specific ribose methylation by box C/D RNA protein complexes, Nature 469 (7331) (2011) 559–563.
[68] S. Nolivos, A.J. Carpousis, B. Clouet-d'Orval, The K-loop, a general feature of the Pyrococcus C/D guide RNAs, is an RNA structural motif related to the K-turn, Nucleic Acids Res. 33 (20) (2005) 6507–6514.
[69] A. Lapinaite, et al., The structure of the box C/D enzyme reveals regulation of RNA methylation, Nature 502 (7472) (2013) 519–523.
[70] W.S. Yip, et al., Box C/D sRNA stem ends act as stabilizing anchors for box C/D di-sRNPs, Nucleic Acids Res. 44 (18) (2016) 8976–8989.
[71] F. Bleichert, et al., A dimeric structure for archaeal box C/D small ribonucleoproteins, Science 325 (5946) (2009) 1384–1387.
[72] K.R. Bower-Phipps, et al., The box C/D sRNP dimeric architecture is conserved across domain Archaea, RNA 18 (8) (2012) 1527–1540.
[73] H. Ghalei, et al., A novel Nop5-sRNA interaction that is required for efficient archaeal box C/D sRNP formation, RNA 16 (12) (2010) 2341–2348.
[74] A. Graziadei, et al., Archaea box C/D enzymes methylate two distinct substrate rRNA sequences with different efficiency, RNA 22 (5) (2016) 764–772.
[75] E.F. Pettersen, et al., UCSF Chimera—a visualization system for exploratory research and analysis, J. Comput. Chem. 25 (13) (2004) 1605–1612.
[76] G.D. Pintilie, et al., Quantitative analysis of cryo-EM density map segmentation by watershed and scale-space filtering, and fitting of structures by alignment to regions, J. Struct. Biol. 170 (3) (2010) 427–438.
[77] G. Qu, et al., The spatial-functional coupling of box C/D and C'/D' RNPs is an evolutionarily conserved feature of the eukaryotic box C/D snoRNP nucleotide modification complex, Mol. Cell. Biol. 31 (2) (2011) 365–374.
[78] E. Tran, et al., Conserved spacing between the box C/D and C'/D' RNPs of the archaeal box C/D sRNP complex is required for efficient 2′-O-methylation of target RNAs, RNA 11 (3) (2005) 285–293.
[79] L.B. Szewczak, et al., Exclusive interaction of the 15.5 kD protein with the terminal box C/D motif of a methylation guide snoRNP, Chem. Biol. 9 (10) (2002) 1095–1107.
[80] N.M. Cahill, et al., Site-specific cross-linking analyses reveal an asymmetric protein distribution for a box C/D snoRNP, EMBO J. 21 (14) (2002) 3816–3828.
[81] N.J. Watkins, A. Dickmanns, R. Luhrmann, Conserved stem II of the box C/D motif is essential for nucleolar localization and is required, along with the 15.5K protein, for the hierarchical assembly of the box C/D snoRNP, Mol. Cell. Biol. 22 (23) (2002) 8342–8352.
[82] M. Kornprobst, et al., Architecture of the 90S Pre-ribosome: a structural view on the birth of the eukaryotic ribosome, Cell 166 (2) (2016) 380–393.
[83] U.T. Meier, The many facets of H/ACA ribonucleoproteins, Chromosoma 114 (1) (2005) 1–14.
[84] S.L. Reichow, et al., The structure and function of small nucleolar ribonucleoproteins, Nucleic Acids Res. 35 (5) (2007) 1452–1464.
[85] Y.T. Yu, U.T. Meier, RNA-guided isomerization of uridine to pseudouridine—pseudouridylation, RNA Biol. 11 (12) (2014) 1483–1494.
[86] P. Ganot, M. Caizergues-Ferrer, T. Kiss, The family of box ACA small nucleolar RNAs is defined by an evolutionarily conserved secondary structure and ubiquitous sequence elements essential for RNA accumulation, Genes Dev. 11 (7) (1997) 941–956.

[87] S. Uliel, et al., Small nucleolar RNAs that guide modification in trypanosomatids: repertoire, targets, genome organisation, and unique functions, Int. J. Parasitol. 34 (4) (2004) 445–454.

[88] A.G. Russell, M.N. Schnare, M.W. Gray, Pseudouridine-guide RNAs and other Cbf5p-associated RNAs in *Euglena gracilis*, RNA 10 (7) (2004) 1034–1046.

[89] S. Muller, et al., Combined in silico and experimental identification of the Pyrococcus abyssi H/ACA sRNAs and their target sites in ribosomal RNAs, Nucleic Acids Res. 36 (8) (2008) 2459–2475.

[90] T.S. Rozhdestvensky, et al., Binding of L7Ae protein to the K-turn of archaeal snoRNAs: a shared RNA binding motif for C/D and H/ACA box snoRNAs in Archaea, Nucleic Acids Res. 31 (3) (2003) 869–877.

[91] D.L. Bernick, et al., Discovery of Pyrobaculum small RNA families with atypical pseudouridine guide RNA features, RNA 18 (3) (2012) 402–411.

[92] J.P. Girard, et al., GAR1 is an essential small nucleolar RNP protein required for pre-rRNA processing in yeast, EMBO J. 11 (2) (1992) 673–682.

[93] A. Henras, et al., Nhp2p and Nop10p are essential for the function of H/ACA snoRNPs, EMBO J. 17 (23) (1998) 7078–7090.

[94] D.L. Lafontaine, et al., The box H + ACA snoRNAs carry Cbf5p, the putative rRNA pseudouridine synthase, Genes Dev. 12 (4) (1998) 527–537.

[95] N.J. Watkins, et al., Cbf5p, a potential pseudouridine synthase, and Nhp2p, a putative RNA-binding protein, are present together with Gar1p in all H BOX/ACA-motif snoRNPs and constitute a common bipartite structure, RNA 4 (12) (1998) 1549–1568.

[96] E. Giordano, et al., Minifly, a Drosophila gene required for ribosome biogenesis, J. Cell Biol. 144 (6) (1999) 1123–1133.

[97] B. Phillips, et al., The Nop60B gene of Drosophila encodes an essential nucleolar protein that functions in yeast, Mol. Gen. Genet. 260 (1) (1998) 20–29.

[98] U.T. Meier, G. Blobel, NAP57, a mammalian nucleolar protein with a putative homolog in yeast and bacteria, J. Cell Biol. 127 (6 Pt. 1) (1994) 1505–1514.

[99] Y. Yang, et al., Conserved composition of mammalian box H/ACA and box C/D small nucleolar ribonucleoprotein particles and their interaction with the common factor Nopp140, Mol. Biol. Cell 11 (2) (2000) 567–577.

[100] N.S. Heiss, et al., X-linked dyskeratosis congenita is caused by mutations in a highly conserved gene with putative nucleolar functions, Nat. Genet. 19 (1) (1998) 32–38.

[101] F. Dragon, V. Pogacic, W. Filipowicz, In vitro assembly of human H/ACA small nucleolar RNPs reveals unique features of U17 and telomerase RNAs, Mol. Cell. Biol. 20 (9) (2000) 3037–3048.

[102] V. Pogacic, F. Dragon, W. Filipowicz, Human H/ACA small nucleolar RNPs and telomerase share evolutionarily conserved proteins NHP2 and NOP10, Mol. Cell. Biol. 20 (23) (2000) 9028–9040.

[103] E.V. Koonin, Pseudouridine synthases: four families of enzymes containing a putative uridine-binding motif also conserved in dUTPases and dCTP deaminases, Nucleic Acids Res. 24 (12) (1996) 2411–2415.

[104] D.L. Baker, et al., RNA-guided RNA modification: functional organization of the archaeal H/ACA RNP, Genes Dev. 19 (10) (2005) 1238–1248.

[105] B. Charpentier, S. Muller, C. Branlant, Reconstitution of archaeal H/ACA small ribonucleoprotein complexes active in pseudouridylation, Nucleic Acids Res. 33 (10) (2005) 3133–3144.

[106] O.A. Youssef, R.M. Terns, M.P. Terns, Dynamic interactions within sub-complexes of the H/ACA pseudouridylation guide RNP, Nucleic Acids Res. 35 (18) (2007) 6196–6206.

[107] T. Hamma, et al., The Cbf5-Nop10 complex is a molecular bracket that organizes box H/ACA RNPs, Nat. Struct. Mol. Biol. 12 (12) (2005) 1101–1107.
[108] X. Manival, et al., Crystal structure determination and site-directed mutagenesis of the Pyrococcus abyssi aCBF5-aNOP10 complex reveal crucial roles of the C-terminal domains of both proteins in H/ACA sRNP activity, Nucleic Acids Res. 34 (3) (2006) 826–839.
[109] R. Rashid, et al., Crystal structure of a Cbf5-Nop10-Gar1 complex and implications in RNA-guided pseudouridylation and dyskeratosis congenita, Mol. Cell 21 (2) (2006) 249–260.
[110] L. Li, K. Ye, Crystal structure of an H/ACA box ribonucleoprotein particle, Nature 443 (7109) (2006) 302–307.
[111] M. Khanna, et al., Structural study of the H/ACA snoRNP components Nop10p and the 3′ hairpin of U65 snoRNA, RNA 12 (1) (2006) 40–52.
[112] S.L. Reichow, G. Varani, Nop10 is a conserved H/ACA snoRNP molecular adaptor, Biochemistry 47 (23) (2008) 6148–6156.
[113] H. Jin, J.P. Loria, P.B. Moore, Solution structure of an rRNA substrate bound to the pseudouridylation pocket of a box H/ACA snoRNA, Mol. Cell 26 (2) (2007) 205–215.
[114] H. Wu, J. Feigon, H/ACA small nucleolar RNA pseudouridylation pockets bind substrate RNA to form three-way junctions that position the target U for modification, Proc. Natl. Acad. Sci. U.S.A. 104 (16) (2007) 6655–6660.
[115] J. Duan, et al., Structural mechanism of substrate RNA recruitment in H/ACA RNA-guided pseudouridine synthase, Mol. Cell 34 (4) (2009) 427–439.
[116] B. Liang, et al., Substrate RNA positioning in the archaeal H/ACA ribonucleoprotein complex, Nat. Struct. Mol. Biol. 14 (12) (2007) 1189–1195.
[117] B. Liang, et al., Structure of a functional ribonucleoprotein pseudouridine synthase bound to a substrate RNA, Nat. Struct. Mol. Biol. 16 (7) (2009) 740–746.
[118] B. Liang, et al., Long-distance placement of substrate RNA by H/ACA proteins, RNA 14 (10) (2008) 2086–2094.
[119] J. Zhou, B. Liang, H. Li, Functional and structural impact of target uridine substitutions on the H/ACA ribonucleoprotein particle pseudouridine synthase, Biochemistry 49 (29) (2010) 6276–6281.
[120] X. Yang, et al., Kinetic and thermodynamic characterization of the reaction pathway of box H/ACA RNA-guided pseudouridine formation, Nucleic Acids Res. 40 (21) (2012) 10925–10936.
[121] P. Wang, et al., Accurate placement of substrate RNA by Gar1 in H/ACA RNA-guided pseudouridylation, Nucleic Acids Res. 43 (15) (2015) 7207–7216.
[122] S. Li, et al., Reconstitution and structural analysis of the yeast box H/ACA RNA-guided pseudouridine synthase, Genes Dev. 25 (22) (2011) 2409–2421.
[123] A.K. Henras, et al., Cbf5p, the putative pseudouridine synthase of H/ACA-type snoRNPs, can form a complex with Gar1p and Nop10p in absence of Nhp2p and box H/ACA snoRNAs, RNA 10 (11) (2004) 1704–1712.
[124] C. Wang, U.T. Meier, Architecture and assembly of mammalian H/ACA small nucleolar and telomerase ribonucleoproteins, EMBO J. 23 (8) (2004) 1857–1867.
[125] A. Henras, et al., Accumulation of H/ACA snoRNPs depends on the integrity of the conserved central domain of the RNA-binding protein Nhp2p, Nucleic Acids Res. 29 (13) (2001) 2733–2746.
[126] B.K. Koo, et al., Structure of H/ACA RNP protein Nhp2p reveals cis/trans isomerization of a conserved proline at the RNA and Nop10 binding interface, J. Mol. Biol. 411 (5) (2011) 927–942.
[127] M.L. Bortolin, P. Ganot, T. Kiss, Elements essential for accumulation and function of small nucleolar RNAs directing site-specific pseudouridylation of ribosomal RNAs, EMBO J. 18 (2) (1999) 457–469.

[128] D.R. Davis, Biophysical and conformational properties of modified nucleosides in RNA (Nuclear magnetic resonance studies), in: H. Grosjean, R. Benne (Eds.), Modification and Editing of RNA, ASM Press, Washington, DC, 1998, pp. 85–102.

[129] D.R. Davis, Stabilization of RNA stacking by pseudouridine, Nucleic Acids Res. 23 (24) (1995) 5020–5026.

[130] R.H. Griffey, et al., 15N-labeled *Escherichia coli* tRNAfMet, tRNAGlu, tRNATyr, and tRNAPhe. Double resonance and two-dimensional NMR of N1-labeled pseudouridine, J. Biol. Chem. 260 (17) (1985) 9734–9741.

[131] S. Roy, et al., Nitrogen-15-labeled yeast tRNAPhe: double and two-dimensional heteronuclear NMR of guanosine and uracil ring NH groups, Biochemistry 23 (19) (1984) 4395–4400.

[132] G. Kawai, et al., Conformational rigidity of specific pyrimidine residues in tRNA arises from posttranscriptional modifications that enhance steric interaction between the base and the 2′-hydroxyl group, Biochemistry 31 (4) (1992) 1040–1046.

[133] W.A. Decatur, M.J. Fournier, rRNA modifications and ribosome function, Trends Biochem. Sci. 27 (7) (2002) 344–351.

[134] A. Ben-Shem, et al., The structure of the eukaryotic ribosome at 3.0 Å resolution, Science 334 (6062) (2011) 1524–1529.

[135] Y.S. Polikanov, T.A. Steitz, C.A. Innis, A proton wire to couple aminoacyl-tRNA accommodation and peptide-bond formation on the ribosome, Nat. Struct. Mol. Biol. 21 (9) (2014) 787–793.

[136] Y. Zebarjadian, et al., Point mutations in yeast CBF5 can abolish in vivo pseudouridylation of rRNA, Mol. Cell. Biol. 19 (11) (1999) 7461–7472.

[137] T.H. King, et al., Ribosome structure and activity are altered in cells lacking snoRNPs that form pseudouridines in the peptidyl transferase center, Mol. Cell 11 (2) (2003) 425–435.

[138] R. Parker, et al., Genetic analysis of small nuclear RNAs in *Saccharomyces cerevisiae*: viable sextuple mutant, Mol. Cell. Biol. 8 (8) (1988) 3150–3159.

[139] L.H. Qu, et al., Seven novel methylation guide small nucleolar RNAs are processed from a common polycistronic transcript by Rat1p and RNase III in yeast, Mol. Cell. Biol. 19 (2) (1999) 1144–1158.

[140] D.A. Samarsky, A.G. Balakin, M.J. Fournier, Characterization of three new snRNAs from *Saccharomyces cerevisiae*: snR34, snR35 and snR36, Nucleic Acids Res. 23 (13) (1995) 2548–2554.

[141] X.H. Liang, Q. Liu, M.J. Fournier, Loss of rRNA modifications in the decoding center of the ribosome impairs translation and strongly delays pre-rRNA processing, RNA 15 (9) (2009) 1716–1728.

[142] B. Meyer, et al., The Bowen-Conradi syndrome protein Nep1 (Emg1) has a dual role in eukaryotic ribosome biogenesis, as an essential assembly factor and in the methylation of Psi1191 in yeast 18S rRNA, Nucleic Acids Res. 39 (4) (2011) 1526–1537.

[143] J.P. Wurm, et al., The ribosome assembly factor Nep1 responsible for Bowen-Conradi syndrome is a pseudouridine-N1-specific methyltransferase, Nucleic Acids Res. 38 (7) (2010) 2387–2398.

[144] B. Meyer, et al., Ribosome biogenesis factor Tsr3 is the aminocarboxypropyl transferase responsible for 18S rRNA hypermodification in yeast and humans, Nucleic Acids Res. 44 (9) (2016) 4304–4316.

[145] G. Badis, M. Fromont-Racine, A. Jacquier, A snoRNA that guides the two most conserved pseudouridine modifications within rRNA confers a growth advantage in yeast, RNA 9 (7) (2003) 771–779.

[146] A. Baudin-Baillieu, et al., Nucleotide modifications in three functionally important regions of the *Saccharomyces cerevisiae* ribosome affect translation accuracy, Nucleic Acids Res. 37 (22) (2009) 7665–7677.

[147] X.H. Liang, et al., Strong dependence between functional domains in a dual-function snoRNA infers coupling of rRNA processing and modification events, Nucleic Acids Res. 38 (10) (2010) 3376–3387.
[148] D. Tollervey, A yeast small nuclear RNA is required for normal processing of pre-ribosomal RNA, EMBO J. 6 (13) (1987) 4169–4175.
[149] D. Tollervey, C. Guthrie, Deletion of a yeast small nuclear RNA gene impairs growth, EMBO J. 4 (13B) (1985) 3873–3878.
[150] X.H. Liang, Q. Liu, M.J. Fournier, rRNA modifications in an intersubunit bridge of the ribosome strongly affect both ribosome biogenesis and activity, Mol. Cell 28 (6) (2007) 965–977.
[151] D. Piekna-Przybylska, et al., Ribosome performance is enhanced by a rich cluster of pseudouridines in the A-site finger region of the large subunit, J. Biol. Chem. 283 (38) (2008) 26026–26036.
[152] P.F. Swann, The effect of ethionine on ribonucleic acid synthesis in rat liver, Biochem. J. 150 (3) (1975) 335–344.
[153] M.H. Vaughan Jr., et al., The effects of methionine deprivation on ribosome synthesis in HeLa cells, Proc. Natl. Acad. Sci. U.S.A. 58 (4) (1967) 1527–1534.
[154] S.F. Wolf, D. Schlessinger, Nuclear metabolism of ribosomal RNA in growing, methionine-limited, and ethionine-treated HeLa cells, Biochemistry 16 (12) (1977) 2783–2791.
[155] M. Caboche, J.P. Bachellerie, RNA methylation and control of eukaryotic RNA biosynthesis. Effects of cycloleucine, a specific inhibitor of methylation, on ribosomal RNA maturation, Eur. J. Biochem. 74 (1) (1977) 19–29.
[156] K. Newton, et al., Fibrillarin is essential for early development and required for accumulation of an intron-encoded small nucleolar RNA in the mouse, Mol. Cell. Biol. 23 (23) (2003) 8519–8527.
[157] D. Ruggero, et al., Dyskeratosis congenita and cancer in mice deficient in ribosomal RNA modification, Science 299 (5604) (2003) 259–262.
[158] Y. Mochizuki, et al., Mouse dyskerin mutations affect accumulation of telomerase RNA and small nucleolar RNA, telomerase activity, and ribosomal RNA processing, Proc. Natl. Acad. Sci. U.S.A. 101 (29) (2004) 10756–10761.
[159] A. Yoon, et al., Impaired control of IRES-mediated translation in X-linked dyskeratosis congenita, Science 312 (5775) (2006) 902–906.
[160] C. Bellodi, N. Kopmar, D. Ruggero, Deregulation of oncogene-induced senescence and p53 translational control in X-linked dyskeratosis congenita, EMBO J. 29 (11) (2010) 1865–1876.
[161] K. Jack, et al., rRNA pseudouridylation defects affect ribosomal ligand binding and translational fidelity from yeast to human cells, Mol. Cell 44 (4) (2011) 660–666.
[162] L. Rocchi, et al., Dyskerin depletion increases VEGF mRNA internal ribosome entry site-mediated translation, Nucleic Acids Res. 41 (17) (2013) 8308–8318.
[163] M. Penzo, et al., Human ribosomes from cells with reduced dyskerin levels are intrinsically altered in translation, FASEB J. 29 (8) (2015) 3472–3482.
[164] C. Bellodi, et al., H/ACA small RNA dysfunctions in disease reveal key roles for noncoding RNA modifications in hematopoietic stem cell differentiation, Cell Rep. 3 (5) (2013) 1493–1502.
[165] S. Higa-Nakamine, et al., Loss of ribosomal RNA modification causes developmental defects in zebrafish, Nucleic Acids Res. 40 (1) (2012) 391–398.
[166] M. Nicoloso, et al., Intron-encoded, antisense small nucleolar RNAs: the characterization of nine novel species points to their direct role as guides for the 2′-O-ribose methylation of rRNAs, J. Mol. Biol. 260 (2) (1996) 178–195.
[167] J. Lee, et al., Rpl13a small nucleolar RNAs regulate systemic glucose metabolism, J. Clin. Invest. 126 (12) (2016) 4616–4625.

[168] S. Schwartz, et al., Transcriptome-wide mapping reveals widespread dynamic-regulated pseudouridylation of ncRNA and mRNA, Cell 159 (1) (2014) 148–162.
[169] M. Taoka, et al., A mass spectrometry-based method for comprehensive quantitative determination of post-transcriptional RNA modifications: the complete chemical structure of Schizosaccharomyces pombe ribosomal RNAs, Nucleic Acids Res. 43 (18) (2015) e115.
[170] M. Buchhaupt, et al., Partial methylation at Am100 in 18S rRNA of baker's yeast reveals ribosome heterogeneity on the level of eukaryotic rRNA modification, PLoS One 9 (2) (2014) e89640.
[171] M. McMahon, A. Contreras, D. Ruggero, Small RNAs with big implications: new insights into H/ACA snoRNA function and their role in human disease, Wiley Interdiscip. Rev. RNA 6 (2) (2015) 173–189.
[172] Y.W. Choi, et al., Identification of differentially expressed genes using annealing control primer-based GeneFishing in human squamous cell cervical carcinoma, Clin. Oncol. (R. Coll. Radiol.) 19 (5) (2007) 308–318.
[173] C.M. Koh, et al., Alterations in nucleolar structure and gene expression programs in prostatic neoplasia are driven by the MYC oncogene, Am. J. Pathol. 178 (4) (2011) 1824–1834.
[174] H. Su, et al., Elevated snoRNA biogenesis is essential in breast cancer, Oncogene 33 (11) (2014) 1348–1358.
[175] S. Belin, et al., Dysregulation of ribosome biogenesis and translational capacity is associated with tumor progression of human breast cancer cells, PLoS One 4 (9) (2009) e7147.
[176] V. Marcel, et al., p53 acts as a safeguard of translational control by regulating fibrillarin and rRNA methylation in cancer, Cancer Cell 24 (3) (2013) 318–330.
[177] L. Montanaro, et al., Dyskerin expression influences the level of ribosomal RNA pseudo-uridylation and telomerase RNA component in human breast cancer, J. Pathol. 210 (1) (2006) 10–18.
[178] S. Xue, M. Barna, Specialized ribosomes: a new frontier in gene regulation and organismal biology, Nat. Rev. Mol. Cell Biol. 13 (6) (2012) 355–369.

CHAPTER SEVEN

Adenosine Deaminases That Act on RNA (ADARs)

Yuru Wang[1], Yuxuan Zheng[1], Peter A. Beal[2]
University of California, Davis, CA, United States
[2]Corresponding author: e-mail address: pabeal@ucdavis.edu

Contents

[1] These authors contributed equally to this work.

The Enzymes, Volume 41
ISSN 1874-6047
http://dx.doi.org/10.1016/bs.enz.2017.03.006

Abstract

Inosine is one of the most common modifications found in human RNAs and the Adenosine Deaminases that act on RNA (ADARs) are the main enzymes responsible for its production. ADARs were first discovered in the 1980s and since then our understanding of ADARs has advanced tremendously. For instance, it is now known that defective ADAR function can cause human diseases. Furthermore, recently solved crystal structures of the human ADAR2 deaminase bound to RNA have provided insights regarding the catalytic and substrate recognition mechanisms. In this chapter, we describe the occurrence of inosine in human RNAs and the newest perspective on the ADAR family of enzymes, including their substrate recognition, catalytic mechanism, regulation as well as the consequences of A-to-I editing, and their relation to human diseases.

1. INTRODUCTION

RNA structure is often altered after transcription via enzymatic reactions that modify the component nucleosides. Methylation, acetylation, transglycosylation, and deamination are some of the transformations that occur on nascent RNAs [1–4]. Modified nucleosides in RNA have received considerable attention recently because of new discoveries related to their abundance in mRNA [5,6], their mechanism of incorporation and removal [5–7] and biological function [7–9]. Indeed, the term "epitranscriptomics" has been coined to refer to the study of the biochemical features of a transcriptome not genetically encoded in the ribonucleotide sequence [10,11].

A commonly occurring modified nucleoside in human RNA is inosine (I), the deamination product of adenosine (A) (Fig. 1A) [12]. Inosine presents a pattern of hydrogen bonding sites on its Watson–Crick edge similar to that of guanosine (G). This allows I to preferentially base pair with cytidine (C), even though it lacks G's 2-amino group (Fig. 1B). Thus, adenosine

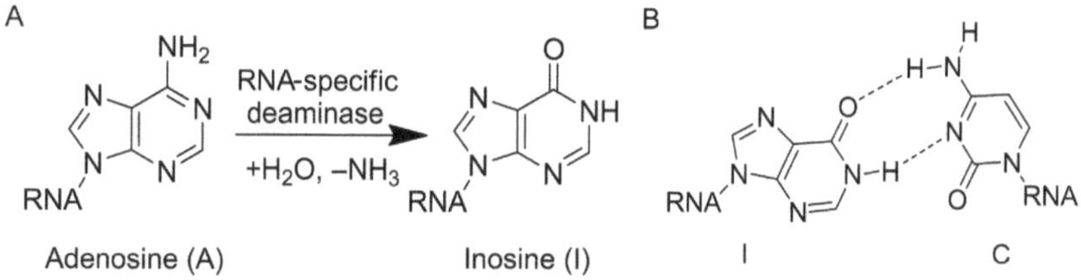

Fig. 1 (A) Adenosine (A) to inosine (I) conversion in RNA. (B) I—C base pair.

deamination changes the base pairing properties at the site of reaction from a nucleoside (A) that pairs selectively with uridine (U) to one that pairs selectively with C. This change can have a profound effect on RNA structure and function. For instance, deamination of adenosine at specific locations in codons (e.g., UAG stop codon) can change codon meaning (e.g., to UGG tryptophan codon). Indeed, since coding properties of the RNA can be altered by this reaction, adenosine deamination is a type of RNA editing [13]. However, adenosine deamination can modulate RNA properties in ways that are not directly related to an RNA's coding potential. For instance, the A-to-I reaction can alter RNA stability by, for instance, converting an A•C mismatch to the more stable I—C pair or by converting an A–U pair to the less stable I•U mismatch. Modulation of RNA properties by adenosine deamination is essential in mammals, and dysfunction in the enzymes responsible leads to disease. In this chapter, we discuss where inosine is found in human RNAs and the family of enzymes responsible for most of the inosine present; the Adenosine Deaminases that act on RNA (ADARs).

2. OCCURRENCE OF INOSINE IN RNA

Inosine is widespread among various types of RNAs including transfer RNA (tRNA), ribosomal RNA (rRNA), messenger RNA (mRNA), long noncoding RNA (lncRNA), and microRNA (miRNA). Within these RNAs, inosine can appear in different locations as well. Table 1 lists examples of the occurrence of inosine in various human RNAs. The functional consequence of the inosine modification depends on both the type of RNA and nucleotide position modified. In this section, we discuss inosine occurrence in different RNAs and methods for detection of inosine at specific locations in an RNA.

Table 1 Selected Examples of Inosine in Human RNAs

RNA	Location	Methods Used for Detection	Enzyme Responsible	Functional Consequences	References
human tRNA Ala (AGC)	Anticodon loop, A34 and A37	NGS RNA-seq	ADAT2/ADAT3 (A34); ADAT1 (A37)	Inosine at A34 modulates the number of codons recognized. The function of inosine at A37 in human is unclear	Torres *et al.* [14]
JPX25 (lncRNA)	41 sites	NGS	Unknown	Unknown	Peng *et al.* [15]
Aryl hydrocarbon receptor pre-mRNA	38 sites in 3′ UTR	Sequencing cDNA	ADAR1	Creates miRNA recognition sequence, affects AhR expression level	Nakano *et al.* [16]
GluR B receptor pre-mRNA	Exon 11	Sequencing cDNA	ADAR2	Recoding reduces the $Ca2+$ permeability of the receptor	Sommer *et al.* [17]
Neil1 pre-mRNA	Exon 6	NGS	ADAR1	Recoding produces protein isoforms with different substrate specificity	Li *et al.* [18] and Yeo *et al.* [19]
ADAR2 pre-mRNA	Intron 4	Sequencing cDNA	ADAR1 and ADAR2	Creates an alternative splicing site	Rueter *et al.* [20]
pri-miR-142	Near the Drosha cleavage site	Internal labeling and TLC	ADAR1 and/or ADAR2	Inhibits Drosha cleavage	Yang *et al.* [21]
miR-376 cluster transcripts	miRNA seed region	Sequencing cDNA	ADAR2 (+4 site); ADAR1 (+44 site)	Redirects silencing target	Kawahara *et al.* [23]

NrCAM pre-mRNA	Intron	Inosine-specific cleavage of RNA	Unknown	Unknown	Morse *et al.* [24]
CFLAR transcript	3′ UTR	Searching EST database	Unknown	Destabilizes the dsRNA	Levanon *et al.* [25]
KCNA1 transcript	Exon 1	Comparative genomics	Unknown	Potentially alters the inactivation kinetics of K+ channels	Hoopengardner *et al.* [26] and Eaholtz *et al.* [26a]
MAPK8 transcript	5′ UTR	NGS	Unknown	Unknown	Li *et al.* [18]
IGFBP7 transcript	Exon 1	Searching SNP database	Unknown	Potentially influences the heparin binding and proteolytic processing of the protein	Gommans *et al.* [27]
LNP pre-mRNA	3′ UTR	ICE-seq	Unknown	Unknown	Sakurai *et al.* [28]

2.1 Inosine in tRNA and rRNA

Inosine can be found in three different positions in tRNAs. Position 34 is the first nucleotide position of the anticodon loop, and inosine is found at this position in both eukaryotes and prokaryotes [29,30]. It was believed that inosine would be found in all tRNAs with A encoded at position 34 (ANN tRNAs) in eukaryotes. However, only 8 out of 16 ANN tRNAs were shown to contain inosine at this position above the detection limit in a human cell line. This includes the tRNAs with AGC, ACG, AAT, AAG, AGG, AGA, AGT, and AAC anticodons [14]. In eukaryotes, the heterodimer of adenosine deaminase acting on tRNA 2 (ADAT2) and ADAT3 is responsible for the generation of inosine in this position [31]. In prokaryotes, inosine was only found in $tRNA^{Arg}$ (ACG) [30], and the adenosine to inosine conversion is accomplished by the homodimer of tRNA adenosine deaminase A (TadA) [32]. Inosine is also found at position 37 in yeast $tRNA^{Ala}$ (AGC) [33]. It is believed that ADAT1 performs the conversion at this position [34]. However, inosine is not the final product of modification at this position. Inosine at this position is further modified by tRNA methyltransferase 5 (Trm5) to generate 1-methylinosine [35]. Similarly, 1-methylinosine is also found at position 57 in all tRNAs except $tRNA^{His}$ in prokaryotes [36]. However, it appears that the methylation occurs before the deamination for this site [37]. The biological function of these inosine modifications in tRNAs is not fully understood. More is known about inosine at position 34 because of its location in the anticodon loop. Inosine at this position allows recognition of multiple codons through wobble pairing, whereas adenosine at this position can only pair with uridine [38]. However, other studies suggest the first anticodon position has a complicated effect on codon recognition, and inosine at this position actually restricts the number of codons that can be recognized [39,40].

Reports of inosine in rRNA are rare [13]. 2′-*O*-Methylinosine was found in rRNA from *Crithidia fasciculata* [41]. Recently, inosine was detected in both mRNA and rRNA in the mitochondria of *Diplonema papillatum* through comparative analysis of transcriptome and genome data [42]. The A-to-G differences between genome and transcriptome sequences were tested biochemically to confirm that these differences arise from A-to-I deamination. The biological functions and the enzymes generating the inosine in rRNA are unclear, but it has been suggested the modification could destabilize rRNA secondary and tertiary structure by changing base pairing patterns.

2.2 Inosine in Alu Elements

The human genome is rich in repetitive sequences originating from transposable elements. The most common of these repetitive sequences is the approximately 280 nt Alu element, believed to arise from the 7SL RNA gene encoding the RNA component of the signal recognition particle [43]. Because Alu insertions often occur as inverted repeats, transcribed RNAs form nearly perfect duplexes of approximately 300 bp. This dsRNA structure provides a nearly ideal-binding surface for the ADAR enzymes responsible for A-to-I modification reactions (see below). A-to-I editing sites were first discovered in Alu elements in the early 2000s [25]. At that time, thousands of A-to-I sites were identified in Alu sequences from various tissue samples. The majority of these editing sites were clustered within introns and UTRs [44,45]. In a more recent study using ultradeep sequencing techniques, nearly all the adenosines in the tested Alu repeats were found to be converted to inosine, albeit most are at very low levels (<1%) [46]. With this information, it was estimated that over one hundred million A-to-I modification sites could be found in the Alus within human genome.

2.3 Inosine in mRNA

Inosine is found in untranslated and coding regions of mRNAs. Depending on the location of these inosines, the modifications have different effects on mRNA function. In this section, we discuss examples of inosine found in different regions of mRNA in detail.

2.3.1 5′ and 3′ UTRs

Inosine is more commonly found in the mRNA's 3′ UTR than in the 5′ UTR, likely because these regions harbor more duplex structure, which is important for the binding of the ADAR enzymes responsible for inosine production (see below). It was previously thought that highly edited mRNAs were retained in the nucleus, and the modified regions were removed by cleavage before export to the cytoplasm [47,48]. However, a later study showed that inosine-containing 3′ UTRs are present on translating ribosomes and suggested that A-to-I modification in the 3′ UTR might play a role in translation regulation under certain conditions [49]. Recently, editing in 3′ UTRs was shown to regulate target gene expression by creating miRNA recognition sites and protein-binding sites [16,50]. Inosine has also been found in 5′ UTRs, such as in the 5′ UTRs of HIV RNAs [51]. These editing events alter dsRNA structure and can potentially affect viral replication [51]. Alu repeat elements are common in 3′ UTRs and are hotspots for A-to-I modification. The inosine modifications found in Alu elements within 3′-UTRs could regulate gene expression by causing 3′-UTRs removal, creating alternative polyadenylation sites, and affecting nuclear retention time of mRNAs [52].

2.3.2 Coding Regions

There are numerous reported examples of inosine in the coding regions of mRNAs [53]. Adenosine deamination in codons can change codon meaning (i.e., cause "recoding") but not all A-to-I sites in exons result in recoding. Transcripts found in cells of the nervous system have multiple A-to-I sites in exons, particularly neurotransmitter receptor mRNAs. Ionotropic glutamate receptor subtype GluR B was one of the earliest transcripts identified bearing inosine in a coding region. GluR B inosine sites were first detected in rat and later were found in other mammals [17]. A CAG codon on exon 11 is converted to CIG at the Q/R site, and an AGA codon on exon 13 is converted IGA at the R/G site. In addition to the GluR B receptor, the mRNA for the 2C subtype of serotonin receptor ($5HT_{2C}R$) is also a substrate for A-to-I conversion [54]. Five editing sites, named A–E sites, were identified in exon 3 of the mRNA [54]. Editing at these sites can alter the amino acid identity in this region of the encoded protein. Considering all possibilities, 32 different mRNA species and 24 protein isoforms can be generated from A-to-I editing at these sites. Seven major protein isoforms and 11 different mRNA species have been found in mammals [55,56]. Inosine has also been detected at exonic sites in mRNAs from various viruses, such as hepatitis delta virus [57–59]. In this example, a region encoding hepatitis delta antigen (HDAg) contains an A-to-I editing site [60]. Editing converts an UAG (stop codon) in the antigenome to an UIG (tryptophan) allowing the expression of an additional 19–20 amino acids at the C-terminus of HDAg to generate HDAg-L.

With the development of next-generation sequencing (NGS) techniques, investigators now can compare whole-genome sequences to whole-transcriptome sequences to identify A-to-G changes since inosine is copied during replication as guanosine. In the first attempt to identify genome-wide inosine sites within exonic regions, several hundred human RNA editing sites were detected within seven different human tissues [18]. With stringent thresholds, 55 A-to-I editing sites were identified within exons, and 38 of these sites were shown to be recoding sites. Editing sites identified through NGS have stimulated additional studies of the edited transcripts and the consequences of the recoding events, such as for Nei endonuclease VIII-like 1 (NEIL1), filamin β (FLNB), cyclin I (CCNI), and ornithine decarboxylase antizyme inhibitor 1 (AZIN1) [19,61–63].

2.3.3 Introns

Inosine is also found in introns. In fact, introns have one of the highest inosine contents among different mRNA elements [15]. Regulation of splicing is one of the consequences of inosine within introns. Interestingly, multiple inosines are found in the intron 4 region of the human ADAR2 mRNA [20]. These A-to-I modifications can remove branch sites, create alternative splice sites, and destroy splice sites [20,64]. It is now believed that RNA editing and splicing are not isolated events but rather coupled events that regulate each other [65]. Like 3′ UTRs, intronic regions of mRNA are also rich in Alu repeat elements. Alu and inverted Alu repeat elements in the intronic regions are substrates for A-to-I modifications. An illustrative example is the A-to-I modification of human nuclear prelamin A recognition factor transcript [66]. The intronic region between exon 7 and exon 9 contains two Alu elements in reverse orientation. Exonization of exon 8 exlusively depends on A-to-I editing of this intronic region.

2.4 Inosine in lncRNA

The term lncRNA is given to the group of RNA transcripts longer than 200 nucleotides that are not translated into protein. lncRNAs bear modified nucleosides including inosine [67]. A recent review reported over 300,000 lncRNA A-to-I editing sites found in the publicly available databases [67]. Most A-to-I editing sites were identified through NGS methods. Though many of the editing sites were confirmed by various studies, the biological consequences of A-to-I changes in lncRNAs remain unclear.

2.5 Inosine in miRNA

miRNAs constitute a group of short, noncoding RNAs involved in regulation of gene expression via the Argonaute proteins [68]. Inosine has been identified in mature miRNAs and in miRNA biosynthetic precursors (pri-mRNA, pre-miRNA) [69]. MiRNAs are initially transcribed as pri-mRNAs and later processed by the RNAse III-like endonucleases Drosha and Dicer to generate mature miRNAs [70]. The miRNA precursors have substantial duplex structure which allow for ADAR-catalyzed A-to-I modification [22,71]. It is believed that most A-to-I modifications in miRNA occur at the pri-miRNA stage. For instance, A-to-I sites are found in pri-miRNA-142, and Drosha cleavage is inhibited by inosine present near the Drosha processing site [21]. In the case of miRNA-151, inosine modifications affect Dicer/TRBP processing of the pre-miRNA; therefore, A-to-I-modified pre-miRNA-151 is enriched in the cytoplasm [22]. Also, inosine has been detected in the miRNA-376 cluster [23]. These A-to-I modifications are located close to the mature miRNA seed region, which is important for target recognition. In fact, inosine in the miRNA-376 cluster can redirect these mature miRNAs to silence new targets [23].

2.6 Methods for Detecting Inosine in RNA

The examples described earlier required methods for detection of inosine in RNA. In this section we review different methods used for detecting the presence and location of inosine in specific RNA sequences.

2.6.1 Historical Methods

The first detection of inosine as a product of an enzymatic modification in dsRNA was achieved in 1988 [72]. In the experiment, dsRNAs were internally labeled at one of the four nucleotides (A, U, C, G). After incubating with cellular extracts, the RNAs were digested with a nuclease and then separated by two-dimensional thin layer chromatography. With corresponding labeled free mononucleotides used as controls, a spot that shifted differently from unmodified AMP was detected and shown to be IMP. This method was later used as a way to study A-to-I editing activities on dsRNAs in vitro [73,74]. Similar methods were used in early studies of A-to-I conversion in tRNAs [33].

A hallmark of an A-to-I RNA editing event is an A-to-G transition when comparing genomic and cDNA sequences since inosine is read as a guanosine by reverse transcriptases. The early discoveries of A-to-I editing events in mRNA were largely serendipitous, and the editing targets encoded proteins with important roles in the nervous system [17,54,75]. Analysis of sequences surrounding the early A-to-I editing sites revealed that the exonic sequences neighboring the modification sites are highly conserved between species. Using evolutionary conservation as a potential signature for A-to-I RNA editing (i.e., comparative genomics), Reenan and coworkers identified editing substrates in a broad range of tissues, opening up possible roles for RNA editing outside the nervous system [76]. A biochemical method using inosine-specific cleavage of RNA by RNAse T1 after treatment with glyoxal was also applied to identify A-to-I editing substrates in

Caenorhabditis elegans and humans [24,26]. New editing sites in the noncoding regions of RNA were identified with this method, raising the possibility that A-to-I editing may be more prevalent in the noncoding regions than in the coding regions of mRNA. As a large-scale database of expressed sequence tags (EST) became publicly available in the early 2000s, a method aligning mRNA sequences from the EST database to the genome attracted great interest. Some constraints based on prior knowledge of RNA editing were applied to the computational pipelines, such as searching for clustered editing sites and editing sites within intramolecular dsRNA structures [25,45]. Studies using this method revealed that A-to-I editing in humans primarily occurs in noncoding regions of the RNA, typically in the Alu repeats [25,44,45,77]. In addition, some A-to-I sites have been identified from the database of single-nucleotide polymorphisms (SNPs) [27,78].

2.6.2 Next-Generation Sequencing

The early 2000s also saw the development of NGS technology. In 2009, a method combining NGS and target capture of computationally predicted editing sites was applied to unbiased detection of A-to-I editing sites [18]. Several hundred human A-to-I sites were detected, which expanded the known nonrepetitive A-to-I editing sites repertoire by around 10-fold. Recently, the development of whole-transcriptome sequencing technology (RNA-seq) allowed researchers to map the transcriptome sequence to the genome sequence from the same individual to identify new editing sites in a high-throughput manner. This method allows for de novo identification of editing sites, without the need of prior knowledge related to RNA editing features. Numerous groups have utilized RNA-seq and genome sequencing to study RNA editing leading to a vastly grown list of editing sites [14,46,79]. In addition, computational strategies to predict RNA editing sites using publically available RNA-seq databases in the absence of relevant genomic sequences were also developed [80]. A recent work in 2014 analyzing large-scale RNA-seq data and ultra-deep sequencing estimated that there are over 100 million human Alu RNA editing sites, located in the majority of human genes [46].

The outputs of the methods described earlier inevitably include false positives due to contaminating guanosine signals derived from SNPs, unfavorable amplification of pseudogenes and sequencing errors. In order to differentiate inosine from contaminating guanosine, a chemical biology method called inosine chemical erasing (ICE) was developed [81]. Briefly, in this method RNA is cyanoethylated to label inosines but not guanosines.

When the RNA is subjected to reverse transcription, a cyanoethylated inosine produces a stop. The method was then combined with deep sequencing (ICE-seq) to identify RNA editing sites on a large scale [28,82]. One study utilizing ICE-seq for mapping A-to-I editing sites in human brain transcriptome found around 20,000 novel sites in around 1200 edited mRNAs not known previously, among which 66 novel sites are in coding regions and 41 of which cause recoding [28].

2.6.3 Databases of A-to-I Sites

There are several existing databases of known A-to-I editing sites, including DARNED, RADAR, e23D, REDIportal, and a human-specific inosinome atlas [83–87]. These databases summarize the A-to-I sites identified thus far and offer searching features to identify RNA editing sites in a particular gene as well as provide detailed information including functional consequence of editing and editing levels in a variety of tissues.

3. ADENOSINE DEAMINASES THAT ACT ON RNA (ADARs)

As can be appreciated from the discussion earlier, inosine is widespread in the human transcriptome. The vast majority of inosine sites in human RNAs are products of Adenosine Deaminases that act on RNA (ADARs) [88]. These enzymes catalyze the hydrolytic deamination of adenosine and require duplex structure in the substrate for reaction. The conversion of A-to-I in an RNA duplex changes an A–U base pair to the less stable I•U mismatch. Indeed, ADARs were first discovered as RNA duplex unwinding enzymes [72,89,90]. In humans, two active ADARs are known: ADAR1 and ADAR2 (ADARB1). A third member of the ADAR family (ADAR3/ADARB2) is expressed but not catalytically active [91]. The *ADAR1* gene is located on chromosome 1 in the human genome. There are two abundant isoforms of ADAR1. ADAR1 p150 is induced by interferon, and ADAR1 p110 is constitutively expressed [92]. An upstream interferon inducible promoter allows the generation of the p150 isoform and p110 is generated from multiple promoters [93]. Alternative splicing results in the generation of additional ADAR1 isoforms [94]. *ADAR2* is located on human chromosome 21. There is one abundant form of ADAR2 expressed in human. However, alternative splicing can create multiple ADAR2 spliced isoforms. Several ADAR2 splice variants have been identified with altered enzymatic activity [20,95]. The *ADAR3* gene is located on chromosome 10 and may have originated from an *ADAR2* gene duplication, given the fact that ADAR2 and ADAR3 are similar in sequence and domain organization [91,96]. Even though ADAR3 is catalytically inactive, it has the capability to bind both dsRNA and ssRNA [91]. It has been proposed that ADAR3 could modulate ADAR editing by binding to potential substrates without editing them [91].

3.1 ADAR Domains

Human ADARs are modular with multiple independently folded domains, each with a specialized function (Fig. 2). Various number of double-stranded RNA-binding domains (dsRBDs) are located near the N-terminus, and one catalytic deaminase domain is located near the C-terminus of each ADAR. ADAR1 also has Z-DNA/RNA-binding domains, and ADAR3 has an arginine/lysine-rich domain (R-domain) at their N-terminus. In this section, we describe the structure and function of the individual ADAR domains.

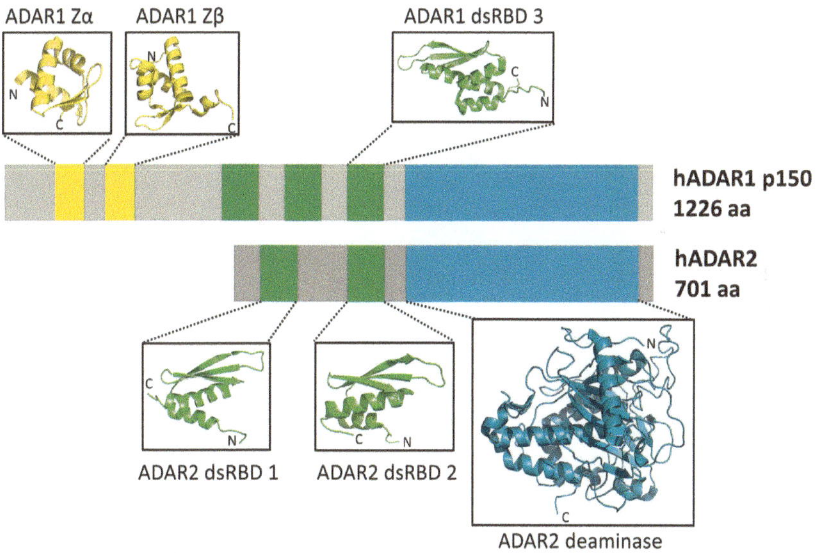

Fig. 2 Domain maps and structures of the active human ADARs: hADAR1 and hADAR2. Human ADAR1 is expressed as a 150-kDa protein (p150, shown) and as a 110-kDa protein, p110. ADAR1 Z-DNA/RNA-binding domains are colored in *yellow*. Double-stranded RNA-binding domains (dsRBDs) are shown in *green*. Deaminase domains are colored *blue*. Human ADAR1 Z alpha domain crystal structure (PDB 2GXB), human ADAR1 Z beta domain crystal structure (PDB 1XMK), human ADAR2 dsRBD1 NMR structure (PDB 2L3C), human ADAR2 dsRBD2 NMR structure (PDB 2L3K), human ADAR1 dsRBD3 NMR structure (PDB 2MDR), and human ADAR2 catalytic domain crystal structure (PDB 5HP3) are listed below and above domain maps.

3.1.1 dsRBDs

The dsRBD is an approximately 65 amino acid protein domain with the major function of binding to double-stranded RNA [97–99]. The motif was first identified in the protein staufen from *Drosophila* and the RNA-binding protein A from *Xenopus laevis* (Xlrbpa), and was later shown to exist in proteins spanning a broad range of functions, including RNA localization, cleavage, editing, and protein phosphorylation [97,100–103]. The domain has an αβββα topology, in which the two alpha helices lie on one face of a three-stranded antiparallel beta sheet [104–106]. The NMR structure of the second dsRBD of human ADAR2 shows that this domain interacts with 12bp of dsRNA contacting two minor grooves and the phosphodiesters at the opening of the intervening major groove (Fig. 3) [107]. This and other structures of dsRBDs in complex with RNA indicate the dsRBD recognizes a near A-form helix in a sequence-independent manner, since the majority of dsRBD–RNA interactions are direct contacts with the 2′-hydroxyl groups and direct or water-mediated contacts with oxygen atoms of the phosphodiester backbone [108,109]. Residues that are important for the conformation of the motif or for potential interaction with RNA have been

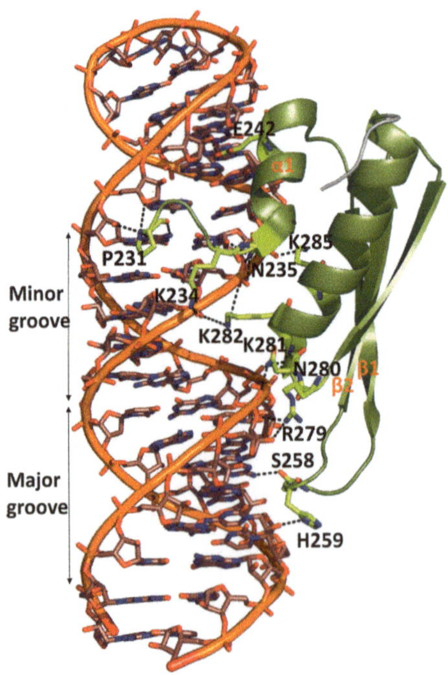

Fig. 3 NMR structure of rat ADAR2 dsRBD2 bound to GluR B R/G site-derived RNA [107].

identified and are highly conserved among dsRBDs from different proteins [105]. In spite of these similarities among dsRBDs, it appears that different dsRBD-containing proteins have intrinsic-binding patterns to suit their different biological functions arising from their individual set of dsRBDs [102]. For example, the dsRBDs of ADAR2 bind to an editing substrate differently from the dsRBDs of the RNA-dependent protein kinase (PKR) [110]. When the two dsRBDs of ADAR2 are replaced by dsRBDs of PKR, the editing efficiency was found to be severely diminished or abolished [110].

3.1.2 Deaminase Domains

The approximately 400 amino acid deaminase domain of human ADAR2 has been characterized by X-ray crystallography [111]. This domain adopts a roughly spherical 40 Å diameter structure comprising 10 α helices, 11 β strands, and multiple connecting loops (Fig. 4). The active site, as indicated by an ordered zinc ion that coordinates a water molecule, is buried deeply in a pocket of the protein. Residues H394, C451, and C516 coordinate the zinc ion to lock it at its position, while E396 hydrogen bonds with the water molecule coordinated by the zinc (Fig. 4A). The ADAR deaminase motif shares little sequence similarities with adenosine deaminase (ADA), an enzyme that carries out deamination of the free adenosine nucleoside

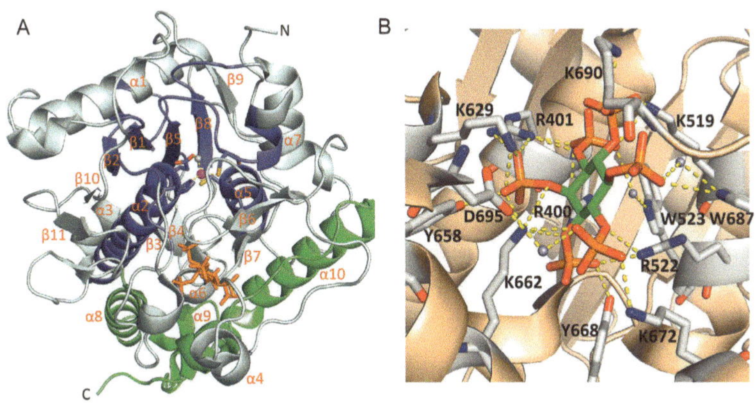

Fig. 4 (A) Structure of isolated human ADAR2 deaminase domain. *Magenta sphere*, zinc ion; *gray sphere*, water molecule; *orange*, IP_6; *blue*, deaminase motif; *green*, C-terminal elements coordinating IP_6. Side chains for zinc-coordinating residues (H394, C451, and C516) and the proton transfer residue (E396) are shown. (B) Close-up view of IP_6-binding site in ADAR2 deaminase domain. The contacts between conserved residues and IP_6 are shown.

[112]. However, ADARs are closely related by sequence to the cytidine deaminase (CDA) family [101]. In agreement with this, the geometry of the active site of ADAR2 deaminase domain is different from that of ADA but is similar to CDAs (and TadA). However, there is a difference between ADAR2 and CDA/TadA in the arrangement of the two cysteine residues coordinating the zinc, indicating the evolutionary divergence between ADARs and CDA/TadA. Superimposition of the ADAR2, CDA, and TadA structures revealed a conserved deaminase motif spanning 77 residues comprising two helices (α2 and α5) and four strands (β1, β2, β5, and β8) as well as the connecting loops (Fig. 4), while other residues do not have a structural equivalence in CDA, further emphasizing the evolutionary separation between these enzymes [111]. Overlapping CDA in complex with the cytidine analogue zebularine onto the ADAR2 deaminase structure suggests a steric clash between an ADAR2 loop and the zebularine [111]. This explains why ADAR2 can catalyze adenosine deamination, but not cytidine deamination.

The ADAR2 deaminase domain also has an inositol hexakisphosphate (IP_6) molecule buried in a cavity formed mainly by C-terminal elements (Fig. 4B) [111]. IP_6 is the phosphorylated derivative of myo-inositol and is generated in the phospholipase C signal transduction pathway [113]. Phospholipase C cleaves the phospholipid phosphatidylinositol 4,5-bisphosphate (PIP_2) to generate inositol 1,4,5-trisphosphate (IP_3), which is then phosphorylated by a series of kinases to make the cellular inositol phosphate derivative IP_6 [114]. IP_6 is implicated in many cellular functions including RNA export, DNA repair, endocytosis, and hormone signaling [113,115]. The IP_6-binding cavity in ADAR2 is extremely basic, where IP_6 binds arginine and lysine residues as well as tryptophan and tyrosine residues (Fig. 4B). Some of these residues directly contact IP_6, while others contact IP_6 via ordered water molecules. Most of the IP_6-binding residues are invariant among ADARs, implying that ADAR1 and ADAR3 also contain a binding pocket for IP_6. In addition, these residues are also conserved in the ADAT1 family of tRNA-modifying enzymes [111]. However, the ADAT2 family (to which TadA belongs) does not have an IP_6-binding pocket.

IP_6 is required for ADAR2 activity, as an *Saccharomyces cerevisiae* strain which lacks the ability to produce IP_6 generated inactive human ADAR2 upon exogenous protein expression [111]. Similar experiments have been performed for ADAT1 showing that IP_6 is also required for ADAT1 activity. Since IP_6 is deeply buried in the ADAR2 deaminase domain, it is believed to be important for ADAR2 activity by stabilizing the protein fold. In addition to the structural requirement, IP_6 is thought to possibly play a role in modulating catalytic efficiency by indirectly coordinating the zinc ion. Specifically, two of the IP_6 phosphate groups, which approach the zinc ion within 10 Å, initiate a chain of hydrogen bonding interactions involving residues K519, D392, and K483. Residue K483, with its ordered side chain, further contacts C516, which is one of the zinc coordinating residues. IP_6 is also found to be bound in other proteins with effects that include facilitating formation of protein trimers, stabilizing the protein structures and inducing allosteric activation [116–120]. In most of these cases, IP_6 is bound to pockets formed near the protein surfaces, unlike ADAR2's deeply buried IP_6. In addition, binding of IP_6 to these proteins appears to induce conformational reorganization, indicating that IP_6 may not be required for the nascent folding of the proteins but is required for modulating the protein structure (and function). In ADARs, however, IP_6 is in an almost completely buried environment and forms an extensive array of hydrogen bonds with conserved ADAR side chains and, thus, seems to be indispensable for the nascent folding of ADAR2.

The deaminase domains for the different ADARs are highly similar in sequence. Indeed, the amino acid sequences of the approximately 400 residue human ADAR1 and ADAR2 deaminase domains are 39% identical and 59% similar. However, a surface loop involved in RNA recognition differs substantially between ADAR1 and ADAR2 [121]. Structural elements present in the catalytic domains that differ between the two ADARs, such as this loop, are likely responsible for observed differences in substrate selectivity (see below).

3.1.3 Z Domains

ADAR1 has additional nucleic acid-binding domains near its N-terminus, its Z domains. ADAR1 has two Z domains, Zα and Zβ, but only Zα can bind to Z-DNA [122–124]. The crystal structure of ADAR1 Z-DNA-binding domain was reported in the late 1990s [124]. One possible role of the Zα domain is to direct ADAR1 to the actively transcribing DNA to allow efficient editing on the RNA before splicing. Also, the binding of ADAR1 Zα domain to the Z-DNA in a promoter region can regulate gene expression [125]. Unexpectedly, the ADAR1 Zα domain was shown to bind to the c-myc promoter G-quadruplex formed in $NHEIII_1$ region, suggesting a role of Zα domain in regulating G-quadruplex-mediated gene expression [126]. The ADAR1 Zα domain can also bind to Z-RNA [122,127]. It has been shown that sequences promoting the formation of Z-RNA in dsRNA substrates facilitated editing by ADAR1, indicating a direct role of the Zα domain in RNA editing [127]. ADAR1 p110 lacks the Zα domain and only contains the Zβ domain. The Zβ domain is not able to bind to Z-DNA/RNA due to a lack of some crucial residues for binding but is believed to be involved in ADAR dimerization [128]. The presence or absence of a Zα domain in ADAR1 isoforms influences their subcellular localization due to the existence of a nuclear export signal (NES) in the Zα domain (see below).

3.2 Cellular Localization and Tissue-Specific Expression of ADARs

In mammalian cells, ADARs are typically located in the nucleus with some exceptions. The two isoforms of ADAR1, p110 and p150, have different subcellular targeting. The constitutively expressed ADAR1 p110 is primary located in the nucleolus but can shuttle between the nucleolus and nucleoplasm [129]. However, the interferon-induced form, ADAR1 p150, is primarily located in the cytoplasm, but can shuttle between the nucleus and the cytoplasm [130]. ADAR1 nuclear localization is caused by a nuclear localization signal (NLS) in the third dsRBD, and the cytoplasm localization is caused by a NES in the Zα domain [130–132]. ADAR1 p110 lacks the Zα domain; therefore, it only locates to the nucleus. The subcellular localization is likely related to the immune properties of this protein (see below) [133,134]. Like ADAR1 p110, ADAR2 is located mostly in the nucleolus and is able to shuttle between nucleolus and nucleoplasm [129,135]. The nuclear localization of ADAR2 is imparted by sequence found in the N-terminus of the protein, especially in the dsRBD region [129]. The subcellular localization may be a mechanism to control RNA editing activity for these enzymes (see below) [135]. The R-domain of ADAR3 also acts as a NLS [91,136]. ADAR1 and ADAR2 can be found in almost all tissues in mammals. However, ADAR expression levels are generally higher in brain [137]. Interestingly, ADAR3, unlike ADAR1 and ADAR2, is only expressed in the brain.

3.3 ADARs in Different Organisms

The ADAR family of proteins is found in most of metazoans [138–140]. Most species have more than one type of ADAR protein. If classified based on domain makeup, ADARs can be placed into three categories: ADAR1-like, ADAR2-like, and ADAD-like (ADAR-related proteins) [141]. There is at most one ADAR1-like ADAR present in most species. However, there are up to three versions of ADAR2-like proteins and two versions of ADAD-like proteins found in one species (*Mnemiopsis leidyi*) [141]. ADARs are not found in several nonmetazoan eukaryotes, such as yeast, fungi, plants, and choanoflagellates.

3.4 The ADAR-Related Proteins

Several families of enzymes are closely related to ADARs, including ADADs (ADAR-related proteins), ADATs (adenosine deaminases that act on tRNA), and APOBECs (apolipoprotein B mRNA-editing enzyme catalytic polypeptide). ADADs are ADAR-related proteins that have similar modular structure compared with ADARs [142]. Typically, ADADs possess only one dsRBD, while ADARs have more than one dsRBD. Two ADADs are expressed in human and both are catalytically inactive; however, they retain RNA-binding ability [142]. The ADAT family of proteins catalyze adenosine deamination in tRNA wobble pair positions [31,143]. ADATs are found in all eukaryotes. Similar to ADARs, ADATs have one deaminase domain, but dsRBDs are missing in the ADATs. Also, the region C-terminal to the deaminase domain is shorter than in the ADAR catalytic domains. It is believed that ADARs evolved from ADATs by incorporation of dsRBDs [144]. The APOBEC family is able to perform C-to-U deamination on both single-stranded (ss) DNA and RNA substrates [145]. This kind of editing was first found on apolipoprotein B mRNA [146]. ADARs, ADATs, and APOBECs all employ a similar catalytic mechanism (see below). A water is activated by a coordinated zinc ion for nucleophilic attack at the carbon atom bonded to the amino group that departs during deamination. Indeed, ADARs, ADATs, and APOBECs all belong to the CDA super family of enzymes. Sequence analysis and structural studies have shown that key features of the catalytic domains of ADARs, ADATs, and APOBECs are highly conserved [111,147–149].

4. ADAR SUBSTRATE RECOGNITION

ADARs have multiple domains capable of binding RNA substrates. ADAR-reactive adenosines are located in duplex structures consistent with the presence of dsRBDs in the ADAR proteins. However, the ADAR deaminase domains also require duplex structure [150]. No high-resolution structures of full-length ADARs bound to RNA have been reported to date. However, there is a wealth of knowledge in the literature on dsRBD–RNA interactions, including NMR structures of human ADAR2 dsRBDs bound to RNA [107]. In addition, our laboratory recently reported crystal structures of the human ADAR2 deaminase domain bound to RNA providing a detailed picture of ADAR interactions near an editing site [121]. In this section, we describe our current understanding of ADAR-substrate recognition with a particular focus on the reported high-resolution structures of dsRBD–dsRNA and deaminase domain–dsRNA complexes.

4.1 dsRBD–RNA Recognition

DsRBDs are important for ADAR editing by providing high binding affinity to target RNAs. In addition, the dsRBDs were shown to contribute to editing specificity by selectively binding to certain duplex structures found in substrate RNAs [73]. In many cases, deleting dsRBDs results in decreased editing activity [111,151]. However, in some cases, dsRBDs are not required for editing [152,153]. For instance, an RNA substrate modeling the glioma factor 1 (Gli1) recoding site was edited by the isolated ADAR2 catalytic domain nearly as efficiently as by the full-length enzyme [152]. Also, certain sites in the $5HT_{2C}R$ substrate are more efficiently edited by the full-length ADARs, while other sites are edited better by the isolated ADAR deaminase domains [151]. Thus, the role of dsRBDs in editing reactions catalyzed by ADARs can vary depending on the substrates and editing sites.

ADARs contain various numbers of dsRBDs. ADAR1 contains three whereas ADAR2 and ADAR3 each contains two. For ADAR1, the three dsRBDs are not functionally equivalent. Deletion of the first or the third dsRBD dramatically decreases the catalytic activity, whereas the second dsRBD seems to be dispensable [74]. In addition, ADAR1 dsRBD1 has a unique RNA-dependent cytoplasmic anchoring function, and the dsRBD3 of ADAR1 harbors a dual function acting as both a RNA-binding domain and a NLS [131,132]. Studies have also shown that individual dsRBDs within ADAR1 can recognize different RNAs in vivo [102]. The two dsRBDs of ADAR2 were suggested to serve different roles as well. Poulsen *et al.* reported that editing relies more on a functional dsRBD2, while dsRBD1 is more important for protein dimerization [154]. In agreement with this observation, the N-terminus and dsRBD1 of ADAR2 were reported to be dispensable for enzymatic activity, and a truncated construct composed of dsRBD2 and the C-terminal deaminase domain was shown to be sufficient to edit a 15-bp RNA substrate [155]. Adding the N-terminus and dsRBD1 *in trans* inhibited editing on the 15-bp substrate. This study implies that there is a minimum length requirement on editing substrates in order to accommodate both of ADAR2's dsRBDs.

The NMR structures of isolated dsRBDs of ADAR2 were reported in 2006 [106]. An NMR chemical shift perturbation study identified the protein and RNA surfaces involved in complex formation with a model of the GluR B R/G site hairpin stem. On basis of these studies, a model of the dsRBD:RNA complex was established, suggesting that dsRBD1 recognizes a conserved pentaloop in the RNA, whereas dsRBD2 recognizes two bulged bases adjacent to the editing site [106]. This demonstrated RNA structure-dependent recognition by the ADAR2 dsRBDs [106]. Later, the solution structures were solved for complexes of ADAR2's dsRBDs and short RNA fragments derived from GluR B R/G-site RNA [107]. From these structures, a model was generated for the dsRBDs bound to a 71-nt GluR B R/G-site RNA. These studies indicated that dsRBD1 and dsRBD2 similarly recognize the RNA helix at two consecutive minor grooves, via helix $\alpha1$ and $\beta1$–$\beta2$ loop, respectively. Although most of the contacts observed were sequence-independent, surprisingly, the structures suggested that the helix $\alpha2$ and the $\beta1$–$\beta2$ loop of both dsRBDs achieved sequence-specific contacts with their respective contacting minor grooves (Fig. 3). Functional groups involved in these sequence-specific contacts were shown to be important for binding affinity and editing activity of ADAR2. Furthermore, this study suggested that the 3′ nearest neighbor preference of ADAR2 originates from a contact between S258 of dsRBD2 and the amino group on the G 3′ to the editing site (Fig. 3). However, a feature of the ADAR2 deaminase domain/RNA complex recently solved by X-ray crystallography appears to explain the 3′ G preference for ADAR2 (see below) [121]. Nevertheless, NMR studies of the ADAR2 dsRBDs bound to RNA challenged the conventional opinion that dsRBDs recognize RNA in a shape-specific rather than a sequence-specific manner.

4.2 Deaminase Domain–RNA Recognition

Our understanding of ADAR/substrate interactions advanced significantly recently with a report of crystal structures of human ADAR2 deaminase domain bound to substrate RNAs bearing an analog of an intermediate in the adenosine deamination reaction [121]. These structures showed that

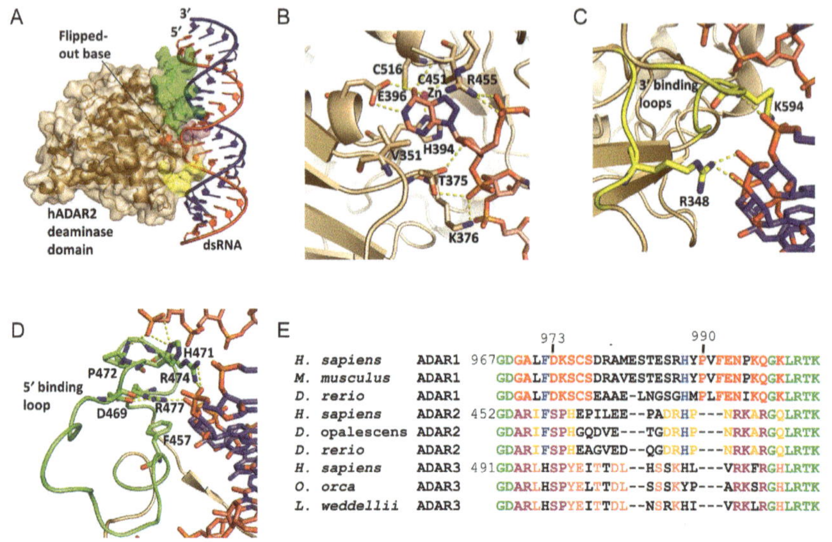

Fig. 5 A. Structure of hADAR2 deaminase domain bound to dsRNA. The binding surfaces are colored in *green*, 5′ binding loop; *purple*, base-flipping loop; *yellow*, 3′ binding loops. (B) Close-up view of active site interactions with reaction intermediate analog 8-azanebularine hydrate. (C) Close-up view of 3′ binding loops interactions with RNA. (D) Close-up view of 5′ binding loop interactions with RNA. (E) Sequence alignment of ADAR1s, ADAR2s, and ADAR3s from different organisms. The conservation patterns are labeled in different *colors*. *Green*, conserved in all ADARs; *blue*, conserved in both ADAR1s and ADAR2s; *purple*, conserved in both ADAR2s and ADAR3s; *red*, conserved in ADAR1s; *yellow*, conserved in ADAR2s; *pink*, conserved in ADAR3s.

the deaminase domain binds duplex RNAs on one face over approximately 20 base pairs with three main contact surfaces. The RNA is contacted at the minor groove near the editing site and at the two adjacent major grooves. The binding site is asymmetric with respect to the reactive adenosine mapping to 10 bp on the 5′ side and 8 bp on the 3′ side of the editing site (Fig. 5).

4.2.1 Base Flipping and Active Site Contacts

In the zinc-containing active site, the side chain of E396 hydrogen bonds to N1 and O6 of the substrate base (a purine analogue in the crystals), which is consistent with the proposed role of E396 in mediating proton transfer to and from N1 of the substrate during the deamination reaction (Figs. 5 and 6) (see Section 5). The site of nucleophilic attack during the ADAR reaction (C6 of adenine) lies deep in the major groove of a dsRNA substrate. In order to access the target adenosine, ADARs use a base-flipping

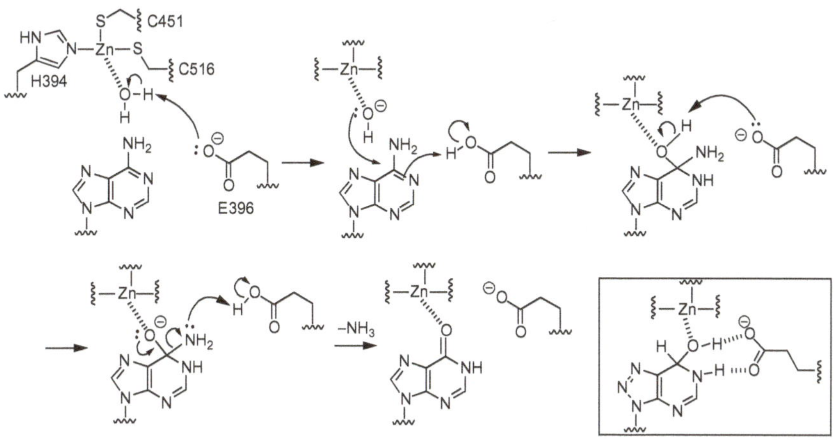

Fig. 6 Mechanism for ADAR-catalyzed adenosine deamination. *Box*: Interactions with hydrated 8-azanebularine revealed in substrate bound ADAR2 catalytic domain crystal structure [121].

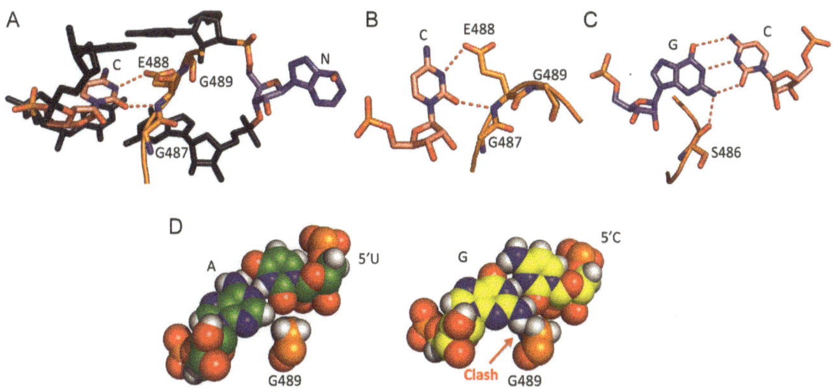

Fig. 7 Interaction between human ADAR2 and its substrate near the editing site. (A) E488 occupies the void left behind when the editing site nucleotide (N) flips out into the active site. (B) Hydrogen bonding interaction between the E488 and the orphan C is shown with *red dashes*. (C) The 2-amino group of guanosine 3′ to the editing site is hydrogen bonded to S486 at its main chain carbonyl group; hydrogen bonding interaction is shown with *red dashes*. (D) Space-filling model showing close approach of 5′ nearest neighbor base pair and G489. A C–G pair at this position could clash with G489 (PDB 5HP3).

mechanism to position the target adenosine in the zinc-containing active site pocket. ADAR base flipping involves a loop of the protein that bears E488 (Fig. 7). This loop approaches the RNA duplex from the minor groove side at the editing site, with the side chain of E488 penetrating the helix, occupying the space vacated by the flipped-out base and contacting the orphaned base located on the complementary strand. The contact between E488 and the orphan base explains ADAR's preference for the opposite base as discussed in more detail later. E488 is flanked by two glycine residues in the loop, which provide conformational flexibility for E488 to penetrate into the RNA helix. Indeed, screening of a library of ADAR1 clones sequence randomized at the codons for one of the glycine residues flanking E1008 in ADAR1 (E488 in ADAR2) showed that only residues with small side chains are allowed at that location [156]. Although crystal structures of ADAR2 in complex with RNA substrates have clearly revealed base flipping and E488 penetration into the RNA helix, the endpoint structures provide limited information about whether the infiltrating residue activated the flipping of the adenosine, or it merely occupies the space vacated by the flipped adenosine to stabilize the flipped out state. Further studies are required to clarify fully the individual steps of the ADAR base-flipping pathway.

A nucleoside analog that has a high propensity to form a covalent hydrate was used to replace the adenosine at the editing site in order to trap protein–RNA complexes (see Section 5). The flipped out base is bound in the active site as a covalent hydrate, where it is contacted by six residues (V351, T375, K376, E396, C451, and R455) (Fig. 5B). The main chain carbonyl of T375 contacts the 2′-OH of the flipped nucleotide. The T375 side chain and the K376 side chain contact the 3′ phosphodiester group of the edited nucleotide, while the R455 side chain contacts its 5′ phosphodiester group. These interactions likely help to position the flipped nucleotide by anchoring the two phosphodiester linkages flanking the edited base. A random mutagenesis/screening study varied the position 375 residue and showed that large residues are not favored at this position [157]. This is consistent with the observation that T375 is in close proximity to the flipped nucleotide and the RNA backbone. Interestingly, mutation of R455 to an alanine does not substantially influence the enzymatic activity of ADAR2, indicating that the interaction of R455 with the RNA backbone is not essential [158,159]. Besides E396, V351 and C451 directly contact the edited base with V351 utilizing its side chain for the interaction while C451 uses its main chain.

4.2.2 The 5′ and 3′ Binding Loops

The interactions with RNA regions distal to the editing site involve the 5′ and 3′ binding loops of ADAR2, named for their RNA contact positions relative to the editing site (Fig. 5C and D). The RNA contact region 3′ from the editing site is bound by two loops of ADAR2: loop 347–352 and loop

584–597 (Fig. 5C). In loop 584–597, residue K594 makes contact with the RNA through its main chain and its side chain inserts into the duplex RNA major groove. This residue is conserved among all ADARs. In loop 347–352, residue R348 binds the RNA via a side chain interaction. This residue is conserved in ADAR2 and ADAR3 but does not seem to have an obvious counterpart in ADAR1 based on sequence alignment. This interaction, thus, may not exist in an ADAR1–RNA complex. Both the K594A and R348A mutants of the ADAR2 deaminase domain have reduced editing activity compared to the wild-type enzyme [121].

The 5′ binding loop of ADAR2 spans amino acids 454–477 and is unstructured in the RNA-free protein structure [111], but becomes ordered upon binding to RNA (Fig. 5D). This loop contacts the RNA mainly by interacting with the phosphodiester backbone. In this loop, 18 residues are highly conserved in ADAR2 enzymes from different organisms indicating an important role for the loop in ADAR2 function (Fig. 5E). Consistent with the conservation pattern, the regions composed of the conserved residues are close to the RNA with some residues making direct contact to the RNA, while the variable residues are farther from the RNA. A high-throughput mutagenesis/phenotypic screening assay for ADAR activity revealed that 6 out the 18 conserved residues are particularly important for editing activity [159]. These six residues provide essential structural elements supporting the fold of the loop and key RNA-binding functional groups for interacting with the RNA. Other conserved residues in this loop region were shown to be tolerant to different mutations. Since they are highly conserved among ADAR2s from different organisms, they are likely important for other reasons, such as processing RNAs that are different from the type used in the screening experiment or binding to proteins that regulate ADAR function in vivo [160–163]. Interestingly, sequences from different ADARs deviate substantially in the 5′ binding loop region (Fig. 5E). ADAR1 and ADAR2 differ in not only the amino acid composition of the loop but also the size of the loop. In both the ADAR1 and ADAR2 loops, the conserved residues are separated into two subsets by a variable region comprising approximately eight residues. In the ADAR2 loop, the two subsets of conserved residues interact with each other to form the conformation required to bind RNA. It is reasonable to speculate that in the ADAR1 loop, the two subsets of conserved residues also interact through contacts between specific residues and bind RNA in a manner different from the ADAR2 loop. The 5′ binding loop of ADAR3 is of similar length to that of the ADAR2 loop; however, it has a distinct conservation pattern.

The region in the ADAR3 loop corresponding to the ADAR2 variable region is highly conserved at some positions. Thus, the hADAR3 5′ binding loop may adopt a distinct folding pattern and recognize a set of RNAs that are different from those recognized by ADAR1 and ADAR2. Indeed, it is tempting to speculate that this loop contributes to functional differences between ADARs. Further experiments, such as swapping the loops between the different ADARs, are necessary to shed light on this hypothesis.

5. CATALYTIC MECHANISM

ADARs catalyze a hydrolytic deamination reaction on dsRNA substrates similar to that catalyzed by ADA and CDA on nucleoside substrates [164,165]. In analogy to the mechanisms of ADA and CDA, a mechanism for ADAR2 has been proposed (Fig. 6) [4,166]. The zinc ion is chelated by H394, C451, and C516 and is also bound to a water molecule. Once the substrate adenosine is flipped into the ADAR2 active site, E396 (as glutamate) deprotonates the water molecule bound to zinc. The resulting hydroxide then attacks the purine at C6 with protonation at N1 via the acid form of E396. These steps form a high energy intermediate with a tetrahedral center at C6. Next, with the participation of E396, a proton is transferred from the newly formed C6 hydroxyl to the adenosine amino group residing on the same carbon. Once protonated, the adenosine amino group can leave as ammonia (NH_3) and at the same time the product inosine is formed.

A key step in the ADAR reaction is the formation of the adenosine covalent hydrate with a tetrahedral center at C6. The nucleoside analog 8-azanebularine (box in Fig. 6), with its relatively high propensity to form a covalent hydrate, is useful as an adenosine replacement in ADAR2 substrates to trap protein–RNA complexes [121,150,166]. Since the 8-azanebularine covalent hydrate is similar in structure to the adenosine deamination transition state, high affinity complexes are formed by the binding of ADAR2 to RNAs bearing this analog [121,150]. Furthermore, once formed in the ADAR2 active site, the 8-azanebularine hydrate cannot proceed forward along the reaction pathway since it lacks a C6 amino group [166]. Indeed, stabilization of human ADAR2 deaminase domain–RNA complexes by replacing the reactive adenosine with 8-azanebularine enabled trapping and crystallization of the complex in the "flipped out" conformation (Fig. 5) [121]. It remains to be seen whether a similar approach can stabilize ADAR1–RNA complexes.

5.1 ADAR Reactivity With Modified Nucleosides in RNA

Nucleoside modifications have been introduced to both the base and the sugar moieties of the editing site adenosine in model RNA substrates [158,166–170]. For example, we have used synthetic RNA substrates containing nucleoside analogs modifying the 2′ position, which showed that ADAR2 cannot accommodate the bulk of a 2′-*O*-methyl group but can react with 2′-deoxy and 2′-fluoro analogs [170]. In addition, by testing the reactivity of ADAR2 with RNAs bearing N6-alkyladenosine at the editing site, it was shown that ADAR2 is sensitive to the size of substituents at C6 and cannot accommodate a substituent larger than an *N*-methyl group at this position [171]. Indeed, N6-methyladenosine (M6A) in RNA is poorly deaminated and N6-ethyladenosine in RNA is not a substrate of human ADAR2 [171]. Also, RNAs containing 8-aza adenosine, 7-deaza adenosine, 8-aza-7-deaza adenosine, and other modified purines have been synthesized and used to test substrate recognition models and probe the mechanistic differences between ADAR1 and ADAR2 [158,169]. These studies contributed to an understanding of the structural requirements in the reactive nucleoside and laid the basis for the base trapping mechanism of ADAR–RNA complexes described earlier.

6. SUBSTRATE SPECIFICITY

Editing reactions catalyzed by ADARs are not strictly sequence specific but do show preferences for adenosines at certain locations in RNA substrates. In addition, ADARs can bind most duplex RNAs beyond a certain length (~15 bp). It is now known that the opposite base, the flanking sequence, and the local secondary structure each play an important role in determining ADAR editing efficiency. Below we discuss each of these factors in detail.

6.1 Opposite Base

The base opposite the editing site adenosine affects editing efficiency. It is known that ADARs prefer adenosine in an A•C mismatch. In fact, previous studies have tested all four-possible opposite bases indicating that ADARs efficiently edit adenosine in A•C mismatches and A–U pairs, but A•A,

A•G mismatches are not preferred [19,172,173]. This preference appears to arise from interactions with the "intercalating" residue E488 that occupies the space vacated by the flipped out adenosine (Fig. 7A and B). When a purine base is present at the opposite position, E488 will clash with these bases. In addition, E488 directly binds with the "orphan" base through hydrogen bonding when the editing site adenosine is flipped out of the helix [121]. While not yet proven, the preference for an A•C mismatch over an A–U pair may arise from differences in base-flipping propensity, since E488 can hydrogen bond with either a uridine or a cytidine [121].

6.2 Flanking Sequence

ADARs preferentially edit adenosines with specific 5′ and 3′ nearest neighbor nucleotides. Human ADAR1 and ADAR2 have similar preferences for the 5′ nearest neighbor (U > A > C > G) [151]. As for the 3′ nearest neighbor, ADAR1 was reported to have a preference of G > C ≈ A > U with ADAR2 having a preference of C ≈ G ≈ A > U [151]. With the help of NGS, investigators have recently been able to analyze more editing sites and define the nearest neighbor preferences with large substrate sets [18,44,152]. Sites identified by NGS agree with most of the preferences mentioned earlier except for the ADAR2 3′ nearest neighbor. Many ADAR2 substrates have been identified with guanosine 3′ to the editing site, while both adenosine and uridine are underrepresented at this position [18,152]. In fact, mutating the 3′ nearest neighbor from guanosine to adenosine or uridine significantly decreased editing efficiency for a model ADAR2 substrate [121]. Together these studies suggest that both ADARs prefer guanosine 3′ to an editing site rather than adenosine or uridine. A result of the nearest neighbor preferences is high ADAR reactivity in 5′-UAG-3′ and 5′-AAG-3′ sequences [174,175]. Different structural studies have shed light on the origin of these preferences but also appear to be in conflict [107,121]. In the NMR structure of RNA bound to ADAR2 dsRBD 2, Stefl *et al.* suggested that a dsRBD contact to the 3′ guanine may contribute to the 3′ nearest neighbor preference [107]. However, crystal structures of RNA bound the ADAR2 catalytic domain identified an interaction between the deaminase domain S486 backbone carbonyl group and the 2-amino group of the substrate 3′ nearest neighbor guanosine (Fig. 7C) [121]. The crystal structure also provided insights for the 5′ nearest neighbor preference. The protein cannot readily accommodate an amino group in the minor groove at the 5′ nearest neighbor base pair; therefore,

G–C and C–G base pairs are not favorable adjacent to the editing site on the 5′ side (Fig. 7D).

6.3 Secondary Structure

ADARs are capable of converting A-to-I at millions of sites in the human transcriptome. ADARs have many editing sites because the reaction is not strictly sequence specific, but rather has local sequence preferences and a dependence on duplex secondary structure in the RNA substrate. ADARs have large RNA-binding surfaces (i.e., for human ADAR2: two dsRBDs that each can bind across one face of ~12 bp duplex and a catalytic domain that can bind across one face of ~20 bp duplex). However, not all of the binding surfaces need to contact the RNA for reaction to occur. In addition, perfect duplex structure is also not required for ADAR reactions. Indeed, as described earlier, an A•C mismatch at the editing site is more reactive than an A–U pair. Other helix defects (bulges, hairpin, loop, stems) are also common among ADAR substrates. Some of these structural elements enhance editing at specific sites and others reduce editing [176]. Also, ADAR1 and ADAR2 favor different secondary structures. One illustrative example is the 5-$HT_{2C}R$ pre-mRNA, where ADAR1 and ADAR2 selectively edit multiple closely located sites [54]. The secondary structure is important for maintaining the specificity for these sites. Indeed, deletion of an internal loop greatly reduces ADAR2 selectivity over different sites [177]. Tertiary structure may also be important for ADARs to select its substrates. In a large-scale sequencing experiment, investigators identified editing sites within specific distances from each other and positioned on the same side of a helix, suggesting the importance of three-dimensional structure for ADARs selectivity [178].

7. REGULATION

RNA editing levels differ with developmental stage, pathological or carcinogenic conditions, and in a tissue-specific manner [94,179–184]. Some of these changes correlate with changes in ADAR expression levels, but not all do, indicating that other mechanisms must exist for regulation of ADAR activity [185–189]. Indeed, ADARs have been shown to be regulated by posttranscriptional RNA processing (i.e., splicing and auto editing of ADAR pre-mRNA), by posttranslational modifications (e.g., phosphorylation, ubiquitination, and SUMOylation), by protein–protein interactions, by colocalization with substrate RNAs, by competition with substrate RNAs from both proteins and RNAs, as well as by changes in ADAR expression levels. In this section we discuss different mechanisms for regulation of ADAR activity.

7.1 Processing of ADAR pre-mRNA

Multiple natural isoforms of ADARs exist due to induction by different promoters, alternative splicing and self-editing. ADAR1 has three promoter-specific transcripts whose relative expression is tissue-specific [93]. Two of these promoters are constitutive and one of them is interferon-inducible [93]. Alternative splicing can lead to production of ADAR1 variants with different distance between functional domains, which is thought to modulate the way in which different functional domains in ADAR1 interact with dsRNA substrates [190]. For ADAR2, alternative splicing can generate isoforms with different composition in the catalytic domain. One alternative splicing isoform, ADAR2b, contains an insertion of an Alu cassette compared to isoform ADAR2a [191]. ADAR2b exhibited similar specificity to ADAR2a and could efficiently edit sites in the GluR B transcript, with efficiencies of about 50% of ADAR2a. According to the cocrystal structure of ADAR2a with RNA substrates, the Alu insertion is located in the variable region of the 5′ binding loop (Fig. 5E) [121]. This could provide a structural explanation for the editing pattern observed with ADAR2b, as the variable region does not make direct contact to the RNA substrates thus has limited contribution to the specificity and activity. The other two splicing isoforms, ADAR2c and ADAR2d, have a truncated C-terminal alpha-helix 10 compared to ADAR2b and ADAR2a, respectively [192]. Alpha-helix 10 is one of the IP_6 coordinating helices (Fig. 4A). In agreement with the structural and functional importance of IP_6 for ADAR2 editing, these two truncated isoforms exhibited little editing activity.

Other than alternative splicing, ADAR mRNAs are regulated by self-editing which further contributes to the isoform diversity [20] and by binding of miRNAs [193]. A self-editing process of ADAR2 leads to the retention of 47 nt by generating an alternative 3′ splicing acceptor in intron 1 [20]. This leads to a frameshift and production of a truncated protein that lacks the double-stranded RNA-binding motifs and deaminase domain required for ADAR function. In rat, the alternatively spliced variant containing the 47 nt cassette exists in high percentage in the total ADAR2 transcripts; however, the truncated protein is expressed in a very low level [20]. In fact, an alternative translation initiation site downstream of the 47 nt retention region could be used to generate an active protein [20]. Transgenetic mice lacking ADAR2 self-editing activity have a higher expression of ADAR2 accompanied by the corresponding increase in the editing of ADAR2 substrates [194]. Therefore, self-editing is thought to be a feedback mechanism to negatively tune the ADAR activity. Self-editing in dADAR leads to a serine to glycine change at a location within the 5′ binding loop in the catalytic domain, resulting in an isoform exhibiting decreased activity in vitro [139,195]. The serine undergoing self-editing (S458 in human ADAR2) is highly conserved in the 5′ binding loop, consistent with the self-edited isoform showing compromised activity (Fig. 5E). The self-edited isoform may also have an altered substrate specificity given that the 5′ binding loop is implied in controlling substrate specificity. Indeed, the self-edited isoform exhibited a selective negative feedback effect on dADAR function in vivo [196]. However, it was suggested by the authors that the selective reduction of the dADAR function is less likely to result from a change in the target specificity, but rather due to a fine-tuned effect by other cellular factors, including a sequestration of the self-edited isoform into an unknown nuclear subcompartment [196]. Self-editing is also shown to be developmentally and spatially regulated [188], suggesting that it is involved in regulating the RNA editing at different developmental stages and cellular compartments. ADAR function is also regulated by small RNAs. Finally, the ADAR1 transcript is targeted by miR-17 and miR-432 at its 3′-UTR region, which leads to a reduction in ADAR1 expression level [193].

7.2 Posttranslational Modifications

Both ADAR1 and ADAR2 are shown to undergo phosphorylation at multiple sites [162,197]. The phosphorylation of ADAR2 has been shown to be important for proper localization in the nucleus, indicating that phosphorylation may play a central role in the regulation of ADAR enzymatic activity. Other than phosphorylation, ADAR1 is posttranslationally modified by SUMOylation, in which process the small ubiquitin-like modifier is covalently linked to Lys 418 of ADAR1 [198]. This leads to a reduction in the editing activity of ADAR1. In addition, an excess of the excitatory neurotransmitter glutamate can lead to specific cleavage of ADAR2 in cells, which could then lead to diminished editing on certain ADAR2 substrates [199].

7.3 Dimerization and Protein–Protein Interactions

Dimerization is essential for the activity of cytidine deaminases and ADATs such as ADAT2/3 and TadA. There are numerous reports on the role of dimerization in ADAR reactions [200–202]. However, whether ADARs require dimerization for activity is still debated. ADARs have been reported to dimerize in both an RNA-dependent and RNA-independent fashion [154,200,201]. In addition, there are conflicting reports regarding whether purified human ADAR2 exists primarily as a monomer or a dimer in solution [121,202,203]. While the data in the literature clearly supports the conclusion that ADARs are able to self-associate and form oligomeric states, whether this oligomerization is important for editing activity is not as clear. For instance, the human ADAR2 catalytic domain has been crystallized with RNA substrates in a catalytically relevant conformation as a 1:1 complex [121]. However, a truncation mutant of dADAR that maintains the ability to bind RNA but could not form dimers loses its enzymatic activity, suggesting that dimerization is required for proper ADAR activity [200]. It is possible that full-length ADARs require dimerization for activity. However, given the large RNA contact surface of the ADAR catalytic domain and the relatively short substrates that can react with full-length ADARs, it seems unlikely that, if dimerization is necessary, all possible RNA contact surfaces (i.e., four or six dsRBDs and two catalytic domains) could engage an RNA substrate during the reaction.

ADARs have also been reported to form heterodimers. ADAR1 and ADAR2 can form heterodimers in human cells [204,205], which could lead to a downregulation of editing activity. ADAR1 expression is elevated in astrocytes which leads to an interference of ADAR2 activity, probably through ADAR1–ADAR2 heterodimerization [205]. In *Drosophila melanogaster*, dADAR heterodimerizes with an inactive protein that inhibits editing [200].

ADAR function is also regulated by many other proteins via direct physical interaction. For example, in *C. elegans*, ADR-2-binding protein 1 (ADBP-1) physically interacts with ADR-2 and facilitates the localization of ADR-2 to the nucleus [206]. In an adbp-1 mutant, A-to-I RNA editing was not detected, suggesting that ADBP-1 is required for the RNA-editing activity of ADR-2. In another case, *Drosophila* fragile X homolog (dFMR1) interacts with dADAR and loss or overexpression of dFMR1 affects editing efficiency on certain dADAR targets [207]. Moreover, the regulation by Jun N-terminal kinase (JNK) is involved in the glucose-augmented ADAR2 expression [208]. In a recent screen for ADAR activating proteins, several editing enhancers were identified including the DSS1/SHFM1, a component of the proteasome subunit [161]. However, whether it has a direct interaction with ADARs is unknown.

7.4 Colocalization With RNA Substrates

The subcellular localization of both ADAR proteins and RNA substrates is one factor that modulates A-to-I editing. RNA substrates of ADARs are primarily found in the nucleus, which is where most RNA editing occurs [209]. Efficient editing has also been observed in the cytoplasm, in which cases the editing targets are mainly viral RNAs and RNA substrates whose editing site is retained after processing [209,210]. ADAR2 is restricted to the nucleus and is concentrated in the nucleolus. It was proposed that the accumulation of ADAR2 in nucleolus is a mechanism to transiently sequester its activity, and ADAR2 can translocate to the nucleoplasm once specific editing substrates are present [129]. Indeed, an enhanced translocation of endogenous ADAR2 from the nucleolus to the nucleoplasm results in increased editing of endogenous ADAR2 substrates [135]. ADAR1 can shuttle between nucleus and cytoplasm due to the presence of both a NLS and a NES in the protein [132].

7.5 *Trans*-Factors That Bind Substrate RNAs

Various RNA-binding proteins can cooperate with ADARs to regulate RNA processing or alter the accessibility of ADARs to the RNA targets. For example, it was revealed recently that the protein HuR and ADAR1 bind a common set of RNA targets, and these two proteins might regulate RNA processing cooperatively [211]. In *C. elegans*, the deaminase-deficient ADR-1 protein can bind editing substrates through its dsRBDs and bind to the common transcripts with ADR-2 in vivo thereby regulating the editing on cellular substrates globally [50,212]. The inactive protein hADAR3 can act as a competitive inhibitor for the editing of ADAR1 and ADAR2 in vitro by binding to the RNA substrates [91]. It is possible that ADAR3 occupies the same transcripts with the other ADARs thereby alter the accessibility of other ADARs to the substrates.

8. CONSEQUENCE OF A-TO-I EDITING

Since A-to-I RNA editing can change base pairing properties at a specific nucleotide position, this modification reaction can have a profound impact on the function of the target RNA. In this section, we review different functional consequences of A-to-I editing.

8.1 Recoding

Inosine is read as a guanosine by the ribosome during translation. Therefore, if editing occurs in a coding region, a nonsynonymous substitution can be introduced that alters the meaning of a codon. This type of "recoding" often results in protein products with altered biochemical properties [17,19,54]. One of the most well-studied cases of recoding occurs in the transcript for the AMPA-type glutamate receptor subunit B (GluR B). Editing within a glutamine codon (CAG) converts it to an arginine codon (CIG) (i.e., a Q/R site) resulting in a change of calcium permeability of the receptor channel [17]. A single Q-to-R preediting mutation rescues the lethal phenotype of the ADAR2 knockout in mice indicating the importance of this recoding event for ADAR2 function [213]. Editing at a different location in the same transcript converts an arginine codon to one for glycine (i.e., a R/G site). The R/G site recoding causes the channel to recover from ligand desensitization at a different rate [214]. Recoding occurs in dozens of human transcripts but we know the effects on the encoded protein's properties for only a fraction of these. Various studies have identified consequences of recoding for the 5-HT_{2C}R serotonin receptor [54], Gli1 transcription factor [215], AZIN1 regulator of cell proliferation [216], and in the NEIL1 DNA repair enzyme [19] among others [61].

8.2 RNA Structure, Stability, and Turnover

Most A-to-I editing events in human occur in noncoding regions of mRNAs, particularly in Alu repeat elements, which can be found in the UTRs, introns, and lncRNA. ADAR reaction in these regions is believed to modulate the RNA structure and stability [25]. In fact, A-to-I editing was initially discovered as a dsRNA unwinding activity present in *X. laevis* and mammalian cells and was later shown to be covalent modification of the dsRNA [72]. The "unwinding" effect results from an alteration of A–U base pairs to less stable I•U mismatches. However, many natural editing sites lie in A•C mismatches and editing in these cases produces more stable I—C pairs. RNA editing also influences the turnover of RNAs. Wang *et al.* showed that there is an enrichment of HuR-binding sequences, which are implicated in transcript turnover and processing, around editing sites and on ADAR-bound transcripts [211]. These authors further showed that

HuR cooperatively interacts with ADAR1 to regulate RNA processing. Hyper-edited substrates are also prone to be degraded by RNAse III Tudor Staphylococcal Nuclease (Tudor SN) which specifically recognizes and degrades inosine-containing RNAs [217]. Thus, A-to-I editing can regulate gene expression by regulating RNA stability and turnover.

8.3 Splicing

A-to-I editing can take place on pre-mRNA substrates in the nucleus and can influence splicing. An illustrative example is the self-editing of ADAR2, which creates an alternative splice site and produces a truncated protein isoform [20]. Also, editing at the GluR B Q/R and R/G sites were shown to influence the splicing rate [218,219]. A-to-I editing can also convert so-called noncanonical splice sites into canonical ones (e.g., a noncanonical GT–AA splice site in NRK gene is edited to a canonical GT–AI splice site) [220]. A recent study showed that knockdown of ADAR led to changes of splicing patterns for hundreds of genes, indicating an important role of A-to-I editing in modulating splicing [221]. However, the authors suggested that this global alteration in the splicing pattern is less likely to be achieved by direct editing of the splicing motifs, which occur relatively infrequently, but by modulation of *trans*-factors involved in the splicing machinery.

8.4 A-to-I Editing and RNA Interference

There is an interplay between A-to-I editing and RNA interference (RNAi) pathways that involve small RNAs and Argonaute proteins [21,222]. This interplay leads to various outcomes including antagonistic effects on RNAi and modulation of the targeting specificity of small RNAs. ADARs can edit miRNA precursors (pri-miRNA and pre-miRNA) and modulate their expression [223]. This effect is partially due to inosine-containing dsRNA-specific degradation by Tudor-SN [217]. The modulation can also result from the interaction between the edited miRNA precursors with the RNA interference machinery [22]. Editing on certain miRNA precursors can affect Dicer cleavage efficiency and RISC loading leading to a reduction in mature miRNAs [22]. In addition, the competitive binding of ADARs and Dicer to the miRNA precursors is another factor leading to inhibition of miRNA processing [224]. Edited miRNA precursors can produce mature miRNAs with redirected silencing ability, since editing alters the base pairing properties of the miRNA [23,225]. On the other hand, the 3′ untranslated regions (3′-UTRs) of mRNAs, which includes binding sites for miRNAs, are also targets of A-to-I editing [25]. RNA editing in miRNA targets interfere with the posttranscriptional repression by miRNAs by deleting or altering miRNA targeting sites in mRNA [226]. Similarly, RNA editing can create new targeting sites for miRNA in mRNA substrates. One example was the study on the transcript encoding human aryl hydrocarbon receptor (AhR). RNA editing by ADAR1 creates miRNA targeting sites in the AhR mRNA, and knockdown of ADAR1 abolishes miRNA-dependent downregulation of AhR expression [16].

8.5 Innate Immunity and the Antiviral Response

High levels of specific nucleic acid structures in certain locations in human cells signify the presence of a pathogen [227]. For instance, high concentrations of dsRNA in the cytoplasm are considered a hallmark of viral infection [228]. Humans express multiple receptors capable of recognizing different pathogen-associated nucleic acids that, once engaged, trigger an immune response [229]. Typically, these pathways lead to interferon production and interferon-stimulated genes are important for the response. The ADAR1 null is lethal in mice with embryonic death by E12.5, where embryos and derived cells show widespread apotosis and increased levels of interferon [230,231]. Other studies employing different systems have also shown ADARs regulate interferon responses negatively [232,233]. Interestingly, an interferon receptor null mutation can partially rescue ADAR1 null mice embryos and mitochondrial antiviral signaling adaptor protein (MAVS) null mutation can rescue to live birth [234]. More recently, the ADAR1 mutant phenotype was rescued by deletion of melanoma differentiation associated gene 5 (MDA5) without additional abnormal phenotype [235]. In humans, three types of pattern recognition receptors detect oligonucleotides and activate innate immune responses. They are nucleotide-binding oligomerization domain-like receptors (NLRs), RIG-I-like receptors (RLRs), and Toll-like receptors (TLRs) [236]. Both MAVS and MDA5 are part of the RLR detection pathway. Like RIG-I receptor, MDA5 senses dsRNA and initiates signaling cascades via its CARD domain [237]. MAVS is the downstream partner for MDA5 and other RLRs. Phosphorylation of MAVS activates interferon regulatory factor 3 (IRF3) and induces interferon production [238]. Null mutations of MAVS and MDA5 rescuing the ADAR1 null phenotype suggests that a primary function for ADAR1 editing is to prevent sensing of endogenous dsRNA and minimize innate immune responses to these RNAs.

Another study showed ADAR1 can suppress other dsRNA-dependent activities such as stimulation of the PKR and IRF-3 [232,233]. It is also known that ADARs can deaminate viral dsRNAs [137]. Editing can be accomplished by ADAR1 p150 in the cytoplasm and by ADAR1 p110 or ADAR2 in the nucleus [239]. Interestingly, ADARs interaction with viral RNAs can be either antiviral or proviral [239].

9. A-TO-I EDITING AND HUMAN DISEASES

9.1 Neurological Disorders

Transcripts for neurotransmitter receptors constitute an important class of ADAR substrates. Indeed, the abnormal phenotype of ADAR2 null mice can be rescued by mutation of an AMPA-like glutamate receptor into the edited form [213]. Clearly, ADAR editing is important to maintain proper function of this receptor. Over the years, several ADAR-related neurological disorders have been identified. Amyotrophic lateral sclerosis (ALS) is a fatal disease characterized by degeneration of motor neurons. Recent studies found that ADAR2 is significantly downregulated in spinal motor neurons for ALS patients [240]. The cause of the downregulation remains unknown, but this observation suggested that upregulation of ADAR2 editing could be a potential therapeutic strategy. Altered ADAR function has been linked to psychiatric disorders such as schizophrenia and bipolar disorder as ionotropic kainite 2 (GRIK2) mRNA was found under edited in these patients [241]. Also, altered editing of the $5HT_{2c}R$ transcript may be a potential contributor to severe depression and suicide attempts [242]. Prader–Willi syndrome is another condition related to dysregulated ADAR function. Prader–Willi syndrome is characterized by infantile hypotonia, early-childhood onset obesity and other developmental abnormalities. The absence of paternally expressed imprinted genes at 15q11.2–q13 is the cause of Prader–Willi syndrome [243]. The MBII-52 gene is located in this region and MBII-52 functions as a nucleolar C/D RNA [244]. MBII-52 targets the C-site of the $5HT_{2C}R$ mRNA and directs 2′-O-methylation at this site resulting in decreased editing efficiency [245]. Mice expressing only the C site edited form of $5HT_{2C}R$ present Prader–Willi syndrome like symptoms [246]. These results suggest that enhanced ADAR editing at this site may contribute to Prader–Willi syndrome [246]. However, ADAR editing at this site is not the sole cause for Prader–Willi syndrome, since not all characteristic symptoms were found in the transgenic mice.

9.2 Aicardi-Goutieres Syndrome

Aicardi-Goutieres syndrome (AGS) is an autoimmune disorder that can result in physical and intellectual handicap [247]. AGS can be caused by mutations in seven possible genes, including ADAR1 [248]. AGS patients with mutations in any of the seven genes showed a type I interferon signature, which could be used to distinguish AGS patients from healthy individuals [247]. As introduced earlier, ADAR1 acts as a suppressor of type I interferon signaling. It was proposed that inefficient ADAR activity might lead to an increase in immunoreactive dsRNA, and this may be involved in the pathogenesis of AGS [249]. A total of eight amino acid substitutions in ADAR1 have been identified in patients with AGS, including P193A, A870T, I872T, R892H, K999N, G1007R, Y1112F, and D1113H [249]. P193 lies in the Z-DNA-binding domain, indicating that the interferon-inducible isoform p150 is involved in the AGS phenotype. The other residues lie in the deaminase domain of ADAR1. The crystal structures of ADAR2 deaminase domain–RNA complexes provided insights into the possible effects of these mutations. Based on their localization in ADAR2, we have suggested these mutations could be categorized into three groups [250]. The first group of mutations affect residues mapped at the protein–RNA-binding surface, including R892, K999, and G1007. Mutations at these locations could change the activity of ADAR1 enzyme by directly influencing protein–RNA binding interactions. The second group of mutations impact residues A870 and I872, which are located internal of the protein, and are possibly important for the proper conformation of the active site glutamate residue and the IP_6-binding helix α2. A third group of mutations are at Y1112 and D1113, which do not map to the RNA-binding surface or near active site, but are located adjacent to RNA-binding loops. ADAR2–RNA structures suggest that mutations at these two locations may disrupt interactions of the RNA-binding loops with RNA substrates. Interestingly, mutation G1007R is also observed in patients with Dyschromatosis symmetrica hereditaria (DHS) (see below) [249]. However, the phenotypes typical of DHS have not been identified in AGS patients.

9.3 Dyschromatosis Symmetrica Hereditaria

DSH is a skin pigment disorder characterized by a mixture of hyperpigmented and hypopigmented macules on the hands and feet [251]. Mutations in the ADAR1 gene were identified as the cause of DSH. More specifically, loss of the functional ADAR1 p150 is the major cause for

DSH [252]. Hundreds of DSH-related mutations have been identified in the ADAR1 gene among patients [253–255]. Interestingly, the majority of mutations are located within the deaminase domain suggesting that DSH is closely related to the catalytic activity of the enzyme [256]. More recently, a novel splice site mutation was identified in a Japanese DSH patient which causes exon skipping [254]. As indicated earlier, both DSH and AGS are caused by mutations in the ADAR1 gene. A recent case of a DSH patient with AGS symptoms draws a connection between DSH and AGS [257]. It has been suggested that heterozygous ADAR1 mutations may cause DSH, while homozygous as well as compound heterozygous ADAR1 mutations may cause a combination of AGS and DSH [257]. The background skin color might be the reason why DSH symptoms are not reported in Caucasian AGS patients.

9.4 RNA Editing and Cancer

Altered ADAR function has also been linked to various cancers, including brain cancers, human hepatocellular carcinoma (HCC), colorectal cancer, and others [61,258,259]. For example, in patients with HHC, editing of the AZIN1 transcript by ADAR1 p110 is increased due to an upregulation of ADAR1 p110 [216]. This leads to increased levels of two oncoproteins, ornithine decarboxylase and cyclin D1, suggesting that editing on AZIN1 mRNA may drive the pathogenesis of HCC [216]. Also, editing of the Ras homolog family member Q (RHOQ) is enhanced in colorectal cancer promoting invasion potential through deregulation of actin remodeling [258]. Indeed, recent global analyses of cancer transcriptomes have led to the conclusions that A-to-I editing is generally dysregulated in human cancers (typically higher editing levels than in paired normal tissues) and occurs in functionally and clinically relevant sites [260–262]. In addition, increased editing in cancer cells is primarily a result of increases in ADAR1 expression [260]. These studies suggest the possibility of therapeutic effects arising from molecules that reduce A-to-I editing at specific, clinically relevant editing sites and/or lead to a reduction in ADAR1 activity to normal tissue levels [263].

ACKNOWLEDGMENTS

P.A.B. acknowledges the National Institutes of Health for financial support in the form of Grant GM061115. The contents of this publication are solely the responsibility of the authors and do not necessarily represent the official views of NIH.

REFERENCES

[1] K.D. Meyer, S.R. Jaffrey, The dynamic epitranscriptome: N6-methyladenosine and gene expression control, Nat. Rev. Mol. Cell Biol. 15 (5) (2014) 313–326.
[2] M. Taoka, D. Ishikawa, Y. Nobe, H. Ishikawa, Y. Yamauchi, G. Terukina, H. Nakayama, K. Hirota, N. Takahashi, T. Isobe, RNA cytidine acetyltransferase of small-subunit ribosomal RNA: identification of acetylation sites and the responsible acetyltransferase in fission yeast, *Schizosaccharomyces pombe*, PLoS One 9 (11) (2014) e112156.
[3] G.A. Garcia, J.D. Kittendorf, Transglycosylation: a mechanism for RNA modification (and editing?), Bioorg. Chem. 33 (3) (2005) 229–251.
[4] R.A. Goodman, M.R. Macbeth, P.A. Beal, ADAR proteins: structure and catalytic mechanism, in: C.E. Samuel (Ed.), Adenosine Deaminases Acting on RNA (ADARs) and A-to-I Editing, Springer Berlin Heidelberg, Berlin, Heidelberg, 2012, pp. 1–33.
[5] T.M. Carlile, M.F. Rojas-Duran, B. Zinshteyn, H. Shin, K.M. Bartoli, W.V. Gilbert, Pseudouridine profiling reveals regulated mRNA pseudouridylation in yeast and human cells, Nature 515 (7525) (2014) 143–146.
[6] X. Li, X. Xiong, K. Wang, L. Wang, X. Shu, S. Ma, C. Yi, Transcriptome-wide mapping reveals reversible and dynamic N1-methyladenosine methylome, Nat. Chem. Biol. 12 (5) (2016) 311–316.
[7] W.V. Gilbert, T.A. Bell, C. Schaening, Messenger RNA modifications: form, distribution, and function, Science 352 (6292) (2016) 1408–1412.
[8] K. Chen, S. Boxuan, C. He, Nucleic acid modifications in regulation of gene expression, Cell Chem. Biol. 23 (1) (2016) 74–85.
[9] M.A. O'Connell, N.M. Mannion, L.P. Keegan, The epitranscriptome and innate immunity, PLoS Genet. 11 (12) (2015) e1005687.
[10] Y. Saletore, K. Meyer, J. Korlach, I.D. Vilfan, S. Jaffrey, C.E. Mason, The birth of the epitranscriptome: deciphering the function of RNA modifications, Genome Biol. 13 (10) (2012) 175.
[11] B.S. Zhao, I.A. Roundtree, C. He, Post-transcriptional gene regulation by mRNA modifications, Nat. Rev. Mol. Cell Biol. 18 (1) (2017) 31–42.
[12] L.P. Keegan, A. Gallo, M.A. O'Connell, The many roles of an RNA editor, Nat. Rev. Genet. 2 (11) (2001) 869–878.
[13] I. Alseth, B. Dalhus, M. Bjørås, Inosine in DNA and RNA, Curr. Opin. Genet. Dev. 26 (2014) 116–123.
[14] A.G. Torres, D. Piñeyro, M. Rodríguez-Escribà, N. Camacho, O. Reina, A. Saint-Léger, L. Filonava, E. Batlle, L. Ribas de Pouplana, Inosine modifications in human tRNAs are incorporated at the precursor tRNA level, Nucleic Acids Res. 43 (10) (2015) 5145–5157.
[15] Z. Peng, Y. Cheng, B.C.-M. Tan, L. Kang, Z. Tian, Y. Zhu, W. Zhang, Y. Liang, X. Hu, X. Tan, et al., Comprehensive analysis of RNA-seq data reveals extensive RNA editing in a human transcriptome, Nat. Biotechnol. 30 (3) (2012) 253–260.
[16] M. Nakano, T. Fukami, S. Gotoh, M. Takamiya, Y. Aoki, M. Nakajima, RNA editing modulates human hepatic Aryl hydrocarbon receptor expression by creating microRNA recognition sequence, J. Biol. Chem. 291 (2) (2016) 894–903.
[17] B. Sommer, M. Köhler, R. Sprengel, P.H. Seeburg, RNA editing in brain controls a determinant of ion flow in glutamate-gated channels, Cell 67 (1) (1991) 11–19.
[18] J.B. Li, E.Y. Levanon, J.-K. Yoon, J. Aach, B. Xie, E. LeProust, K. Zhang, Y. Gao, G.M. Church, Genome-wide identification of human RNA editing sites by parallel DNA capturing and sequencing, Science 324 (5931) (2009) 1210–1213.
[19] J. Yeo, R.A. Goodman, N.T. Schirle, S.S. David, P.A. Beal, RNA editing changes the lesion specificity for the DNA repair enzyme NEIL1, Proc. Natl. Acad. Sci. U.S.A. 107 (48) (2010) 20715–20719.

[20] S.M. Rueter, T.R. Dawson, R.B. Emeson, Regulation of alternative splicing by RNA editing, Nature 399 (6731) (1999) 75–80.

[21] W. Yang, T.P. Chendrimada, Q. Wang, M. Higuchi, P.H. Seeburg, R. Shiekhattar, K. Nishikura, Modulation of microRNA processing and expression through RNA editing by ADAR deaminases, Nat. Struct. Mol. Biol. 13 (1) (2006) 13–21.

[22] Y. Kawahara, B. Zinshteyn, T.P. Chendrimada, R. Shiekhattar, K. Nishikura, RNA editing of the microRNA-151 precursor blocks cleavage by the dicer–TRBP complex, EMBO Rep. 8 (8) (2007) 763–769.

[23] Y. Kawahara, B. Zinshteyn, P. Sethupathy, H. Iizasa, A.G. Hatzigeorgiou, K. Nishikura, Redirection of silencing targets by adenosine-to-inosine editing of miRNAs, Science 315 (5815) (2007) 1137–1140.

[24] D.P. Morse, P.J. Aruscavage, B.L. Bass, RNA hairpins in noncoding regions of human brain and *Caenorhabditis elegans* mRNA are edited by adenosine deaminases that act on RNA, Proc. Natl. Acad. Sci. U.S.A. 99 (12) (2002) 7906–7911.

[25] E.Y. Levanon, E. Eisenberg, R. Yelin, S. Nemzer, M. Hallegger, R. Shemesh, Z.Y. Fligelman, A. Shoshan, S.R. Pollock, D. Sztybel, et al., Systematic identification of abundant A-to-I editing sites in the human transcriptome, Nat. Biotechnol. 22 (8) (2004) 1001–1005.

[26] D.P. Morse, B.L. Bass, Detection of inosine in messenger RNA by inosine-specific cleavage, Biochemistry 36 (28) (1997) 8429–8434.

[26a] G. Eaholtz, T. Scheuer, W.A. Catterall, Restoration of inactivation and block of open sodium channels by an inactivation gate peptide, Neuron 12 (5) (1994) 1041–1048.

[27] W.M. Gommans, N.E. Tatalias, C.P. Sie, D. Dupuis, N. Vendetti, L. Smith, R. Kaushal, S. Maas, Screening of human SNP database identifies recoding sites of A-to-I RNA editing, RNA 14 (10) (2008) 2074–2085.

[28] M. Sakurai, H. Ueda, T. Yano, S. Okada, H. Terajima, T. Mitsuyama, A. Toyoda, A. Fujiyama, H. Kawabata, T. Suzuki, A biochemical landscape of A-to-I RNA editing in the human brain transcriptome, Genome Res. 24 (2014) 522–534.

[29] R.W. Holley, G.A. Everett, J.T. Madison, A. Zamir, Nucleotide sequences in the yeast alanine transfer Ribonucleic acid, J. Biol. Chem. 240 (5) (1965) 2122–2128.

[30] F. Jühling, M. Mörl, R.K. Hartmann, M. Sprinzl, P.F. Stadler, J. Pütz, tRNAdb 2009: compilation of tRNA sequences and tRNA genes, Nucleic Acids Res. **37** (Suppl. 1) (2008) D159–D162.

[31] A.P. Gerber, W. Keller, An adenosine deaminase that generates inosine at the wobble position of tRNAs, Science 286 (5442) (1999) 1146.

[32] J. Wolf, A.P. Gerber, W. Keller, tadA, an essential tRNA-specific adenosine deaminase from *Escherichia coli*, EMBO J. 21 (14) (2002) 3841–3851.

[33] H. Grosjean, S. Auxilien, F. Constantinesco, C. Simon, Y. Corda, H.F. Becker, D. Foiret, A. Morin, Y.X. Jin, M. Fournier, et al., Enzymatic conversion of adenosine to inosine and to N1-methylinosine in transfer RNAs: a review, Biochimie 78 (6) (1996) 488–501.

[34] A. Gerber, H. Grosjean, T. Melcher, W. Keller, Tad1p, a yeast tRNA-specific adenosine deaminase, is related to the mammalian pre-mRNA editing enzymes ADAR1 and ADAR2, EMBO J. 17 (16) (1998) 4780–4789.

[35] G.R. Björk, K. Jacobsson, K. Nilsson, M.J.O. Johansson, A.S. Byström, O.P. Persson, A primordial tRNA modification required for the evolution of life? EMBO J. 20 (1–2) (2001) 231–239.

[36] M.A. Machnicka, K. Milanowska, O. Osman Oglou, E. Purta, M. Kurkowska, A. Olchowik, W. Januszewski, S. Kalinowski, S. Dunin-Horkawicz, K.M. Rother, et al., MODOMICS: a database of RNA modification pathways—2013 update, Nucleic Acids Res. 41 (D1) (2013) D262–D267.

[37] H. Grosjean, F. Constantinesco, D. Foiret, N. Benachenhou, A novel enzymatic pathway leading to 1-methylinosine modification in Haloferax volcanii tRNA, Nucleic Acids Res. 23 (21) (1995) 4312–4319.
[38] F.H.C. Crick, Codon—anticodon pairing: the wobble hypothesis, J. Mol. Biol. 19 (2) (1966) 548–555.
[39] T. Borén, P. Elias, T. Samuelsson, C. Claesson, M. Barciszewska, C.W. Gehrke, K.C. Kuo, F. Lustig, Undiscriminating codon reading with Adenosine in the wobble position, J. Mol. Biol. 230 (3) (1993) 739–749.
[40] J.F. Curran, Decoding with the A:I wobble pair is inefficient, Nucleic Acids Res. 23 (4) (1995) 683–688.
[41] M.W. Gray, O2'-methylinosine, a constituent of the ribosomal RNA of Crithidia fasciculata, Nucleic Acids Res. 3 (4) (1976) 977–988.
[42] S. Moreira, M. Valach, M. Aoulad-Aissa, C. Otto, G. Burger, Novel modes of RNA editing in mitochondria, Nucleic Acids Res. 44 (10) (2016) 4907–4919.
[43] E. Ullu, C. Tschudi, Alu sequences are processed 7SL RNA genes, Nature 312 (5990) (1984) 171–172.
[44] D.D.Y. Kim, T.T.Y. Kim, T. Walsh, Y. Kobayashi, T.C. Matise, S. Buyske, A. Gabriel, Widespread RNA editing of embedded Alu elements in the human transcriptome, Genome Res. 14 (9) (2004) 1719–1725.
[45] A. Athanasiadis, A. Rich, S. Maas, Widespread A-to-I RNA editing of Alu-containing mRNAs in the human transcriptome, PLoS Biol. 2 (12) (2004) e391.
[46] L. Bazak, A. Haviv, M. Barak, J. Jacob-Hirsch, P. Deng, R. Zhang, F.J. Isaacs, G. Rechavi, J.B. Li, E. Eisenberg, et al., A-to-I RNA editing occurs at over a hundred million genomic sites, located in a majority of human genes, Genome Res. 24 (3) (2014) 365–376.
[47] M. Kumar, G.G. Carmichael, Nuclear antisense RNA induces extensive adenosine modifications and nuclear retention of target transcripts, Proc. Natl. Acad. Sci. U.S.A. 94 (8) (1997) 3542–3547.
[48] K.V. Prasanth, S.G. Prasanth, Z. Xuan, S. Hearn, S.M. Freier, C.F. Bennett, M.Q. Zhang, D.L. Spector, Regulating gene expression through RNA nuclear retention, Cell 123 (2) (2005) 249–263.
[49] H.A. Hundley, A.A. Krauchuk, B.L. Bass, *C. elegans* and *H. sapiens* mRNAs with edited 3′ UTRs are present on polysomes, RNA 14 (10) (2008) 2050–2060.
[50] M.C. Washburn, H.A. Hundley, Trans and cis factors affecting A-to-I RNA editing efficiency of a noncoding editing target in *C. elegans*, RNA 22 (5) (2016) 722–728.
[51] M. Doria, F. Neri, A. Gallo, M.G. Farace, A. Michienzi, Editing of HIV-1 RNA by the double-stranded RNA deaminase ADAR1 stimulates viral infection, Nucleic Acids Res. 37 (17) (2009) 5848–5858.
[52] L.L. Chen, J.N. DeCerbo, G.G. Carmichael, Alu element-mediated gene silencing, EMBO J. 27 (12) (2008) 1694–1705.
[53] A. Mallela, K. Nishikura, A-to-I editing of protein coding and noncoding RNAs, Crit. Rev. Biochem. Mol. Biol. 47 (6) (2012) 493–501.
[54] C.M. Burns, H. Chu, S.M. Rueter, L.K. Hutchinson, H. Canton, E. Sanders-Bush, R.B. Emeson, Regulation of serotonin-2C receptor G-protein coupling by RNA editing, Nature 387 (6630) (1997) 303–308.
[55] C.M. Niswender, S.C. Copeland, K. Herrick-Davis, R.B. Emeson, E. Sanders-Bush, RNA editing of the human serotonin 5-hydroxytryptamine 2C receptor silences constitutive activity, J. Biol. Chem. 274 (14) (1999) 9472–9478.
[56] A.I. Abbas, D.J. Urban, N.H. Jensen, M.S. Farrell, W.K. Kroeze, P. Mieczkowski, Z. Wang, B.L. Roth, Assessing serotonin receptor mRNA editing frequency by a novel ultra high-throughput sequencing method, Nucleic Acids Res. 38 (10) (2010) e118.

[57] G.X. Luo, M. Chao, S.Y. Hsieh, C. Sureau, K. Nishikura, J. Taylor, A specific base transition occurs on replicating hepatitis delta virus RNA, J. Virol. 64 (3) (1990) 1021–1027.
[58] H. Zheng, T.B. Fu, D. Lazinski, J. Taylor, Editing on the genomic RNA of human hepatitis delta virus, J. Virol. 66 (8) (1992) 4693–4697.
[59] J.L. Casey, K.F. Bergmann, T.L. Brown, J.L. Gerin, Structural requirements for RNA editing in hepatitis delta virus: evidence for a uridine-to-cytidine editing mechanism, Proc. Natl. Acad. Sci. U.S.A. 89 (15) (1992) 7149–7153.
[60] S.K. Wong, D.W. Lazinski, Replicating hepatitis delta virus RNA is edited in the nucleus by the small form of ADAR1, Proc. Natl. Acad. Sci. U.S.A. 99 (23) (2002) 15118–15123.
[61] T.H.M. Chan, C.H. Lin, L. Qi, J. Fei, Y. Li, K.J. Yong, M. Liu, Y. Song, R.K.K. Chow, V.H.E. Ng, et al., A disrupted RNA editing balance mediated by ADARs (Adenosine DeAminases that act on RNA) in human hepatocellular carcinoma, Gut 63 (5) (2014) 832–843.
[62] O. Solomon, L. Bazak, E.Y. Levanon, N. Amariglio, R. Unger, G. Rechavi, E. Eyal, Characterizing of functional human coding RNA editing from evolutionary, structural, and dynamic perspectives, Proteins: Struct., Funct., Bioinf. 82 (11) (2014) 3117–3131.
[63] Y. Li, L. Chen, T.H.M. Chan, X.-Y. Guan, Hepatocellular carcinoma: transcriptome diversity regulated by RNA editing, Int. J. Biochem. Cell Biol. 45 (8) (2013) 1843–1848.
[64] L. Valente, K. Nishikura, ADAR gene family and A-to-I RNA editing: diverse roles in posttranscriptional gene regulation, Prog. Nucleic Acid Res. Mol. Biol. 79 (2005) 299–338. Academic Press.
[65] L.E. Rieder, R.A. Reenan, The intricate relationship between RNA structure, editing, and splicing, Semin. Cell Dev. Biol. 23 (3) (2012) 281–288.
[66] G. Lev-Maor, R. Sorek, E.Y. Levanon, N. Paz, E. Eisenberg, G. Ast, RNA-editing-mediated exon evolution, Genome Biol. 8 (2) (2007) R29.
[67] A. Shafik, U. Schumann, M. Evers, T. Sibbritt, T. Preiss, The emerging epitranscriptomics of long noncoding RNAs, Biochim. Biophys. Acta 1859 (1) (2016) 59–70.
[68] A. Azlan, N. Dzaki, G. Azzam, Argonaute: the executor of small RNA function, J. Genet. Genomics 43 (8) (2016) 481–494.
[69] S. Tomaselli, B. Bonamassa, A. Alisi, V. Nobili, F. Locatelli, A. Gallo, ADAR enzyme and miRNA story: a nucleotide that can make the difference, Int. J. Mol. Sci. 14 (11) (2013) 22796–22816.
[70] P. Connerty, A. Ahadi, G. Hutvagner, RNA binding proteins in the miRNA pathway, Int. J. Mol. Sci. 17 (1) (2016) 31.
[71] D.J. Luciano, H. Mirsky, N.J. Vendetti, S. Maas, RNA editing of a miRNA precursor, RNA 10 (8) (2004) 1174–1177.
[72] B.L. Bass, H. Weintraub, An unwinding activity that covalently modifies its double-stranded RNA substrate, Cell 55 (6) (1988) 1089–1098.
[73] O.M. Stephens, B.L. Haudenschild, P.A. Beal, The binding selectivity of ADAR2's dsRBMs contributes to RNA-editing selectivity, Chem. Biol. 11 (9) (2004) 1239–1250.
[74] F. Lai, R. Drakas, K. Nishikura, Mutagenic analysis of double-stranded RNA adenosine deaminase, a candidate enzyme for RNA editing of glutamate-gated ion channel transcripts, J. Biol. Chem. 270 (29) (1995) 17098–17105.
[75] R.A. Reenan, C.J. Hanrahan, B. Ganetzky, The mlenapts RNA helicase mutation in Drosophila results in a splicing catastrophe of the para Na+ channel transcript in a region of RNA editing, Neuron 25 (1) (2000) 139–149.

[76] B. Hoopengardner, T. Bhalla, C. Staber, R. Reenan, Nervous system targets of RNA editing identified by comparative genomics, Science 301 (5634) (2003) 832–836.
[77] M. Blow, P.A. Futreal, R. Wooster, M.R. Stratton, A survey of RNA editing in human brain, Genome Res. 14 (12) (2004) 2379–2387.
[78] E. Eisenberg, K. Adamsky, L. Cohen, N. Amariglio, A. Hirshberg, G. Rechavi, E.Y. Levanon, Identification of RNA editing sites in the SNP database, Nucleic Acids Res. 33 (14) (2005) 4612–4617.
[79] J.H. Bahn, J.H. Lee, G. Li, C. Greer, G. Peng, X. Xiao, Accurate identification of A-to-I RNA editing in human by transcriptome sequencing, Genome Res. 22 (2011) 142–150.
[80] S. Zhu, J.-F. Xiang, T. Chen, L.-L. Chen, L. Yang, Prediction of constitutive A-to-I editing sites from human transcriptomes in the absence of genomic sequences, BMC Genomics 14 (1) (2013) 206.
[81] M. Sakurai, T. Yano, H. Kawabata, H. Ueda, T. Suzuki, Inosine cyanoethylation identifies A-to-I RNA editing sites in the human transcriptome, Nat. Chem. Biol. 6 (10) (2010) 733–740.
[82] T. Suzuki, H. Ueda, S. Okada, M. Sakurai, Transcriptome-wide identification of adenosine-to-inosine editing using the ICE-seq method, Nat. Protocols 10 (5) (2015) 715–732.
[83] E. Picardi, C. Manzari, F. Mastropasqua, I. Aiello, A.M. D'Erchia, G. Pesole, Profiling RNA editing in human tissues: towards the inosinome Atlas, Sci. Rep. 5 (2015) 14941.
[84] G. Ramaswami, J.B. Li, RADAR: a rigorously annotated database of A-to-I RNA editing, Nucleic Acids Res. 42 (D1) (2014) D109–D113.
[85] A. Kiran, P.V. Baranov, DARNED: a DAtabase of RNa EDiting in humans, Bioinformatics 26 (14) (2010) 1772–1776.
[86] O. Solomon, E. Eyal, N. Amariglio, R. Unger, G. Rechavi, e23D: database and visualization of A-to-I RNA editing sites mapped to 3D protein structures, Bioinformatics 32 (14) (2016) 2213–2215.
[87] E. Picardi, A.M. D'Erchia, C. Lo Giudice, G. Pesole, REDIportal: a comprehensive database of A-to-I RNA editing events in humans, Nucleic Acids Res. 45 (D1) (2017) D750–D757.
[88] B.L. Bass, K. Nishikura, W. Keller, P.H. Seeburg, R.B. Emeson, M.A. O'Connell, C.E. Samuel, A. Herbert, A standardized nomenclature for adenosine deaminases that act on RNA, RNA 3 (1997) 947–949.
[89] R.W. Wagner, K. Nishikura, Cell cycle expression of RNA duplex unwindase activity in mammalian cells, Mol. Cell. Biol. 8 (2) (1988) 770–777.
[90] B.L. Bass, H. Weintraub, A developmentally regulated activity that unwinds RNA duplexes, Cell 48 (4) (1987) 607–613.
[91] C.X. Chen, D.S. Cho, Q. Wang, F. Lai, K.C. Carter, K. Nishikura, A third member of the RNA-specific adenosine deaminase gene family, ADAR3, contains both single- and double-stranded RNA binding domains, RNA 6 (5) (2000) 755–767.
[92] J.B. Patterson, C.E. Samuel, Expression and regulation by interferon of a double-stranded-RNA-specific adenosine deaminase from human cells: evidence for two forms of the deaminase, Mol. Cell. Biol. 15 (10) (1995) 5376–5388.
[93] C.X. George, C.E. Samuel, Human RNA-specific adenosine deaminase ADAR1 transcripts possess alternative exon 1 structures that initiate from different promoters, one constitutively active and the other interferon inducible, Proc. Natl. Acad. Sci. U.S.A. 96 (8) (1999) 4621–4626.
[94] S. Lykke-Andersen, S. Piñol-Roma, J. Kjems, Alternative splicing of the ADAR1 transcript in a region that functions either as a 5′-UTR or an ORF, RNA 13 (10) (2007) 1732–1744.

[95] Y. Kawahara, K. Ito, M. Ito, S. Tsuji, S. Kwak, Novel splice variants of human ADAR2 mRNA: skipping of the exon encoding the dsRNA-binding domains, and multiple C-terminal splice sites, Gene 363 (2005) 193–201.
[96] P. Barraud, F.H.-T. Allain, ADAR proteins: double-stranded RNA and Z-DNA binding domains, in: C.E. Samuel (Ed.), Adenosine Deaminases Acting on RNA (ADARs) and A-to-I Editing, Springer Berlin Heidelberg, Berlin, Heidelberg, 2012, pp. 35–60.
[97] D. St Johnston, N.H. Brown, J.G. Gall, M. Jantsch, A conserved double-stranded RNA-binding domain, Proc. Natl. Acad. Sci. U.S.A. 89 (22) (1992) 10979–10983.
[98] T.J. Gibson, J.D. Thompson, Detection of dsRNA-binding domains in RNA helicase A and Drosophila maleless: implications for monomeric RNA helicases, Nucleic Acids Res. 22 (13) (1994) 2552–2556.
[99] K.-Y. Chang, A. Ramos, The double-stranded RNA-binding motif, a versatile macromolecular docking platform, FEBS J. 272 (9) (2005) 2109–2117.
[100] S.J. McCormack, D.C. Thomis, C.E. Samuel, Mechanism of interferon action: identification of a RNA binding domain within the N-terminal region of the human RNA-dependent P1/eIF-2 alpha protein kinase, Virology 188 (1) (1992) 47–56.
[101] U. Kim, Y. Wang, T. Sanford, Y. Zeng, K. Nishikura, Molecular cloning of cDNA for double-stranded RNA adenosine deaminase, a candidate enzyme for nuclear RNA editing, Proc. Natl. Acad. Sci. U.S.A. 91 (24) (1994) 11457–11461.
[102] M. Doyle, M.F. Jantsch, New and old roles of the double-stranded RNA-binding domain, J. Struct. Biol. 140 (1–3) (2002) 147–153.
[103] J. Gan, J.E. Tropea, B.P. Austin, D.L. Court, D.S. Waugh, X. Ji, Structural insight into the mechanism of double-stranded RNA processing by ribonuclease III, Cell 124 (2) (2006) 355–366.
[104] P. Barraud, B.S.E. Heale, M.A. O'Connell, F.H.T. Allain, Solution structure of the N-terminal dsRBD of Drosophila ADAR and interaction studies with RNA, Biochimie 94 (7) (2012) 1499–1509.
[105] M. Bycroft, S. Grunert, A.G. Murzin, M. Proctor, D.S. Johnston, NMR solution structure of a dsRNA binding domain from Drosophila staufen protein reveals homology to the N-terminal domain of ribosomal protein S5, EMBO J. 14 (14) (1995) 3563–3571.
[106] R. Stefl, M. Xu, L. Skrisovska, R.B. Emeson, F.H.T. Allain, Structure and specific RNA binding of ADAR2 double-stranded RNA binding motifs, Structure 14 (2) (2006) 345–355.
[107] R. Stefl, F.C. Oberstrass, J.L. Hood, M. Jourdan, M. Zimmermann, L. Skrisovska, C. Maris, L. Peng, C. Hofr, R.B. Emeson, et al., The solution structure of the ADAR2 dsRBM-RNA complex reveals a sequence-specific readout of the minor groove, Cell 143 (2) (2010) 225–237.
[108] A. Ramos, S. Grünert, J. Adams, D.R. Micklem, M.R. Proctor, S. Freund, M. Bycroft, D.S. Johnston, G. Varani, RNA recognition by a Staufen double-stranded RNA-binding domain, EMBO J. 19 (5) (2000) 997–1009.
[109] H. Wu, A. Henras, G. Chanfreau, J. Feigon, Structural basis for recognition of the AGNN tetraloop RNA fold by the double-stranded RNA-binding domain of Rnt1p RNase III, Proc. Natl. Acad. Sci. U.S.A. 101 (22) (2004) 8307–8312.
[110] Y. Liu, M. Lei, C.E. Samuel, Chimeric double-stranded RNA-specific adenosine deaminase ADAR1 proteins reveal functional selectivity of double-stranded RNA-binding domains from ADAR1 and protein kinase PKR, Proc. Natl. Acad. Sci. U.S.A. 97 (23) (2000) 12541–12546.
[111] M.R. Macbeth, H.L. Schubert, A.P. VanDemark, A.T. Lingam, C.P. Hill, B.L. Bass, Inositol hexakisphosphate is bound in the ADAR2 core and required for RNA editing, Science 309 (5740) (2005) 1534–1539.

[112] Z. Chang, P. Nygaard, A.C. Chinault, R.E. Kellems, Deduced amino acid sequence of *Escherichia coli* adenosine deaminase reveals evolutionarily conserved amino acid residues: implications for catalytic function, Biochemistry 30 (8) (1991) 2273–2280.

[113] R.H. Michell, Inositol derivatives: evolution and functions, Nat. Rev. Mol. Cell Biol. 9 (2) (2008) 151–161.

[114] J.W. Verbsky, S.-C. Chang, M.P. Wilson, Y. Mochizuki, P.W. Majerus, The pathway for the production of inositol hexakisphosphate in human cells, J. Biol. Chem. 280 (3) (2005) 1911–1920.

[115] A.R. Alcazar-Roman, E.J. Tran, S. Guo, S.R. Wente, Inositol hexakisphosphate and Gle1 activate the DEAD-box protein Dbp5 for nuclear mRNA export, Nat. Cell Biol. 8 (7) (2006) 711–716.

[116] A. Arnone, M.F. Perutz, Structure of inositol hexaphosphate-human deoxyhaemoglobin complex, Nature 249 (5452) (1974) 34–36.

[117] B.M. Collins, A.J. McCoy, H.M. Kent, P.R. Evans, D.J. Owen, Molecular architecture and functional model of the endocytic AP2 complex, Cell 109 (4) (2002) 523–535.

[118] S.A.K. Datta, Z. Zhao, P.K. Clark, S. Tarasov, J.N. Alexandratos, S.J. Campbell, M. Kvaratskhelia, J. Lebowitz, A. Rein, Interactions between HIV-1 Gag molecules in solution: an inositol phosphate-mediated switch, J. Mol. Biol. 365 (3) (2007) 799–811.

[119] C.E. Paulsen, J.-P. Armache, Y. Gao, Y. Cheng, D. Julius, Structure of the TRPA1 ion channel suggests regulatory mechanisms, Nature 520 (7548) (2015) 511–517.

[120] R.N. Pruitt, B. Chagot, M. Cover, W.J. Chazin, B. Spiller, D.B. Lacy, Structure-function analysis of inositol hexakisphosphate-induced autoprocessing in clostridium difficile toxin A, J. Biol. Chem. 284 (33) (2009) 21934–21940.

[121] M.M. Matthews, J.M. Thomas, Y. Zheng, K. Tran, K.J. Phelps, A.I. Scott, J. Havel, A.J. Fisher, P.A. Beal, Structures of human ADAR2 bound to dsRNA reveal base-flipping mechanism and basis for site selectivity, Nat. Struct. Mol. Biol. 23 (5) (2016) 426–433.

[122] B.A. Brown, K. Lowenhaupt, C.M. Wilbert, E.B. Hanlon, A. Rich, The Zα domain of the editing enzyme dsRNA adenosine deaminase binds left-handed Z-RNA as well as Z-DNA, Proc. Natl. Acad. Sci. U.S.A. 97 (25) (2000) 13532–13536.

[123] A. Herbert, J. Alfken, Y.-G. Kim, I.S. Mian, K. Nishikura, A. Rich, A Z-DNA binding domain present in the human editing enzyme, double-stranded RNA adenosine deaminase, Proc. Natl. Acad. Sci. U.S.A. 94 (16) (1997) 8421–8426.

[124] T. Schwartz, M.A. Rould, K. Lowenhaupt, A. Herbert, A. Rich, Crystal structure of the Zα domain of the human editing enzyme ADAR1 bound to left-handed Z-DNA, Science 284 (5421) (1999) 1841–1845.

[125] D.-B. Oh, Y.-G. Kim, A. Rich, Z-DNA-binding proteins can act as potent effectors of gene expression in vivo, Proc. Natl. Acad. Sci. U.S.A. 99 (26) (2002) 16666–16671.

[126] H.-J. Kang, T.V.T. Le, K. Kim, J. Hur, K.K. Kim, H.-J. Park, Novel interaction of the Z-DNA binding domain of human ADAR1 with the oncogenic c-Myc promoter G-Quadruplex, J. Mol. Biol. 426 (14) (2014) 2594–2604.

[127] M. Koeris, L. Funke, J. Shrestha, A. Rich, S. Maas, Modulation of ADAR1 editing activity by Z-RNA in vitro, Nucleic Acids Res. 33 (16) (2005) 5362–5370.

[128] A. Athanasiadis, D. Placido, S. Maas, B.A. Brown Ii, K. Lowenhaupt, A. Rich, The crystal structure of the Zβ domain of the RNA-editing enzyme ADAR1 reveals distinct conserved surfaces among Z-domains, J. Mol. Biol. 351 (3) (2005) 496–507.

[129] J.M.P. Desterro, L.P. Keegan, M. Lafarga, M.T. Berciano, M. O'Connell, M. Carmo-Fonseca, Dynamic association of RNA-editing enzymes with the nucleolus, J. Cell Sci. 116 (9) (2003) 1805–1818.

[130] H. Poulsen, J. Nilsson, C.K. Damgaard, J. Egebjerg, J. Kjems, CRM1 mediates the export of ADAR1 through a nuclear export signal within the Z-DNA binding domain, Mol. Cell. Biol. 21 (22) (2001) 7862–7871.
[131] A. Strehblow, M. Hallegger, M.F. Jantsch, Nucleocytoplasmic distribution of human RNA-editing enzyme ADAR1 is modulated by double-stranded RNA-binding domains, a leucine-rich export signal, and a putative dimerization domain, Mol. Biol. Cell 13 (11) (2002) 3822–3835.
[132] C.R. Eckmann, A. Neunteufl, L. Pfaffstetter, M.F. Jantsch, The human but not the xenopus RNA-editing enzyme ADAR1 has an atypical nuclear localization signal and displays the characteristics of a shuttling protein, Mol. Biol. Cell 12 (7) (2001) 1911–1924.
[133] C.X. George, Z. Gan, Y. Liu, C.E. Samuel, Adenosine deaminases acting on RNA, RNA editing, and interferon action, J. Interferon Cytokine Res. 31 (1) (2010) 99–117.
[134] S.K. Ng, R. Weissbach, G.E. Ronson, A.D.J. Scadden, Proteins that contain a functional Z-DNA-binding domain localize to cytoplasmic stress granules, Nucleic Acids Res. 41 (21) (2013) 9786–9799.
[135] C.L. Sansam, K.S. Wells, R.B. Emeson, Modulation of RNA editing by functional nucleolar sequestration of ADAR2, Proc. Natl. Acad. Sci. U.S.A. 100 (24) (2003) 14018–14023.
[136] T. Melcher, S. Maas, A. Herb, R. Sprengel, P.H. Seeburg, M. Higuchi, A mammalian RNA editing enzyme, Nature 379 (6564) (1996) 460–464.
[137] B.L. Bass, RNA editing by adenosine deaminases that act on RNA, Annu. Rev. Biochem. 71 (1) (2002) 817–846.
[138] L.A. Tonkin, L. Saccomanno, D.P. Morse, T. Brodigan, M. Krause, B.L. Bass, RNA editing by ADARs is important for normal behavior in *Caenorhabditis elegans*, EMBO J. 21 (22) (2002) 6025–6035.
[139] M.J. Palladino, L.P. Keegan, M.A. O'Connell, R.A. Reenan, dADAR, a Drosophila double-stranded RNA-specific adenosine deaminase is highly developmentally regulated and is itself a target for RNA editing, RNA 6 (7) (2000) 1004–1018.
[140] M.A. O'Connell, S. Krause, M. Higuchi, J.J. Hsuan, N.F. Totty, A. Jenny, W. Keller, Cloning of cDNAs encoding mammalian double-stranded RNA-specific adenosine deaminase, Mol. Cell. Biol. 15 (3) (1995) 1389–1397.
[141] L.F. Grice, B.M. Degnan, The origin of the ADAR gene family and animal RNA editing, BMC Evol. Biol. 15 (1) (2015) 4.
[142] N. Mannion, F. Arieti, A. Gallo, L.P. Keegan, M.A. O'Connell, New insights into the biological role of mammalian ADARs; the RNA editing proteins, Biomolecules 5 (4) (2015) 2338–2362.
[143] H.C. Losey, A.J. Ruthenburg, G.L. Verdine, Crystal structure of *Staphylococcus aureus* tRNA adenosine deaminase TadA in complex with RNA, Nat. Struct. Mol. Biol. 13 (2) (2006) 153–159.
[144] L.P. Keegan, A. Leroy, D. Sproul, M.A. O'Connell, Adenosine deaminases acting on RNA (ADARs): RNA-editing enzymes, Genome Biol. 5 (2) (2004) 209.
[145] J.D. Salter, R.P. Bennett, H.C. Smith, The APOBEC protein family: united by structure, divergent in function, Trends Biochem. Sci. 41 (7) (2016) 578–594.
[146] J. Greeve, I. Altkemper, J.H. Dieterich, H. Greten, E. Windler, Apolipoprotein B mRNA editing in 12 different mammalian species: hepatic expression is reflected in low concentrations of apoB-containing plasma lipoproteins, J. Lipid Res. 34 (8) (1993) 1367–1383.
[147] A.P. Gerber, W. Keller, RNA editing by base deamination: more enzymes, more targets, new mysteries, Trends Biochem. Sci. 26 (6) (2001) 376–384.

[148] M. Kuratani, R. Ishii, Y. Bessho, R. Fukunaga, T. Sengoku, M. Shirouzu, S.-i. Sekine, S. Yokoyama, Crystal structure of tRNA adenosine deaminase (TadA) from aquifex aeolicus, J. Biol. Chem. 280 (16) (2005) 16002–16008.

[149] C. Prochnow, R. Bransteitter, M.G. Klein, M.F. Goodman, X.S. Chen, The APOBEC-2 crystal structure and functional implications for the deaminase AID, Nature 445 (7126) (2007) 447–451.

[150] K.J. Phelps, K. Tran, T. Eifler, A.I. Erickson, A.J. Fisher, P.A. Beal, Recognition of duplex RNA by the deaminase domain of the RNA editing enzyme ADAR2, Nucleic Acids Res. 43 (2) (2015) 1123–1132.

[151] J.M. Eggington, T. Greene, B.L. Bass, Predicting sites of ADAR editing in double-stranded RNA, Nat. Commun. 2 (2011) 319.

[152] T. Eifler, S. Pokharel, P.A. Beal, RNA-seq analysis identifies a novel set of editing substrates for human ADAR2 present in *Saccharomyces cerevisiae*, Biochemistry 52 (45) (2013) 7857–7869.

[153] A. Herbert, A. Rich, The role of binding domains for dsRNA and Z-DNA in the in vivo editing of minimal substrates by ADAR1, Proc. Natl. Acad. Sci. U.S.A. 98 (21) (2001) 12132–12137.

[154] H. Poulsen, R. Jorgensen, A. Heding, F.C. Nielsen, B. Bonven, J. Egebjerg, Dimerization of ADAR2 is mediated by the double-stranded RNA binding domain, RNA 12 (7) (2006) 1350–1360.

[155] M.R. Macbeth, A.T. Lingam, B.L. Bass, Evidence for auto-inhibition by the N terminus of hADAR2 and activation by dsRNA binding, RNA 10 (10) (2004) 1563–1571.

[156] Y. Wang, J. Havel, P.A. Beal, A phenotypic screen for functional mutants of human adenosine deaminase acting on RNA 1, ACS Chem. Biol. 10 (11) (2015) 2512–2519.

[157] S. Pokharel, P.A. Beal, High-throughput screening for functional adenosine to inosine RNA editing systems, ACS Chem. Biol. 1 (12) (2006) 761–765.

[158] S. Pokharel, P. Jayalath, O. Maydanovych, R.A. Goodman, S.C. Wang, D.J. Tantillo, P.A. Beal, Matching active site and substrate structures for an RNA editing reaction, J. Am. Chem. Soc. 131 (33) (2009) 11882–11891.

[159] Y. Wang, P.A. Beal, Probing RNA recognition by human ADAR2 using a high-throughput mutagenesis method, Nucleic Acids Res. 44 (20) (2016) 9872–9880.

[160] S.N. Deffit, H.A. Hundley, To edit or not to edit: regulation of ADAR editing specificity and efficiency, Wiley Interdiscip. Rev. RNA 7 (1) (2016) 113–127.

[161] W. Garncarz, A. Tariq, C. Handl, O. Pusch, M.F. Jantsch, A high-throughput screen to identify enhancers of ADAR-mediated RNA-editing, RNA Biol. 10 (2) (2013) 192–204.

[162] R. Marcucci, J. Brindle, S. Paro, A. Casadio, S. Hempel, N. Morrice, A. Bisso, L.P. Keegan, G. Del Sal, M.A. O'Connell, Pin1 and WWP2 regulate GluR2 Q/R site RNA editing by ADAR2 with opposing effects, EMBO J. 30 (20) (2011) 4211–4222.

[163] A. Tariq, W. Garncarz, C. Handl, A. Balik, O. Pusch, M.F. Jantsch, RNA-interacting proteins act as site-specific repressors of ADAR2-mediated RNA editing and fluctuate upon neuronal stimulation, Nucleic Acids Res. 41 (4) (2013) 2581–2593.

[164] D.K. Wilson, F.B. Rudolph, F.A. Quiocho, Atomic structure of adenosine deaminase complexed with a transition-state analog: understanding catalysis and immunodeficiency mutations, Science 252 (5010) (1991) 1278–1284.

[165] L. Betts, S. Xiang, S.A. Short, R. Wolfenden, C.W. Carter Jr., Cytidine deaminase. The 2.3 Å structure of an enzyme: transition-state analog complex, J. Mol. Biol. **235** (1994) 635–656.

[166] B.L. Haudenschild, O. Maydanovych, E.A. Véliz, M.R. Macbeth, B.L. Bass, P.A. Beal, A transition state analogue for an RNA-editing reaction, J. Am. Chem. Soc. 126 (36) (2004) 11213–11219.
[167] L.M. Easterwood, E.A. Véliz, P.A. Beal, Demethylation of 6-O-methylinosine by an RNA-editing adenosine deaminase, J. Am. Chem. Soc. 122 (46) (2000) 11537–11538.
[168] O. Maydanovych, P.A. Beal, C6-substituted analogues of 8-azanebularine: probes of an RNA-editing enzyme active site, Org. Lett. 8 (17) (2006) 3753–3756.
[169] R.A. Mizrahi, K.J. Phelps, A.Y. Ching, P.A. Beal, Nucleoside analog studies indicate mechanistic differences between RNA-editing adenosine deaminases, Nucleic Acids Res. 40 (19) (2012) 9825–9835.
[170] H.Y. Yi-Brunozzi, L.M. Easterwood, G.M. Kamilar, P.A. Beal, Synthetic substrate analogs for the RNA-editing adenosine deaminase ADAR-2, Nucleic Acids Res. 27 (14) (1999) 2912–2917.
[171] E.A. Véliz, L.M. Easterwood, P.A. Beal, Substrate analogues for an RNA-editing adenosine deaminase: mechanistic investigation and inhibitor design, J. Am. Chem. Soc. 125 (36) (2003) 10867–10876.
[172] S.K. Wong, S. Sato, D.W. Lazinski, Substrate recognition by ADAR1 and ADAR2, RNA 7 (6) (2001) 846–858.
[173] A.M. Källman, M. Sahlin, M. Öhman, ADAR2 A→I editing: site selectivity and editing efficiency are separate events, Nucleic Acids Res. 31 (16) (2003) 4874–4881.
[174] A.A. Dillman, M.R. Cookson, D. Galter, ADAR2 affects mRNA coding sequence edits with only modest effects on gene expression or splicing in vivo, RNA Biol. 13 (1) (2016) 15–24.
[175] K.A. Lehmann, B.L. Bass, Double-stranded RNA adenosine deaminases ADAR1 and ADAR2 have overlapping specificities, Biochemistry 39 (42) (2000) 12875–12884.
[176] K.A. Lehmann, B.L. Bass, The importance of internal loops within RNA substrates of ADAR11, J. Mol. Biol. 291 (1) (1999) 1–13.
[177] N.T. Schirle, R.A. Goodman, M. Krishnamurthy, P.A. Beal, Selective inhibition of ADAR2-catalyzed editing of the serotonin 2c receptor pre-mRNA by a helix-threading peptide, Bioorg. Med. Chem. Lett. 8 (21) (2010) 4898–4904.
[178] M. Ensterö, C. Daniel, H. Wahlstedt, F. Major, M. Öhman, Recognition and coupling of A-to-I edited sites are determined by the tertiary structure of the RNA, Nucleic Acids Res. 37 (20) (2009) 6916–6926.
[179] Y. Kawahara, K. Ito, H. Sun, M. Ito, I. Kanazawa, S. Kwak, Regulation of glutamate receptor RNA editing and ADAR mRNA expression in developing human normal and Down's syndrome brains, Dev. Brain Res. 148 (1) (2004) 151–155.
[180] J. Ohlson, J.S. Pedersen, D. Haussler, M. Öhman, Editing modifies the GABAA receptor subunit α3, RNA 13 (5) (2007) 698–703.
[181] E.Y. Rula, A.H. Lagrange, M.M. Jacobs, N. Hu, R.L. Macdonald, R.B. Emeson, Developmental modulation of GABA(A) receptor function by RNA editing, J. Neurosci. 28 (24) (2008) 6196–6201.
[182] A. Barbon, F. Fumagalli, L. Caracciolo, L. Madaschi, E. Lesma, C. Mora, S. Carelli, T.A. Slotkin, G. Racagni, A.M. Di Giulio, et al., Acute spinal cord injury persistently reduces R/G RNA editing of AMPA receptors, J. Neurochem. 114 (2) (2010) 397–407.
[183] M.T. Englander, S.C. Dulawa, P. Bhansali, C. Schmauss, How stress and fluoxetine modulate serotonin 2C receptor pre-mRNA editing, J. Neurosci. 25 (3) (2005) 648–651.
[184] M.A. Huntley, M. Lou, L.D. Goldstein, M. Lawrence, G.J.P. Dijkgraaf, J.S. Kaminker, R. Gentleman, Complex regulation of ADAR-mediated RNA-editing across tissues, BMC Genomics 17 (1) (2016) 61.

[185] P.N.T. Hang, M. Tohda, K. Matsumoto, Developmental changes in expression and self-editing of adenosine deaminase type 2 pre-mRNA and mRNA in rat brain and cultured cortical neurons, Neurosci. Res. 61 (4) (2008) 398–403.

[186] M. Venø, J.B. Bramsen, C. Bendixen, F. Panitz, I. Holm, M. Öhman, J. Kjems, Spatio-temporal regulation of ADAR editing during development in porcine neural tissues, RNA Biol. 9 (8) (2012) 1054–1065.

[187] M.M. Jacobs, R.L. Fogg, R.B. Emeson, G.D. Stanwood, ADAR1 and ADAR2 expression and editing activity during forebrain development, Dev. Neurosci. 31 (3) (2009) 223–237.

[188] J.E.C. Jepson, Y.A. Savva, K.A. Jay, R.A. Reenan, Visualizing adenosine-to-inosine RNA editing in the Drosophila nervous system, Nat. Methods 9 (2) (2012) 189–194.

[189] H. Wahlstedt, C. Daniel, M. Enstero, M. Öhman, Large-scale mRNA sequencing determines global regulation of RNA editing during brain development, Genome Res. 19 (6) (2009) 978–986.

[190] Y. Liu, C.X. George, J.B. Patterson, C.E. Samuel, Functionally distinct double-stranded RNA-binding domains associated with alternative splice site variants of the interferon-inducible double-stranded RNA-specific adenosine deaminase, J. Biol. Chem. 272 (7) (1997) 4419–4428.

[191] A. Gerber, M.A. O'Connell, W. Keller, Two forms of human double-stranded RNA-specific editase 1 (hRED1) generated by the insertion of an Alu cassette, RNA 3 (5) (1997) 453–463.

[192] F. Lai, C.X. Chen, K.C. Carter, K. Nishikura, Editing of glutamate receptor B subunit ion channel RNAs by four alternatively spliced DRADA2 double-stranded RNA adenosine deaminases, Mol. Cell. Biol. 17 (5) (1997) 2413–2424.

[193] Y. Nemlich, E. Greenberg, R. Ortenberg, M.J. Besser, I. Barshack, J. Jacob-Hirsch, E. Jacoby, E. Eyal, L. Rivkin, V.G. Prieto, et al., MicroRNA-mediated loss of ADAR1 in metastatic melanoma promotes tumor growth, J. Clin. Investig. 123 (6) (2013) 2703–2718.

[194] Y. Feng, C.L. Sansam, M. Singh, R.B. Emeson, Altered RNA editing in mice lacking ADAR2 autoregulation, Mol. Cell. Biol. 26 (2) (2006) 480–488.

[195] L.P. Keegan, J. Brindle, A. Gallo, A. Leroy, R.A. Reenan, M.A. O'Connell, Tuning of RNA editing by ADAR is required in Drosophila, EMBO J. 24 (12) (2005) 2183–2193.

[196] Y.A. Savva, J.E.C. Jepson, A. Sahin, A.U. Sugden, J.S. Dorsky, L. Alpert, C. Lawrence, R.A. Reenan, Auto-regulatory RNA editing fine-tunes mRNA re-coding and complex behaviour in Drosophila, Nat. Commun. 3 (2012) 790.

[197] J.V. Olsen, M. Vermeulen, A. Santamaria, C. Kumar, M.L. Miller, L.J. Jensen, F. Gnad, J. Cox, T.S. Jensen, E.A. Nigg, et al., Quantitative phosphoproteomics reveals widespread full phosphorylation site occupancy during mitosis, Sci. Signal. **3** (104) (2010) ra3.

[198] J.M.P. Desterro, L.P. Keegan, E. Jaffray, R.T. Hay, M.A. O'Connell, M. Carmo-Fonseca, SUMO-1 modification alters ADAR1 editing activity, Mol. Biol. Cell 16 (11) (2005) 5115–5126.

[199] S.S. Mahajan, K.H. Thai, K. Chen, E. Ziff, Exposure of neurons to excitotoxic levels of glutamate induces cleavage of the RNA editing enzyme, adenosine deaminase acting on RNA 2, and loss of GLUR2 editing, Neuroscience 189 (2011) 305–315.

[200] A. Gallo, L.P. Keegan, G.M. Ring, M.A. O'Connell, An ADAR that edits transcripts encoding ion channel subunits functions as a dimer, EMBO J. 22 (13) (2003) 3421–3430.

[201] L. Valente, K. Nishikura, RNA binding-independent dimerization of adenosine deaminases acting on RNA and dominant negative effects of nonfunctional subunits on dimer functions, J. Biol. Chem. 282 (22) (2007) 16054–16061.

[202] D.-S.C. Cho, W. Yang, J.T. Lee, R. Shiekhattar, J.M. Murray, K. Nishikura, Requirement of dimerization for RNA editing activity of adenosine deaminases acting on RNA, J. Biol. Chem. 278 (19) (2003) 17093–17102.

[203] M.R. Macbeth, B.L. Bass, Large-scale overexpression and purification of ADARs from Saccharomyces cerevisiae for biophysical and biochemical studies, in: M.G. Jonatha (Ed.), Methods in Enzymology, Elsevier, 2007, pp. 319–331. Chapter 15.

[204] K.A. Chilibeck, T. Wu, C. Liang, M.J. Schellenberg, E.M. Gesner, J.M. Lynch, A.M. MacMillan, FRET analysis of in vivo dimerization by RNA-editing enzymes, J. Biol. Chem. 281 (24) (2006) 16530–16535.

[205] C. Cenci, R. Barzotti, F. Galeano, S. Corbelli, R. Rota, L. Massimi, C. Di Rocco, M.A. O'Connell, A. Gallo, Down-regulation of RNA editing in pediatric astrocytomas: ADAR2 editing activity inhibits cell migration and proliferation, J. Biol. Chem. 283 (11) (2008) 7251–7260.

[206] H. Ohta, M. Fujiwara, Y. Ohshima, T. Ishihara, ADBP-1 regulates an ADAR RNA-editing enzyme to antagonize RNA-interference-mediated gene silencing in *Caenorhabditis elegans*, Genetics 180 (2) (2008) 785–796.

[207] B. Bhogal, J.E. Jepson, Y.A. Savva, A.S.R. Pepper, R.A. Reenan, T.A. Jongens, Modulation of dADAR-dependent RNA editing by the Drosophila fragile X mental retardation protein, Nat. Neurosci. 14 (12) (2011) 1517–1524.

[208] L. Yang, P. Huang, F. Li, L. Zhao, Y. Zhang, S. Li, Z. Gan, A. Lin, W. Li, Y. Liu, c-Jun amino-terminal kinase-1 mediates glucose-responsive upregulation of the RNA editing enzyme ADAR2 in pancreatic beta-cells, PLoS One 7 (11) (2012) e48611.

[209] W.H. Liu, C.H. Chen, K.H. Yeh, C.L. Li, Y.J. Wu, et al., ADAR2-mediated editing of miR-214 and miR-122 precursor and antisense RNA transcripts in liver cancers, PLoS One 8 (12) (2013) e81922.

[210] S.K. Wong, S. Sato, D.W. Lazinski, Elevated activity of the large form of ADAR1 in vivo: very efficient RNA editing occurs in the cytoplasm, RNA 9 (5) (2003) 586–598.

[211] I.X. Wang, E. So, J.L. Devlin, Y. Zhao, M. Wu, V.G. Cheung, ADAR regulates RNA editing, transcript stability, and gene expression, Cell Rep. 5 (3) (2013) 849–860.

[212] M.C. Washburn, B. Kakaradov, B. Sundararaman, E. Wheeler, S. Hoon, G.W. Yeo, H.A. Hundley, The dsRBP and inactive editor ADR-1 utilizes dsRNA binding to regulate A-to-I RNA editing across the *C. elegans* transcriptome, Cell Rep. 6 (4) (2014) 599–607.

[213] M. Higuchi, S. Maas, F.N. Single, J. Hartner, A. Rozov, N. Burnashev, D. Feldmeyer, R. Sprengel, P.H. Seeburg, Point mutation in an AMPA receptor gene rescues lethality in mice deficient in the RNA-editing enzyme ADAR2, Nature 406 (6791) (2000) 78–81.

[214] H. Lomeli, J. Mosbacher, T. Melcher, T. Hoger, T.K. Geiger, H. Monyer, M. Higuchi, A. Bach, P. Seeburg, Control of kinetic properties of AMPA receptor channels by nuclear RNA editing, Science 266 (5191) (1994) 1709–1713.

[215] T. Shimokawa, M.F.-U. Rahman, U. Tostar, E. Sonkoly, M. Ståhle, A. Pivarcsi, R. Palaniswamy, P.G. Zaphiropoulos, RNA editing of the GLI1 transcription factor modulates the output of Hedgehog signaling, RNA Biol. 10 (2) (2013) 321–333.

[216] L. Chen, Y. Li, C.H. Lin, T.H.M. Chan, R.K.K. Chow, Y. Song, M. Liu, Y.-F. Yuan, L. Fu, K.L. Kong, et al., Recoding RNA editing of AZIN1 predisposes to hepatocellular carcinoma, Nat. Med. 19 (2) (2013) 209–216.

[217] C.-L. Li, W.-Z. Yang, Y.-P. Chen, H.S. Yuan, Structural and functional insights into human Tudor-SN, a key component linking RNA interference and editing, Nucleic Acids Res. 36 (11) (2008) 3579–3589.

[218] E. Bratt, M. Öhman, Coordination of editing and splicing of glutamate receptor pre-mRNA, RNA 9 (3) (2003) 309–318.

[219] V.K. Schoft, S. Schopoff, M.F. Jantsch, Regulation of glutamate receptor B pre-mRNA splicing by RNA editing, Nucleic Acids Res. 35 (11) (2007) 3723–3732.

[220] G.E. Parada, R. Munita, C.A. Cerda, K. Gysling, A comprehensive survey of non-canonical splice sites in the human transcriptome, Nucleic Acids Res. 42 (16) (2014) 10564–10578.

[221] O. Solomon, S. Oren, M. Safran, N. Deshet-Unger, P. Akiva, J. Jacob-Hirsch, K. Cesarkas, R. Kabesa, N. Amariglio, R. Unger, et al., Global regulation of alternative splicing by adenosine deaminase acting on RNA (ADAR), RNA 19 (5) (2013) 591–604.

[222] K. Nishikura, Editor meets silencer: crosstalk between RNA editing and RNA interference, Nat. Rev. Mol. Cell Biol. 7 (12) (2006) 919–931.

[223] Y. Kawahara, M. Megraw, E. Kreider, H. Iizasa, L. Valente, A.G. Hatzigeorgiou, K. Nishikura, Frequency and fate of microRNA editing in human brain, Nucleic Acids Res. 36 (16) (2008) 5270–5280.

[224] W. Yang, Q. Wang, K.L. Howell, J.T. Lee, D.-S.C. Cho, J.M. Murray, K. Nishikura, ADAR1 RNA deaminase limits short interfering RNA efficacy in mammalian cells, J. Biol. Chem. 280 (5) (2005) 3946–3953.

[225] H. Iizasa, B.-E. Wulff, N.R. Alla, M. Maragkakis, M. Megraw, A. Hatzigeorgiou, D. Iwakiri, K. Takada, A. Wiedmer, L. Showe, et al., Editing of epstein-barr virus-encoded BART6 microRNAs controls their dicer targeting and consequently affects viral latency, J. Biol. Chem. 285 (43) (2010) 33358–33370.

[226] L. Zhang, C.-S. Yang, X. Varelas, S. Monti, Altered RNA editing in 3′ UTR perturbs microRNA-mediated regulation of oncogenes and tumor-suppressors, Sci. Rep. 6 (2016) 23226.

[227] B. Webster, S. Assil, M. Dreux, Cell-cell sensing of viral infection by plasmacytoid dendritic cells, J. Virol. 90 (22) (2016) 10050–10053.

[228] M.P. Gantier, B.R.G. Williams, The response of mammalian cells to double-stranded RNA, Cytokine Growth Factor Rev. 18 (5–6) (2007) 363–371.

[229] L. Alexopoulou, A.C. Holt, R. Medzhitov, R.A. Flavell, Recognition of double-stranded RNA and activation of NF-[kappa]B by Toll-like receptor 3, Nature 413 (6857) (2001) 732–738.

[230] Q. Wang, M. Miyakoda, W. Yang, J. Khillan, D.L. Stachura, M.J. Weiss, K. Nishikura, Stress-induced apoptosis associated with null mutation of ADAR1 RNA editing deaminase gene, J. Biol. Chem. 279 (6) (2004) 4952–4961.

[231] J.C. Hartner, C.R. Walkley, J. Lu, S.H. Orkin, ADAR1 is essential for the maintenance of hematopoiesis and suppression of interferon signaling, Nat. Immunol. 10 (1) (2009) 109–115.

[232] A.M. Toth, Z. Li, R. Cattaneo, C.E. Samuel, RNA-specific adenosine deaminase ADAR1 suppresses measles virus-induced apoptosis and activation of protein kinase PKR, J. Biol. Chem. 284 (43) (2009) 29350–29356.

[233] Z. Li, K.C. Wolff, C.E. Samuel, RNA adenosine deaminase ADAR1 deficiency leads to increased activation of protein kinase PKR and reduced vesicular stomatitis virus growth following interferon treatment, Virology 396 (2) (2010) 316–322.

[234] N.M. Mannion, S.M. Greenwood, R. Young, S. Cox, J. Brindle, D. Read, C. Nellåker, C. Vesely, C.P. Ponting, P.J. McLaughlin, et al., The RNA-editing enzyme ADAR1 controls innate immune responses to RNA, Cell Rep. **9** (4) (2014) 1482–1494.

[235] B.J. Liddicoat, R. Piskol, A.M. Chalk, G. Ramaswami, M. Higuchi, J.C. Hartner, J.B. Li, P.H. Seeburg, C.R. Walkley, RNA editing by ADAR1 prevents MDA5 sensing of endogenous dsRNA as nonself, Science 349 (6252) (2015) 1115–1120.

[236] T. Kawai, S. Akira, The roles of TLRs, RLRs and NLRs in pathogen recognition ARTICLE, Int. Immunol. 21 (4) (2009) 317–337.
[237] O. Takeuchi, S. Akira, MDA5/RIG-I and virus recognition, Curr. Opin. Immunol. 20 (1) (2008) 17–22.
[238] S. Liu, X. Cai, J. Wu, Q. Cong, X. Chen, T. Li, F. Du, J. Ren, Y.-T. Wu, N.V. Grishin, et al., Phosphorylation of innate immune adaptor proteins MAVS, STING, and TRIF induces IRF3 activation, Science **347** (6227) (2015) aaa2630.
[239] C.E. Samuel, Adenosine deaminases acting on RNA (ADARs) are both antiviral and proviral, Virology 411 (2) (2011) 180–193.
[240] T. Hideyama, T. Yamashita, H. Aizawa, S. Tsuji, A. Kakita, H. Takahashi, S. Kwak, Profound downregulation of the RNA editing enzyme ADAR2 in ALS spinal motor neurons, Neurobiol. Dis. 45 (3) (2012) 1121–1128.
[241] G. Silberberg, D. Lundin, R. Navon, M. Öhman, Deregulation of the A-to-I RNA editing mechanism in psychiatric disorders, Hum. Mol. Genet. 21 (2) (2012) 311–321.
[242] J. Karanović, S. Šviković, M. Pantović, S. Durica, G. Brajušković, A. Damjanović, V. Jovanović, M. Ivković, S. Romac, D. Savić Pavićević, Joint effect of ADARB1 gene, HTR2C gene and stressful life events on suicide attempt risk in patients with major psychiatric disorders, World J. Biol. Psychiatry 16 (4) (2015) 261–271.
[243] S.B. Cassidy, S. Schwartz, J.L. Miller, D.J. Driscoll, Prader-Willi syndrome, Genet. Med. 14 (1) (2012) 10–26.
[244] B. Rogelj, Brain-specific small nucleolar RNAs, J. Mol. Neurosci. 28 (2) (2006) 103–109.
[245] P. Vitali, E. Basyuk, E. Le Meur, E. Bertrand, F. Muscatelli, J. Cavaillé, A. Huttenhofer, ADAR2-mediated editing of RNA substrates in the nucleolus is inhibited by C/D small nucleolar RNAs, J. Cell Biol. 169 (5) (2005) 745–753.
[246] M.V. Morabito, A.I. Abbas, J.L. Hood, R.A. Kesterson, M.M. Jacobs, D.S. Kump, D.L. Hachey, B.L. Roth, R.B. Emeson, Mice with altered serotonin 2C receptor RNA editing display characteristics of Prader–Willi syndrome, Neurobiol. Dis. 39 (2) (2010) 169–180.
[247] G.I. Rice, G.M.A. Forte, M. Szynkiewicz, D.S. Chase, A. Aeby, M.S. Abdel-Hamid, S. Ackroyd, R. Allcock, K.M. Bailey, U. Balottin, et al., Assessment of interferon-related biomarkers in Aicardi-Goutières syndrome associated with mutations in TREX1, RNASEH2A, RNASEH2B, RNASEH2C, SAMHD1, and ADAR: a case-control study, Lancet Neurol. 12 (12) (2013) 1159–1169.
[248] Y.J. Crow, D.S. Chase, J. Lowenstein Schmidt, M. Szynkiewicz, G.M.A. Forte, H.L. Gornall, A. Oojageer, B. Anderson, A. Pizzino, G. Helman, et al., Characterization of human disease phenotypes associated with mutations in TREX1, RNASEH2A, RNASEH2B, RNASEH2C, SAMHD1, ADAR, and IFIH1, Am. J. Med. Genet. A 167 (2) (2015) 296–312.
[249] G.I. Rice, P.R. Kasher, G.M.A. Forte, N.M. Mannion, S.M. Greenwood, M. Szynkiewicz, J.E. Dickerson, S.S. Bhaskar, M. Zampini, T.A. Briggs, et al., Mutations in ADAR1 cause Aicardi-Goutieres syndrome associated with a type I interferon signature, Nat. Genet. 44 (11) (2012) 1243–1248.
[250] A.J. Fisher, P.A. Beal, Effects of Aicardi-Goutières syndrome mutations predicted from ADAR-RNA structures, RNA Biol. 14 (2016) 164–170.
[251] Y. Miyamura, T. Suzuki, M. Kono, K. Inagaki, S. Ito, N. Suzuki, Y. Tomita, Mutations of the RNA-specific adenosine deaminase gene (DSRAD) are involved in dyschromatosis symmetrica hereditaria, Am. J. Hum. Genet. 73 (3) (2003) 693–699.
[252] N. Suzuki, T. Suzuki, K. Inagaki, S. Ito, M. Kono, T. Horikawa, S. Fujiwara, A. Ishiko, K. Matsunaga, Y. Aoyama, et al., Ten novel mutations of the ADAR1 gene

in Japanese patients with dyschromatosis symmetrica hereditaria, J. Invest. Dermatol. 127 (2) (2007) 309–311.
[253] Y.B. Lee, S.B. Lee, S.J. Kim, S.M. Park, H.R. Ko, A frameshift mutation in the ADAR gene in a Korean family with dyschromatosis symmetrica hereditaria, Eur. J. Dermatol. 24 (6) (2014) 693–695.
[254] O. Ansai, Y. Shigehara, A. Ito, R. Abe, Y. Shimomura, A novel splice site mutation in the ADAR gene leading to exon skipping and dyschromatosis symmetrica hereditaria in a Japanese patient, Clin. Exp. Dermatol. 41 (8) (2016) 933–934.
[255] M.L. Lai, L.J. Yang, X.H. Zhu, M. Li, A novel mutation of the DSRAD gene in a Chinese family with dyschromatosis symmetrica hereditaria, Genet. Mol. Res. 11 (2) (2012) 1731–1737.
[256] M. Li, L. Yang, C. Li, C. Jin, M. Lai, G. Zhang, Y. Hu, J. Ji, Z. Yao, Mutational spectrum of the ADAR1 gene in dyschromatosis symmetrica hereditaria, Arch. Dermatol. Res. 302 (6) (2010) 469–476.
[257] M. Kono, F. Matsumoto, Y. Suzuki, M. Suganuma, H. Saitsu, Y. Ito, S. Fujiwara, S. Moriwaki, K. Matsumoto, N. Matsumoto, et al., Dyschromatosis symmetrica hereditaria and Aicardi-Goutières syndrome 6 are phenotypic variants caused by ADAR1 mutations, J. Invest. Dermatol. 136 (4) (2016) 875–878.
[258] S.-W. Han, H.-P. Kim, J.-Y. Shin, E.-G. Jeong, W.-C. Lee, K.Y. Kim, S.Y. Park, D.-W. Lee, J.-K. Won, S.-Y. Jeong, et al., RNA editing in RHOQ promotes invasion potential in colorectal cancer, J. Exp. Med. 211 (4) (2014) 613–621.
[259] D. Dominissini, S. Moshitch-Moshkovitz, N. Amariglio, G. Rechavi, Adenosine-to-inosine RNA editing meets cancer, Carcinogenesis 32 (11) (2011) 1569–1577.
[260] D. Fumagalli, D. Gacquer, F. Rothé, A. Lefort, F. Libert, D. Brown, N. Kheddoumi, A. Shlien, T. Konopka, R. Salgado, et al., Principles governing A-to-I RNA editing in the breast cancer transcriptome, Cell Rep. 13 (2) (2015) 277–289.
[261] L. Han, L. Diao, S. Yu, X. Xu, J. Li, R. Zhang, Y. Yang, H.M.J. Werner, A.K. Eterovic, Y. Yuan, et al., The genomic landscape and clinical relevance of A-to-I RNA editing in human cancers, Cancer Cell 28 (4) (2015) 515–528.
[262] N. Paz-Yaacov, L. Bazak, I. Buchumenski, H.T. Porath, M. Danan-Gotthold, B.A. Knisbacher, E. Eisenberg, E.Y. Levanon, Elevated RNA editing activity is a major contributor to transcriptomic diversity in tumors, Cell Rep. 13 (2) (2015) 267–276.
[263] V. Rayon-Estrada, F.N. Papavasiliou, D. Harjanto, RNA editing dynamically rewrites the cancer code, Trends Cancer 1 (4) (2015) 211–212.

CHAPTER EIGHT

Reading the Epitranscriptome: New Techniques and Perspectives

Lee E. Vandivier*,†, Brian D. Gregory*,†,‡,1
*University of Pennsylvania, Philadelphia, PA, United States
†Cell and Molecular Biology Graduate Program, University of Pennsylvania, Philadelphia, PA, United States
‡Genomics and Computational Biology Graduate Program, University of Pennsylvania, Philadelphia, PA, United States
[1]Corresponding author: e-mail address: bdgregor@sas.upenn.edu

Contents

Abstract

Ribonucleotides can be covalently modified with over 100 known chemical moieties, greatly expanding the potential coding and regulatory repertoire of RNA. Recent advances in applying high-throughput sequencing to the study of RNA modifications have generated transcriptome-wide modification maps and demonstrated that modifications are abundant features of multiple classes of RNAs, including messenger RNAs.

The Enzymes, Volume 41
ISSN 1874-6047
http://dx.doi.org/10.1016/bs.enz.2017.03.004

In turn, new regulatory functions for RNA modifications have been elucidated. Here, we review both targeted and global methods for surveying RNA modification, with a focus on how transcriptome-wide methods have helped expand the understanding of modification-mediated regulation of the transcriptome.

1. INTRODUCTION

RNA chemical modifications can decorate nearly every known class of RNA, across all kingdoms of life and viruses. To date, over 100 classes of posttranscriptional modifications have been characterized [1–4], each of which can alter the chemical properties of their respective nucleotides, leading to altered base pairing and structural conformation and differential association with RNA-binding proteins (RBPs). In turn, modifications can regulate nearly every step of the RNA lifecycle, from transcription [5] and maturation [6,7] to export [8,9], translation [10,11], and stability [12–15]. Thus, establishing robust methods to survey modifications across the transcriptome is a critical component of understanding posttranscriptional regulation.

Until recently, the majority of nucleotide-resolution RNA modifications studies were limited to highly abundant and predominantly noncoding species like transfer RNAs (tRNAs) [16], ribosomal RNAs (rRNAs) [17], and small nuclear RNAs (snRNAs) [18] since their methods of detection required large amounts of highly pure RNA [16,19,20]. As a result, tRNAs are still the most thoroughly characterized of any RNA class, and their modifications remain a gold standard for measuring the true discovery rate of new techniques. However, progress in marrying biochemical techniques with high-throughput sequencing has yielded rapid advances in the understanding of both the form and function of RNA modifications, particularly in messenger RNAs (mRNAs) and long noncoding RNAs (lncRNAs). These mRNA and lncRNA modifications, often referred to as the "epitranscriptome" [21,22], are now known to encompass N^6-methyladenosine (m^6A), N^1-methyladenosine (m^1A), 5-methylcytosine (m^5C), pseudouridine (ψ), and 2′-O-methylation of ribose (2′OMe) (Fig. 1), and likely contain additional modification types that can be detected but not unambiguously defined, such as modified guanosines, not including the 7-methylguanosine (7mG) cap [23,24]. Here, we review the best-characterized transcriptome modifications, though additional types likely exist.

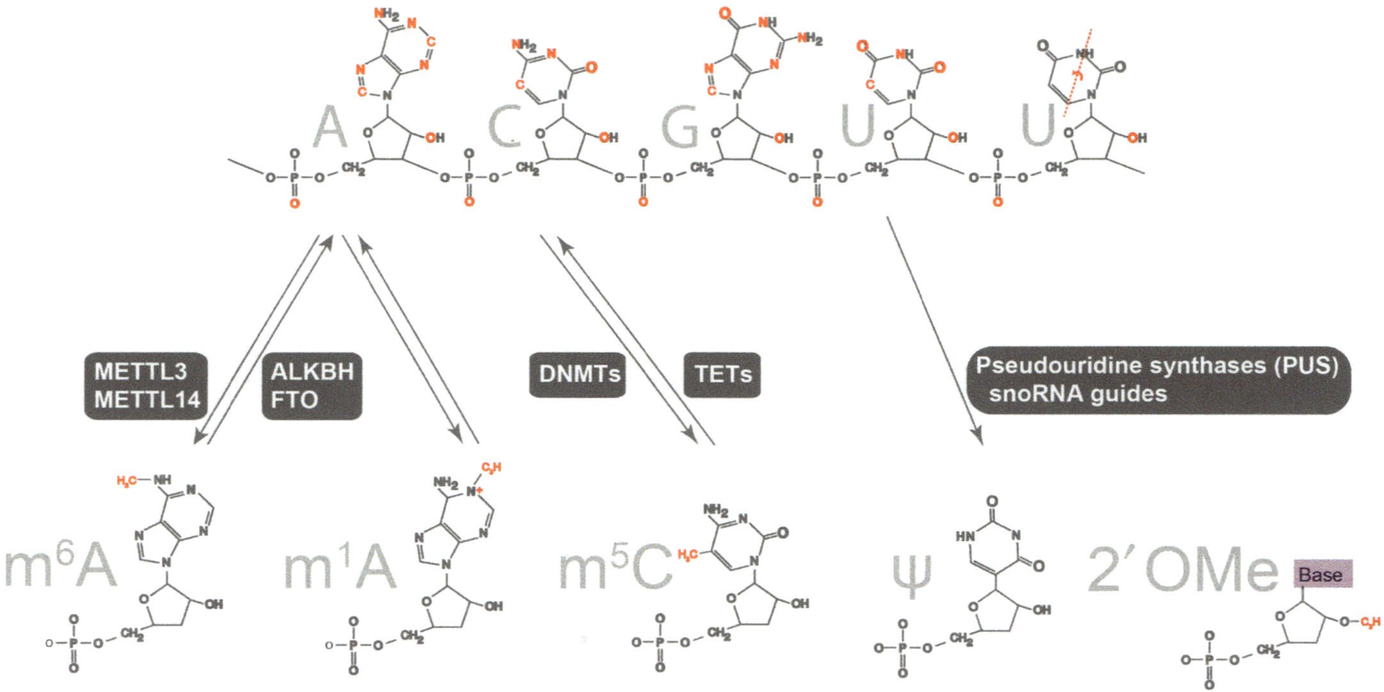

Fig. 1 RNA harbors multiple potential modifications, though only five have been mapped to mRNAs. Unmodified ribonucleotides are shown in the panel above. All atoms corresponding to known sites of modification are labeled in *red*, in addition to uridine's axis of isomerization to form pseudouridine (*top right*). Known mRNA modifications are shown in the panel below and include N^6-methyladenosine (m^6A), N^1-methyladenosine (m^1A), pseudouridine (ψ), 5-methylcytosine (m^5C), and 2′OMe. *Black boxes* between the panels denote writer enzymes known to catalyze the formation of these modifications, and eraser enzymes known to catalyze their removal.

1.1 N^6-Methyladenosine

Of these transcriptome modifications, m^6A (Fig. 1) is the most abundant and well-studied chemical mark and has been reviewed extensively [4,25–30]. Methylation outside the 7mG cap was first detected in mammalian mRNAs through measuring the incorporation of radiolabeled methyl groups from ^{3}H-methylmethionine [31,32] and was later attributed to m^6A specifically through various chromatographic methods such as electrophoresis, thin-layer chromatography, and high-performance liquid chromatography [33,34]. Subsequent studies demonstrated that m^6A is a widespread feature across viruses [35–37] and mRNAs from bacteria [38], actively meiotic yeast [39], and plants like *Arabidopsis* [40], maize [41], wheat [42], and oat [43]. m^6A has likewise been detected in archea [44], though archaeal mRNAs have yet to been assayed. Like DNA methylation, RNA m^6A tends to occur in a specific sequence context [45–47]. Targeted mutation studies and in vitro analysis of methyltransferases indicated a general motif of RRACH (R is either G or A, and H is A, C, or U) that is largely consistent across multiple organisms [38,45,47–50], hinting at broad conservation of the machinery that deposits RNA m^6A. Nonetheless, nonconsensus m^6A is known to exist in rRNA [1,2,51], which could either result from yet to be discovered m^6A writers or from recognition of m^6A secondary structure motifs [51]. Across transcripts, m^6A tends to occur near stop codons, long introns, and 3′ UTRs [21,48,52]. In plants, additional mRNA enrichment is observed at the start codon [48]. In addition, like all modifications m^6A is not mutually exclusive with other chemical marks and has recently been demonstrated to cooccur with 2′OMe (Section 1.5) to form m^6A_m. This newly discovered chemical mark was primarily found in the adenosine proximal to the 7mG cap and was found to enhance mRNA stability by inhibiting binding of DECAPPING PROTEIN 2 (DCP2) [15].

m^6A is also the best example of a complete epitranscriptomic regulatory system, as it possesses known writers (methyltransferase complexes), readers (RBPs), and erasers (demethylases) [30]. The first characterized m^6A writer was the mammalian methyltransferase METTL3 [53,54], which was later shown to function as a heterodimer with its catalytically active paralog METTL14 [13,55,56], alongside cofactors such as the splicing regulator Wilms tumor 1-associated protein (WTAP) [13,55,56] and KIAA1429 [57]. WTAP is required for recruitment of METTL3 and METTL14 to nuclear speckles, which are foci of pre-mRNA splicing and maturation factors [56]. Accordingly, transcriptome-wide studies in mammals have shown its enrichment in long exons [21,49], and alternatively spliced exons and introns [49], consistent with a role in splicing [58]. m^6A has also been shown to be deposited in pre-mRNAs, and loss of methylation has been shown to cause a buildup of nuclear transcripts [59], suggesting m^6A is written in the nucleus. However, loss of the plant homolog of WTAP (FIP37) has been shown to have no significant effects on splicing [60], and no evidence in other organisms has yet to suggest an effect of m^6A on splicing outside of mammals. Nonetheless, METTL3 and WTAP are broadly conserved across yeast [61], plants [40], and nonmammalian animals [62], and disruption of these m^6A writers leads to a broad range of phenotypes such as loss of stem cell differentiation in mammals [63,64], developmental defects in plants [40,60], impaired gametogenesis in *Drosophila melanogaster* [62], and impaired meiosis in yeast [50], indicating that m^6A is an ancient and physiologically relevant RNA regulatory feature.

Currently, m^6A is one of the few posttranscriptional modifications with direct evidence of in vivo reversibility. m^6A erasers include the AlkB family proteins fat mass and obesity-associated protein (FTO) [65] and AlkB homolog 5 (ALKBH5) [9]. FTO is known to catalyze oxidative demethylation, analogous to the TET DNA demethylases [28,30], and FTO has been recently shown to more efficiently demethylate m^6A_m [15]. Intriguingly, the oxidative demethylation intermediates N^6-hydroxymethyladenosine (hm^6A) and N^6-formyladenosine (f^6A) have been observed in vivo in mRNAs [66], though their function remains to be elucidated. Like m^6A writers, both mammalian FTO and ALKBH5 have been shown to function primarily in the nucleus [9,65], and FTO localizes to the nuclear speckles [65]. Consequently, their genetic disruption leads to altered patterns of splicing [6,9] and increased mRNA export [9]. Conversely, loss of METTL3 inhibits RNA export [8].

m^6A is recognized either through "direct readers" like the YTH domain-containing proteins (YTHDs) that contain dedicated aromatic methyladenosine-binding pockets [67,68], or through "indirect readers" such as HNRNPC that directly favor single-stranded RNA, and are recruited via m^6A-induced relaxation of secondary structure [26,69]. m^6A primarily alters base stacking and is not present on the Watson–Crick base pairing edge. Nonetheless, it has been estimated to reduce the stability of duplex RNA by 0.5–1.7 kcal/mol, while in fact stabilizing single-stranded RNA [70]. YTHDs include YTHDC1, which has been shown to enhance exon inclusion via recruitment of SRSF3 and blocking of SRSF10 [7], and can associate with *Xist* m^6A to facilitate X chromosome silencing [5]. YTHDF1 increases translation through recruiting initiation factors [10], and YTHDF2 binding destabilizes mRNAs through localization to processing bodies (P-bodies) [14] and recruitment of the CCR4-NOT deadenylases [12]. Among the indirect readers of m^6A, HNRNPC and HNRNPA2B1 affect splicing [69,71], and HNRNPA2B1 also facilitates miRNA stem–loop processing through recruitment of the microprocessor complex [71].

In addition to its biochemical reversibility, m^6A is also notable in being dynamic across time, development, and stress. For instance, m^6A is a feature of known circadian clock transcripts, and reduction of m^6A via METTL3 knockdown slows nuclear export, leading to a longer circadian period [8]. Additionally, m^6A can be rapidly upregulated in response to stress, leading to enhanced and even cap-independent translation [10,72], which can for instance allow for translation of heat shock proteins in spite of widespread translational repression [72]. m^6A thus illustrates the potential of posttranscriptional chemical modifications to direct rapid and widespread posttranscriptional regulation of mRNAs. Its effects on enhancing export, increasing translation, and promoting RNA decay have led to its proposal as a "fast track" mark that speeds up the RNA lifecycle and reduces the time needed to respond to physiological stimuli [26].

1.2 N^1-Methyladenosine

Like m^6A, m^1A (Fig. 1) is a widespread transcriptome mark that is known to disrupt RNA secondary structure in both coding and noncoding RNAs [73,74]. However, m^1A was only recently shown to be present in mRNAs [75,76], so most of what is known comes from studies of tRNAs and rRNAs. For instance, methyltransferases have been defined for tRNAs [77], though an eraser, ALKBH3, has recently been characterized to direct demethylation both in vitro and in vivo for m^1A [76]. In mRNAs, m^1A has been observed to cluster around the start codon, including noncanonical starts [75,76], as well as the most upstream splice site [75]. Similarly, we have observed strong enrichment of m^1A in ribosome footprints over total polyadenylated RNA, through methods inferring modifications from reverse transcriptase (RT) errors. Thus, it is very likely that m^1A plays a role in facilitating translation, though its precise mechanism has yet to be elucidated.

1.3 Pseudouridine (ψ)

Pseudouridine (Fig. 1) is prevalent in rRNAs [78], tRNAs [16], and snRNAs [79,80] and is the most abundant modification in total cellular RNA [81], which enabled its early detection as the "fifth nucleotide" of RNA [82]. Pseudouridine has been studied primarily through chromatographic methods and primer extension assays, and can be mapped with single-nucleotide resolution. Pseudouridine is formed through isomerization of uracil such that the ribose C1′ binds to uracil C5 instead of N1. In turn, the more accessible N1 is free to form additional hydrogen bonds, leading to pseudouridine stabilizing RNA secondary structure and increasing RNA rigidity [83,84], even though its Watson–Crick edge remains identical. The unique structural properties of pseudouridine contribute to the folding of tRNAs and rRNA, and recent studies indicate that pseudouridylation can also affect mRNA coding potential. For instance, this modification has been found to result in readthrough at stop codons [85,86]. Given its strong conservation in tRNAs and rRNAs, it is not that surprising that pseudouridine is found across all kingdoms of life, including endosymbionts [87].

Pseudouridine writers are termed the pseudouridine synthases (PUSs) and are known to function via two different mechanisms. The RNA-dependent pathway involves the formation of a ribonucleoprotein (RNP) complex containing a PUS, cofactors, and box H/ACA small nucleolar RNAs (snoRNAs). The snoRNAs act as guides that recognize targets with sequence complementarity, thus directing pseudouridylation in a site-specific manner [88,89]. Alternatively, the RNA-independent pathway relies upon direct recognition of targets by PUS complexes [90,91], often at conserved structural or sequence motifs. RNA-independent pseudouridylation of noncoding RNAs tends to occur within paired structures and has been shown to be base pairing-dependent [91,92]. Notably, these paired structures resemble snoRNAs, which are likewise highly paired [93]. In addition, certain PUS enzymes have been shown to target specific sequence motifs [94,95], and recent transcriptome-wide analyses of coding mRNAs have confirmed these motifs [96–98]. Thus, there is evidence that coding and noncoding RNAs share the same pseudouridine writers.

While there is no known mechanism by which pseudouridine is reversed, this chemical mark is still known to be dynamic across development and stress. For instance, inducible pseudouridylation has been observed upon rapamycin treatment [99], heat stress [96,98,100], nutrient deprivation [97,98,100], and oxidative stress [98]. Differential pseudouridylation has also been observed within telomerase RNA in Dyskeratosis Congenita cells [96]. The precise regulatory outcomes of these changes in mRNA pseudouridylation has yet to be clearly defined, though it has been speculated that pseudouridylation stabilizes secondary structures to alter translation efficiency, RNA localization, and RNA stability [96,97].

1.4 5-Methylcytosine and 5-Hydroxymethylcytosine

Early studies using ^{3}H-methylmethionine radiolabeling coupled with various chromatographic methods (analogous to those used to define m^{6}A) demonstrated that m^{5}C (Fig. 1) can mark mRNAs [33], tRNAs [101], and viral RNAs [102]. Nonetheless, this chemical mark has primarily been studied as the characteristic mark of DNA methylation, and has not been extensively characterized in mRNAs until recently [103–105]. As in DNA, RNA m^{5}C can be readily detected at single-nucleotide resolution through bisulfite conversion [100,101]. Additional techniques rely upon antibody pulldown, or upon cytidine analogues that remain bound to their methyltransferases [103,105], and will be covered in Section 3.1.

Known m^{5}C writers were first characterized through their methylation of tRNAs and include the yeast tRNA:m^{5}C methyltransferase (Trm4) [101], its animal homolog NOP2/SUN RNA methyltransferase family member 2 (Nsun2) [106], and the tRNA aspartic acid methyltransferase Dnmt2, which is conserved across plants and animals [107]. Loss of Dnmt2 leads to reduced stress tolerance, in part through leading to an increase in stress-induced tRNA cleavage [108]. Similarly, loss of Nsun2 leads to an increase in aberrant vault RNA cleavage [109], suggesting that m^{5}C may have a general role in protecting RNAs from cleavage. Loss of Nsun2 has also been linked to developmental disability [110] and to impaired male germ cell differentiation [111].

m^{5}C is also known to be reversible, and the oxidative demethylation intermediates 5-hydroxymethylcytosine (hm^{5}C) and 5-formylcytosine (f^{5}C) have been observed in vivo [112–114]. The ten-eleven (TET) family demethylases were previously known to direct DNA demethylation and have been shown to be necessary [112,113] and sufficient [113] to direct the formation of hm^{5}C in RNA, and thus comprise the first set of known RNA m^{5}C erasers. hm^{5}C has not been studied extensively but is known to be enriched among polysome-associated RNA, suggesting a role in facilitating or demarcating active translation [112]. Additionally, loss of Drosophila

Tet (dTet) has been shown to both reduce hm^5C and disrupt brain development [112].

To date, the precise function of RNA m^5C in coding RNAs is still unclear, though transcriptome-wide experiments have shown it to be enriched in the UTRs [104]. Notably, some m^5C marks in introns have been shown to reside in regions with homology to tRNAs, though it is not clear whether these regulate pre-mRNAs directly [103].

1.5 2′-O-Methylation of Ribose

The first studies to characterize mRNA methylation also detected 2′OMe (Fig. 1), which can modify any ribonucleotide (A_m, C_m, G_m, U_m) [31]. 2′OMe was first mapped in rRNAs and is deposited on specific bases through complementary base interactions of box C/D snoRNAs [115]. 2′OMe is also characterized as a plant-specific marker that stabilizes smRNAs and is added by the HEN1 methyltransferase [116,117]. Very little is known about the function of 2′OMe in mRNAs, other than that it has the potential to inhibit adenosine deamination [118] and has recently been found to increase mRNA stability by inhibiting decapping when marking cap-proximal sites of m^6A (m^6A_m) [15]. Given the availability of targeted [119] and high-throughput [120,121] methods for mapping 2′OMe, the function of 2′OMe in the epitranscriptome remains an open but approachable question.

2. TARGETED METHODS FOR DETECTING RNA MODIFICATIONS

Targeted approaches for detecting RNA modifications can be broadly grouped into methods (1) of direct sequencing via nucleotide labeling and chromatography, (2) using mass spectrometry, (3) detecting the stalling and termination of RT, and (4) directly measuring changes in base pairing. Given that nearly all high-throughput techniques for detecting modified ribonucleotides are based upon these existing techniques, we begin by reviewing these targeted techniques for detecting specific modifications.

2.1 Direct Sequencing

The earliest methods for detecting RNA modifications relied upon various one- and two-dimensional chromatographic methods such as high-performance liquid chromatography, electrophoresis, or thin layer chromatography to resolve modified ribonucleotides based upon changes in their migration properties. While these methods lacked nucleotide resolution, they were able to detect global patterns of nucleotide modification (Fig. 2A) and provide some of the earliest evidence of RNA modification [31,82].

Single-nucleotide modification mapping was enabled by applying the chromatographic methods to direct RNA sequencing, in which a purified and homogenous RNA sample is fragmented to produce a "ladder" of fragments, followed by 5′ radiolabeling. These fragments are resolved at base resolution on a gel, followed by separation in an additional dimension through electrophoresis or salt gradients of the terminal radiolabeled nucleotides [16,19,20] (Fig. 2B). The first dimension of electrophoresis thus mapped nucleotide position, while the second dimension determined modification status. While powerful and able to produce high quality data, this method is laborious and requires milligram quantities of highly pure RNA, restricting its use to highly abundant RNAs.

2.2 Direct Sequencing With SCARLET

Interestingly, these mapping techniques have been revived through their combination with techniques to purify less abundant species of RNA, in a technique known as "site-specific cleavage and radioactive-labeling followed by ligation-assisted extraction and thin-layer chromatography" (SCARLET) [51]. This technique involves site-specific cleavage of a target RNA through the use of a DNA probe and RNase H, followed by 5′ radiolabeling and splint ligation to a DNA primer (Fig. 2C). The sample is then digested exhaustively with RNases, leaving only the DNA primer bound to the radiolabeled mononucleotide. This purified ribonucleotide of interest can then be analyzed with methods such as thin layer chromatography [51].

2.3 Direct Sequencing With Mass Spectrometry

Mass spectrometry has become a powerful method for detecting covalent modifications in both proteins and nucleic acids, and new interest has emerged regarding its application to the direct characterization of RNA modifications [123–126]. Unlike chromatographic methods, mass spectrometry can in principle detect any modification that causes a change in mass, and can even be applied to mass-neutral modifications like pseudouridine through treatment with the pseudouridine-specific adduct CMC [127]. Mass spectrometry can also find new modifications in an unbiased manner by defining mass shifts without any a priori knowledge of the modification structure. As such, mass spectrometry was critical in the initial

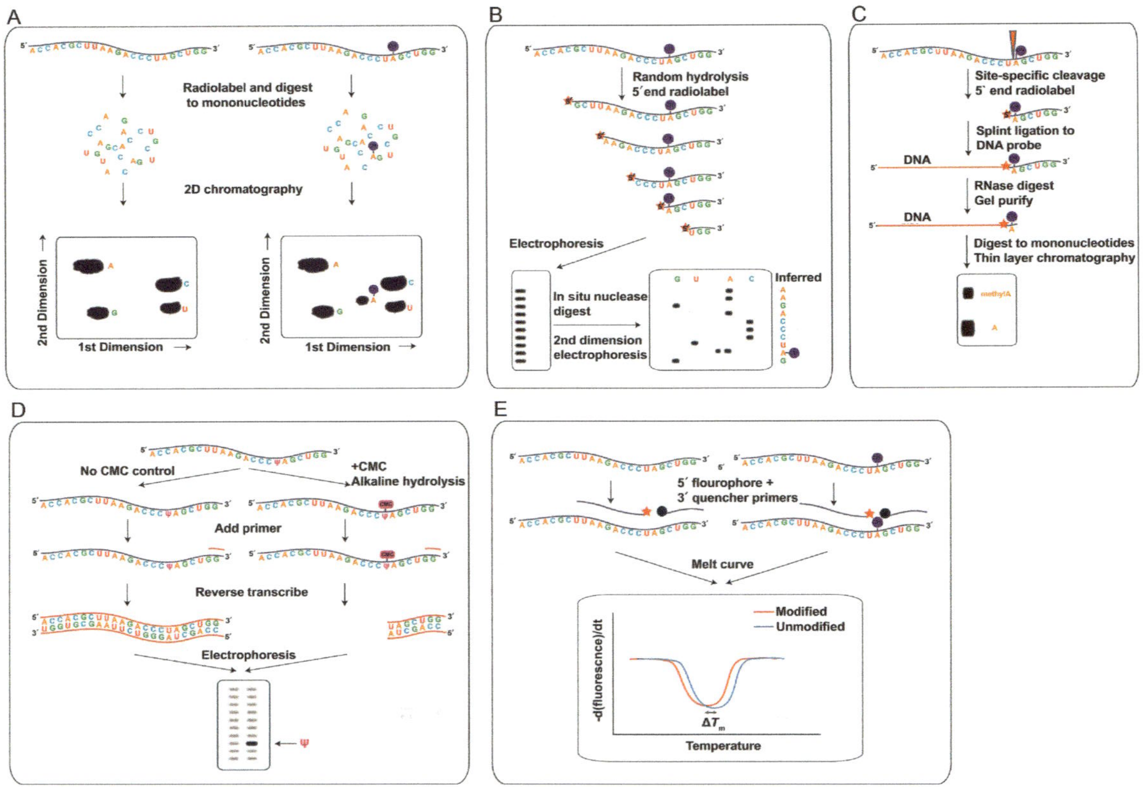

Fig. 2 See legend on next page.

characterization of most known modifications [125] and has now begun to uncover novel modifications such as tRNA geranylation [128], cyclic N^6-threonylcarbamoyladenosine (ct^{6}A) [129], and even new combinations of known and novel modifications [128].

As with proteins, RNA mass spectrometry can be classified into "bottom-up" and "top-down" approaches which analyze fragmented and whole RNA, respectively (analogous to analyzing peptides vs whole proteins) [123]. Bottom-up methods involve prior treatment of RNA with nucleases to produce smaller oligonucleotides, followed by mass spectrometry to look for unexpected m/z peaks that can only be explained via chemical modifications. RNA maps are then constructed "bottom-up" from the aggregate of data from these oligonucleotide fragments [123]. In contrast, "top-down" approaches involve direct mass spectrometry of intact, pure RNAs and are conceptually analogous to classical direct sequencing methods insofar as they provide high quality, albeit labor-intensive data [123,130]. Notably, mass spectrometry has been applied to mapping modifications across large pools of tRNAs [131,132], and in principle could be applied to mapping pools of mRNAs, though not without significant

Fig. 2 Targeted methods for mapping RNA modifications. (A) Bulk mapping of RNA base modification through complete nuclease digestion followed by two-dimensional chromatography. Methylated adenosine migrates differently than other nucleotides, and thus appears as a fifth dot. (B) Direct sequencing of RNA modifications through random hydrolysis to form a "ladder" of fragments differing in size by one nucleotide. Several representative fragments are shown. Fragments are first radiolabeled and separated by gel electrophoresis. Fragments are then digested to single nucleotides and separated again by electrophoresis. Radiolabelled 5′ mononucleotides can be visualized by their different migration patterns. (C) "Site-specific cleavage and radioactive-labeling followed by ligation-assisted extraction and thin-layer chromatography" (SCARLET) [51]. Single nucleotides are radiolabeled by the following sequence of treatments, (1) site-specific cleavage, (2) 5′ end radiolabeling, (3) splint ligation to a DNA probe, and (4) complete RNase digestion to remove all except for the 5′-most labeled ribonucleotide. After purifying the DNA-ligated ribonucleotide of interest, the sample is digested to mononucleotides and analyzed by thin layer chromatography (TLC). Relative levels of nucleotide modification can be inferred by the density of observed bands. (D) A RT-based method involving labeling of pseudouridines with CMC, followed by reverse transcription and observation of stalling sites. Stalls enriched in +CMC over the −CMC control are inferred to be pseudouridylated (*darker band*). (E) High-resolution melting (HRM) analysis of a modified site, using a pair of probes, respectively, 3′ labeled with the quencher BHQ1 and 5′ labeled with the fluorophore FAM [122]. Modifications alter interactions of the transcript and probe, which are observed as a shift in melting temperature.

methodological and computational challenges. Moreover, there exist methods analogous to stable isotope labeling with amino acids in cell culture, in which two samples can be labeled with different nitrogen isotopes, and thus compared on the same mass spectrometry run. For instance, comparative analysis of RNA digests allows analogous sample–sample comparisons by using different isotopes of oxygen [133].

2.4 Reverse Transcriptase-Based Methods

RNA modifications often induce alterations in the ability of RNA to base pair and interact with RBPs. In turn, RNA processing enzymes behave differentially upon encountering a modified base. One such enzyme, RT, has been particularly useful in experimental detection of modifications. When RT encounters a modified ribonucleotide during its first round of DNA synthesis, it stalls, leading either to base misincorporation or termination. Base misincorporation is the basis of global techniques such as high-throughput annotation of modified ribonucleotides (HAMR) [23,24], to be covered in Section 3.3. Transcription termination has enabled the development of primer extension, a technique that has allowed for base-resolution mapping of modifications.

Primer extension involves adding an oligo with homology to an a priori defined region, followed by extension with an RNA-dependent polymerase such as RT (Fig. 2D). Thus, one of the major advantages of primer extension is that it can target transcripts in a heterogeneous pool of RNA, in contrast to approaches like direct sequencing that require large volumes of high purity RNA [134]. Thus, primer extension is an ideal approach for studying less abundant RNAs and helped enable the mapping of mRNA modifications. The first primer extension assays were often coupled to gel-based dideoxynucleotide sequencing, in which aberrant RT termination events could be inferred to result from modified ribonucleotides [135,136]. However, distinguishing modification-induced stalling (signal) from normal variation in RT's movement (noise) is difficult [134].

Later iterations overcame these difficulties through incorporating reagents that specifically target modified ribonucleotides and lead either to large adducts or RNA cleavage. Modifications are inferred from RT stalls that are enriched upon addition of the reagent (Fig. 2D). For instance, pseudouridine is known to preferentially react with the adduct *N*-cyclohexyl-*N*′-(2-morpholinoethyl) carbodiimide metho-*p*-toluenesulfonate (CMC), which forms adducts that induce stalls in RT [137]. CMC initially labels all uridines, while alkaline hydrolysis only removes CMC from unmodified uridines. The development of CMC treatment enabled the rapid survey of pseudouridines across rRNAs from all kingdoms of life [87] and is now the basis for a variety of high-throughput pseudouridine sequencing methods [96–98,138]. Notably, defining modification sites based upon reagent-induced stalling of RT is more reliable than using absolute abundance of stalling and is also able to predict modification identity [134]. Analogously, recent techniques have also relied upon comparing RT stalling in the presence or absence of modification eraser proteins to define sites of m^1A [76].

2.5 High-Resolution Melting

Like RT-based methods, high-resolution melting (HRM) detects modifications through their altered base pairing properties. HRM relies upon detecting changes in the association between oligonucleotide probes and a RNA region of interest, and was first developed as a rapid method for screening DNA polymorphisms [139,140]. It was later applied toward detecting DNA methylation [141,142] and has recently been optimized for assaying covalent RNA modifications [122]. In principle, HRM can be performed with any existing set of PCR probes that cover a putative modification site, but its sensitivity has been increased through the use of fluorescently labeled probes. For instance, one method for HRM analysis in RNA utilizes a pair of probes, one 3′ labeled with the quencher BHQ1 and the other 5′ labeled with the fluorophore FAM [122] (Fig. 2E). The probes are designed such that they hybridize adjacently to their target RNA, bringing the fluorophore in close proximity with the quencher. Putative modifications lie near the boundary of these two primers, and modifications that disrupt primer binding should thus alter the distance between fluorophore and quencher, leading to shifts in the observed melting temperature [122] (Fig. 2E).

3. GLOBAL METHODS FOR DETECTING RNA MODIFICATIONS

Recent advances in merging existing biochemical techniques with high-throughput sequencing have enabled more rapid progress in the study of covalent RNA modifications. Here, we review methods that couple sequencing to (1) antibody pulldown and (2) chemical conversion and chemical adducts. Moreover, we describe (3) in silico methods for detecting modifications from RT errors manifest in high-throughput RNA sequencing data and (4) give a brief introduction to high-throughput single-molecule sequencing, a technique with the potential to read both transcriptomic and epitranscriptomic information simultaneously.

3.1 Antibody-Based Methods

Reliable antibodies have been raised against modified ribonucleotide epitopes, including m^6A, m^1A, pseudouridine, m^5C, and hm^5C. In turn, they have enabled the development of RNA immunoprecipitation (RIP)-based sequencing methods that allow unbiased surveys of these modifications across a transcriptome of interest (Fig. 3A). Antibody-based methods were first used to map m^6A via methyl RIP-seq (MeRIP-seq) [21] and m^6A-seq [49], m^1A via m^1A-seq [75] and m^1A-ID-seq [76] and m^5C [111] and hm^5C via methyl and hydroxymethyl RIP-seq (meRIP-seq and hMeRIP-seq), respectively [112]. Some of these methods involve simple pulldown and sequencing, and are directly analogous to chromatin IP (ChIP) and RBP crosslinking and IP (CLIP), drawing upon similar experimental and computation protocols. Others layer on additional chemical treatment and RT-based detection methods, using antibodies primarily to purify out an informative RNA subpool.

Simple pulldown methods have been used extensively and successfully to map m^6A [21,48,49,57,63,64,72] and more recently hm^5C [112]. For this approach, RNA is first fragmented to a suitable size range, before purification with bead-linked antibodies (Fig. 3A). A related method involves use "suicide inhibitor" nucleotide analogues such as 5-azacytidine, which irreversibly bind its methyltransferase [105]. Modified regions can thus be pulled down through IP of a writer protein. In both methods, RNA fragments are then sequenced and compared to the sequenced reads from a control library composed of input RNA or RNA immunoprecipitated with an antibody isotype control. Sites of RNA modification are inferred from sequencing read peaks in the modification-specific antibody pulldown as compared to the background control, and thus a major drawback of most antibody-based methods is that modification sites cannot be defined with single-nucleotide resolution (Fig. 3A).

Nonetheless, several experimental and computational approaches have been taken to improve resolution. The simplest approach is to infer modification sites from the presence of consensus motifs within the identified sequence read peaks, although this assumption is vulnerable to false negatives of modification at nonconsensus sites, and false positives when multiple

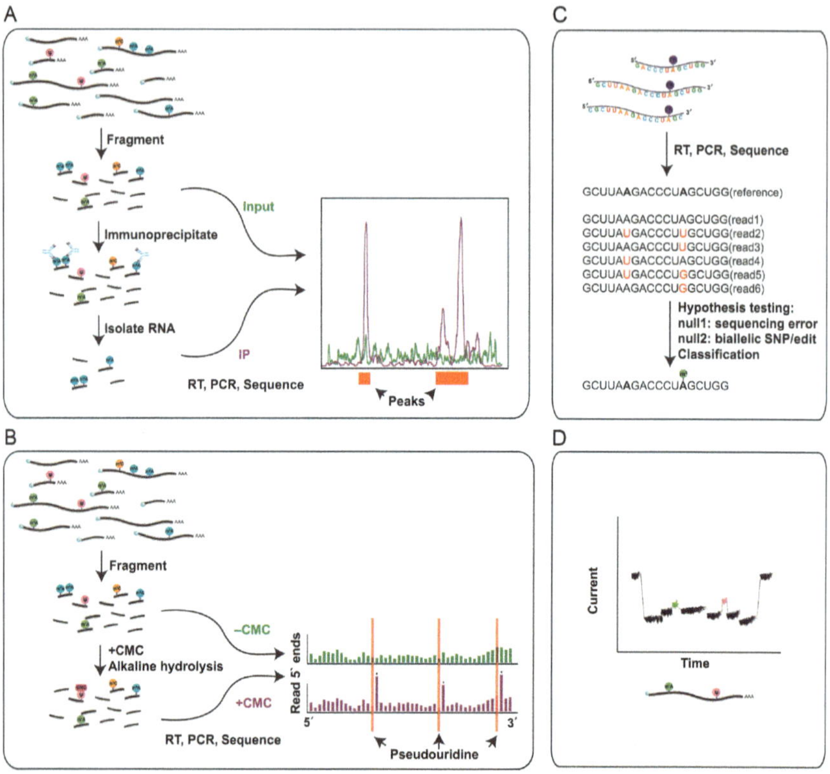

Fig. 3 High-throughput methods for detecting RNA modifications. (A) Antibody-based approaches, which rely upon antibodies recognizing modified ribonucleotide epitopes (shown in figure) or epitopes from modification writer proteins. Immunoprecipitated (IP) fragments are sequenced and compared to an input (shown in figure) or isotype control library, from which peaks of significant IP enrichment are calculated. (B) Reverse transcriptase (RT)-based approaches, such as those involving CMC treatment to induce RT stalling at pseudouridines. CMC-treated and -untreated libraries are sequenced, and significant enrichment of RT stalls indicates the presence of a pseudouridine one base upstream (*red lines*). (C) In silico determination of modifications using the high-throughput annotation of modified ribonucleotides (HAMR) pipeline [23]. Observed mismatches in sequencing data are tabulated, and sites are tested against null hypotheses that (1) sequencing error explains the pattern of mismatches, and (2) biallelic genotypes explain the pattern of mismatches. Sites inferred to be modifications are then classified using machine learning trained on known tRNA modifications. (D) An example of high-throughput direct using the SMRT system (Pacific Biosciences) [143], in which modified ribonucleotides pass more slowly through RT and hence present as spikes in the current (*pink and green lines*).

consensus sequences exist in the same peak. More elaborate inferences might also take into account secondary structure, given for instance that m^6A disrupts structure and tends to occur in single-stranded RNA regions [50]. Another approach is to reduce fragment size in order to reduce peak width and improve density [50], akin to the strategy of treating ChIP samples with exonucleases in the ChIP-exo approach [144]. Alternatively, a more direct approach incorporates crosslinking into the RIP protocol in order to define modification sites based upon crosslinking-induced mutations [145]. This technique was first applied to m^6A mapping and adapts the method called cross-linking and immunoprecipitation sequencing (CLIP-seq) that was first developed to map RBP–RNA interactions [146] and is thus called m^6A individual nucleotide CLIP [145].

Another approach that bypasses this resolution limit is to couple antibody pulldown to chemical modification and assays of RT stalling. For instance, both m^1A-seq [75] and m^1A-ID-seq [76] first utilize anti-m^1A antibodies to pull down methylated RNA fragments. Unlike m^6A, m^1A affects the Watson–Crick base pairing edge and causes RT stalling, which can be used to infer the location of these modifications. To unambiguously define m^1A-induced stalling events, both methods involve comparison to an m^1A-depleted control library, prepared either through in vitro addition of demethylases (m^1A-ID-seq) [76] or through inducing Dimroth rearrangements in which m^1A isomerizes to m^6A and thus no longer blocks RT [75].

Like most high-throughput techniques, it is important to note that the set of called modification sites from these techniques is also dependent upon downstream computational methods. This has been particularly well illustrated in a systematic comparison of meRIP-seq and m^6A-seq, which applied algorithms from one study to another and showed systematic changes in the number of called m^6A sites [22]. For instance, rarefaction analyses indicated that both techniques were nowhere close to saturation for called peaks, indicating there is a strong dependence in peak calling on read depth [22]. Overlap between datasets could likewise be greatly improved simply by filtering based on read depth [22]. Thus, great care should be taken when comparing peaks called between two conditions to ensure that differences are not simply dependent on technical differences such as read coverage, or when comparing data generated with different peak-calling algorithms.

3.2 Chemical-Based Methods

Combining existing compounds that specifically target or exclude modified ribonucleotides with high-throughput sequencing has yielded powerful, single-nucleotide resolution techniques for determining the location of modification sites (Fig. 3). For instance, bisulfite conversion has been used extensively in mapping DNA epigenetic m^5C and has recently been applied to mapping the same modification in RNA [103,104]. In bisulfite sequencing, unmodified cytosines are converted to inosine, while m^5C is unchanged. Thus, every read gives information regarding the modification status of its cytidines, allowing global and quantitative detection of m^5C. However, resulting reads are also less complex, and global mapability is reduced, leading to potential false negatives.

Other chemical compounds that specifically target modified ribonucleotides have been used with success, particularly in mapping pseudouridine. CMC treatment has been used to develop at least four different protocols for global detection of pseudouridine sites, including Pseudo-Seq [97], ψ-seq [96], PSI-seq [138], and CeU-seq [98]. RT stalling and overall read coverage are then compared in the presence or absence of CMC. As with m^6A sequencing methods, the major difference between these methods is in how data are postprocessed, and how single-nucleotide peaks are defined. Similarly, systematic comparison between these methods shows relatively different results, in part depending on the type of peak-calling methods used [147]. For instance, CeU-seq involves calculation of a ratio of 5′ sequencing read end (i.e., RT stall sites) coverage to total read coverage [98], while Pseudo-seq normalizes 5′ sequencing read ends to all 5′ read ends in a surrounding window [97].

3.3 In Silico Methods

Even in the absence of additional chemical adducts, chemical modifications that lie at the Watson–Crick base pairing edge will interfere with base pairing and alter the behavior of RNA-dependent polymerases such as RT. Given that most high-throughput RNA sequencing methods rely upon RT for sequencing library preparation, it follows that the presence of modified ribonucleotides will lead to apparent mismatches from the expected sequence. In fact, this was observed when comparing mismatches in small RNA sequencing data to known tRNA-modified bases [148]. This logic was applied to studying whole transcriptome modifications through HAMR, a novel in silico method for retrospective detection of RNA modifications from any RNA sequencing dataset that does not involve mutation-induced techniques such as crosslinking [23] (Fig. 3). This technique likewise recapitulated existing tRNA modifications and found characteristic mismatch profiles for different modification types, allowing nearest neighbor-based prediction of novel modification identity [23].

Like patterns of RT stalling, patterns of mismatch can be quite messy and lead to artifacts if not properly controlled. HAMR for instance requires multiple steps to ensure that a set of observed mismatches is not due to sequencing error, alignment algorithm error, or single-nucleotide polymorphisms. To account for sequencing error, only reads with high quality score (less than 1/1000 probability of sequencing error) are considered, and bases are only retained if they have significantly more mismatches than expected by sequencing error alone (binomial test). Remaining bases are then tested to ensure that no biallelic genotype can explain the observed pattern of mismatches (ensemble of binomial tests), ruling out RNA editing or polymorphism [23]. As a result, HAMR is limited to diploid and haploid organisms. Moreover, this relatively high bar for modification calling results in a method with low false-positive but high false-negative rates, and HAMR is consistently far from saturation (full census of the genome) even at very high read coverage, a problem it shares with other methods like m6A IP. To address this, we normalize total HAMR modifications to the total number of bases with sufficient read coverage for HAMR to have the statistical power to call a modification site, thereby allowing us to determine what proportion of these "accessible bases" are modified [24].

Nonetheless, HAMR has the advantage of being able to probe modifications retrospectively and can be readily applied to existing data and in meta-analyses. Moreover, it can be applied to specialized library types (such as global mapping of uncapped transcripts [149,150] to survey uncapped, degrading RNAs) that would not normally be amenable to IP or chemical treatment. This has enabled novel observations such as the strong enrichment of modifications in actively degrading transcripts [24]. Moreover, HAMR can survey multiple modification subtypes simultaneously, so long as they affect the Watson–Crick base pairing edge. For instance, it is the only high-throughput technique that can currently detect modified guanosines in the body of mRNAs [23,24].

3.4 High-Throughput Direct Sequencing

Emerging single-molecule sequencing techniques has the potential to survey both sequence and modification data at the same time, providing unbiased views of both the transcriptome and epitranscriptome and allowing direct quantitation of modification abundance unlike most current high-throughput techniques. Current technologies are based upon linking protein enzymes or pore structures to a nanoscale matrix, through which the passage of nucleic acids can be visualized optically or via electric current (Fig. 3D). For instance, Pacific Biosciences' single-molecule, real-time (SMRT) platform links enzymes such as DNA or RT to optical structures called zero-mode waveguides, allowing real-time tracking of the enzyme's incorporation of fluorescently labeled nucleotides [143]. Alternatively, Oxford Nanopore Technologies approach avoids sequencing by synthesis and directly measures the current induced by movement of nucleic acid through a protein pore.

While able to sequence single molecules, these methods are still very prone to noise and sequencing error (approximately 10% compared to <0.1% for Illumina Hi-Seq) and have yet to fully mature [151]. Nonetheless, single-molecule sequencing has successfully been applied to detecting modifications in both DNA and RNA. For instance, SMRT sequencing with DNA polymerase detected m^5C and hm^5C in DNA [152–154], and SMRT sequencing with RT detected m^6A in RNA [22,155]. Moreover, highly structured tRNAs have been successfully applied to nanopores [156], suggesting that single-molecule sequencing will likewise be applicable to highly structured noncoding RNAs. While single-molecule sequencing is still in its infancy, if successfully developed, it has the potential to revolutionize the study of RNA modifications.

4. PERSPECTIVES

Here, we highlight the strengths and pitfalls and high-throughput detection of modified ribonucleotides, and suggest ways to improve their sensitivity, reduce artifacts, and increase spatial and temporal resolution.

4.1 New Insights From High-Throughput Techniques

Coupling existing biochemical techniques to high-throughput sequencing has enabled massively parallel analysis of RNA modification sites and has in particular led to rapid advances in our understanding of the modifications marking mRNAs and long noncoding RNAs. For instance, global patterns emerged such as enrichment of m^6A around the stop codon and long introns, hinting at roles for this modification in translation and splicing. In turn, more detailed mechanistic studies uncovered m^6A reader proteins that directly interact with splicing [7] and translation machinery [10]. Similarly, m^6A in plants and m^1A in animals have been shown to mark the start codon [48,52,76], though their respective mechanisms of action have yet to be uncovered. High-throughput techniques have thus proven to be invaluable methods for generating testable hypotheses as to the function of previous uncharacterized RNA modifications.

Additionally, the relative ease of replicating sequencing-based approaches across different treatment conditions has spurred numerous studies into the dynamics of modification across time, development, and stress. For instance, m^6A has been shown to specifically mark transcripts during meiosis in yeast [39] and play a role in marking circadian transcripts in animals [8]. m^6A has likewise been shown to be upregulated upon stress to increase translation [10,72], and similar stress experiments showed that m^1A [75,76] and pseudouridine [97,98] are dynamic across various types of stress conditions such as heat shock and starvation.

4.2 Avoiding Artifacts and Improving Techniques

Most current high-throughput sequencing technologies rely upon reverse transcription and PCR amplification, and are thus subject to biases inherent in these approaches [157]. For instance, certain loci may be systematically over- or underrepresented based on G/C content, homopolymer runs, or secondary structure. These biases can result in read duplications that result from PCR as opposed to actual biological abundance, and as such methods for detecting modifications often choose to ignore duplicate reads, at the expense of potentially losing reliable data [147]. One proposed solution is labeled molecular barcoding, which incorporates random barcodes at the ends of every single RNA fragment that make up a high-throughput sequencing library. Identical reads with nonidentical barcodes can thus be inferred to derive from different RNA molecules as opposed to PCR artifacts [147]. Direct RNA sequencing with nanopore platforms or mass spectrometry may also soon become a viable means to bypass amplification artifacts.

Sequencing artifacts are likewise of concern, in particular for in silico methods that infer modifications from sequencing errors alone. For instance, untemplated nucleotide addition during adaptor ligation could present as false mismatches in sequencing data, and as such we recommend ignoring the terminal read nucleotides when running the HAMR pipeline [23]. Reads with splice junctions near their termini are likewise a potential source of error, as introducing terminal mismatches may produce a higher quality score by introducing a long gap corresponding to an intron. As such, we also recommend clipping strings of terminal mismatches from reads before HAMR analysis. Another potential method to increase the proportion of true mismatches involves improved methods of sequencing such as circular sequencing [158], which reduce the overall error rate in high-throughput RNA sequencing.

Another major problem with most current high-throughput modification detection techniques is their lack of stoichiometric quantitation, that is to say what proportion of transcripts are modified at a specific base. Modification sites are called from an aggregate of read data, so a low frequency modification on an abundant transcript may thus have the same probability of detection as a high frequency modification on a rare transcript. Modifications may not mark all copies of a transcript equally, and likewise apparent clusters of modifications could in fact be ensembles of singly modified transcripts. Establishing methods to reliably quantify stoichiometry are thus integral to fully understanding modification function, though currently only bisulfite conversion is able to achieve this. Single-molecule methods also have the potential to map modification stoichiometry and cooccurrence on each transcript.

Additional advances in the understanding of RNA modification could come through the development of in vivo visualization techniques. For instance, fluorescent in situ hybridization has been applied to distinguish single-nucleotide polymorphisms in DNA [159], and similar techniques could be applied to distinguish modified from unmodified ribonucleotides, assuming the modification alters base-pairing. In turn, the dynamics of RNA modification across time and space could be tracked in single cells using such an approach.

5. CONCLUDING REMARKS

The advent of massively parallel methods for mapping modified ribonucleotides, in conjunction with highly accurate techniques for validating modification status at single bases, has renewed interested in the study of RNA modification and enabled rapid advances in the understanding of how these chemical marks regulate the RNA lifecycle. Applying deep sequencing to the study of modifications has also helped to demonstrate that modifications are dynamic in that they vary across time and development, are induced by stress and developmental cues, and can even be actively erased. In turn, the "epitranscriptome" is now recognized as an integral carrier of information across all kingdoms and life and viruses. Going forward, improved high-throughput techniques will help increase spatial resolution, provide stoichiometric data, and potentially detect novel mRNA modifications transcriptome-wide.

REFERENCES

[1] M.A. Machnicka, et al., MODOMICS: a database of RNA modification pathways—2013 update, Nucleic Acids Res. 41 (Suppl. 1) (2013) D262–D267.

[2] S. Dunin-Horkawicz, et al., MODOMICS: a database of RNA modification pathways, Nucleic Acids Res. 34 (Suppl. 1) (2006) D145–D149.

[3] P.A. Limbach, P.F. Crain, J.A. McCloskey, Summary: the modified nucleosides of RNA, Nucleic Acids Res. 22 (12) (1994) 2183–2196.

[4] W.A. Cantara, et al., The RNA modification database, RNAMDB: 2011 update, Nucleic Acids Res. 39 (Suppl. 1) (2011) D195–D201.

[5] D.P. Patil, et al., m6A RNA methylation promotes XIST-mediated transcriptional repression, Nature 537 (7620) (2016) 369–373.

[6] X. Zhao, et al., FTO-dependent demethylation of N6-methyladenosine regulates mRNA splicing and is required for adipogenesis, Cell Res. 24 (12) (2014) 1403–1419.

[7] W. Xiao, et al., Nuclear m6A reader YTHDC1 regulates mRNA splicing, Mol. Cell 61 (4) (2016) 507–519.

[8] J.-M. Fustin, et al., RNA-methylation-dependent RNA processing controls the speed of the circadian clock, Cell 155 (4) (2013) 793–806.

[9] G. Zheng, et al., ALKBH5 is a mammalian RNA demethylase that impacts RNA metabolism and mouse fertility, Mol. Cell 49 (1) (2013) 18–29.

[10] X. Wang, et al., N6-methyladenosine modulates messenger RNA translation efficiency, Cell 161 (6) (2015) 1388–1399.

[11] J. Choi, et al., N6-methyladenosine in mRNA disrupts tRNA selection and translation-elongation dynamics, Nat. Struct. Mol. Biol. 23 (2) (2016) 110–115.

[12] H. Du, et al., YTHDF2 destabilizes m6A-containing RNA through direct recruitment of the CCR4–NOT deadenylase complex, Nat. Commun. 7 (2016) 12626.

[13] Y. Wang, et al., N6-methyladenosine modification destabilizes developmental regulators in embryonic stem cells, Nat. Cell Biol. 16 (2) (2014) 191–198.

[14] X. Wang, et al., N6-methyladenosine-dependent regulation of messenger RNA stability, Nature 505 (7481) (2014) 117–120.

[15] J. Mauer, et al., Reversible methylation of m6Am in the 5′ cap controls mRNA stability, Nature 541 (7637) (2017) 371–375.

[16] M. Sprinzl, K.S. Vassilenko, Compilation of tRNA sequences and sequences of tRNA genes, Nucleic Acids Res. 33 (Suppl. 1) (2005) D139–D140.

[17] J.A. McCloskey, J. Rozenski, The small subunit rRNA modification database, Nucleic Acids Res. 33 (Suppl. 1) (2005) D135–D138.

[18] S. Massenet, A. Mougin, C. Branlant, Posttranscriptional modifications in the U small nuclear RNAs, in: H. Grosjean, R. Benne (Eds.), Modification and Editing of RNA, ASM Press, Washington, DC, 1998, pp. 201–227.

[19] Y. Tanaka, T.A. Dyer, G.G. Brownlee, An improved direct RNA sequence method; its application to Vida faba 5.8S ribosomal RNA, Nucleic Acids Res. 8 (6) (1980) 1259–1272.

[20] R.C. Gupta, K. Randerath, Rapid print-readout technique for sequencing of RNA's containing modified nucleotides, Nucleic Acids Res. 6 (11) (1979) 3443–3458.

[21] K.D. Meyer, et al., Comprehensive analysis of mRNA methylation reveals enrichment in 3′ UTRs and near stop codons, Cell 149 (7) (2012) 1635–1646.
[22] Y. Saletore, et al., The birth of the Epitranscriptome: deciphering the function of RNA modifications, Genome Biol. 13 (2012) 175.
[23] P. Ryvkin, et al., HAMR: high-throughput annotation of modified ribonucleotides, RNA 19 (12) (2013) 1684–1692.
[24] L.E. Vandivier, et al., Chemical modifications mark alternatively spliced and uncapped messenger RNAs in Arabidopsis, Plant Cell 27 (11) (2015) 3024–3037.
[25] S. Schwartz, Cracking the epitranscriptome, RNA 22 (2) (2016) 169–174.
[26] B.S. Zhao, I.A. Roundtree, C. He, Post-transcriptional gene regulation by mRNA modifications, Nat. Rev. Mol. Cell Biol. 18 (1) (2017) 31–42.
[27] I.A. Roundtree, C. He, RNA epigenetics—chemical messages for posttranscriptional gene regulation, Curr. Opin. Chem. Biol. 30 (2016) 46–51.
[28] G. Jia, Y. Fu, C. He, Reversible RNA adenosine methylation in biological regulation, Trends Genet. 29 (2) (2013) 108–115.
[29] K. Chen, B.S. Zhao, C. He, Nucleic acid modifications in regulation of gene expression, Cell Chem. Biol. 23 (1) (2016) 74–85.
[30] Y. Fu, D. Dominissini, G. Rechavi, C. He, Gene expression regulation mediated through reversible m6A RNA methylation, Nat. Rev. Genet. 15 (5) (2014) 293–306.
[31] R. Desrosiers, K. Friderici, F. Rottman, Identification of methylated nucleosides in messenger RNA from Novikoff hepatoma cells, Proc. Natl. Acad. Sci. U.S.A. 71 (10) (1974) 3971–3975.
[32] R.P. Perry, D.E. Kelley, Existence of methylated messenger RNA in mouse L cells, Cell 1 (1) (1974) 37–42.
[33] D.T. Dubin, R.H. Taylor, The methylation state of poly A-containing-messenger RNA from cultured hamster cells, Nucleic Acids Res. 2 (10) (1975) 1653–1668.
[34] R.P. Perry, D.E. Kelley, K. Friderici, F. Rottman, The methylated constituents of L cell messenger RNA: evidence for an unusual cluster at the 5′ terminus, Cell 4 (4) (1975) 387–394.
[35] D. Canaani, C. Kahana, S. Lavi, Y. Groner, Identification and mapping of N6-methyladenosine containing sequences in simian virus 40 RNA, Nucleic Acids Res. 6 (8) (1979) 2879–2899.
[36] K. Beemon, J. Keith, Localization of N6-methyladenosine in the Rous sarcoma virus genome, J. Mol. Biol. 113 (1) (1977) 165–179.
[37] Y. Furuichi, A.J. Shatkin, E. Stavnezer, J.M. Bishop, Blocked, methylated 5′-terminal sequence in avian sarcoma virus RNA, Nature 257 (5527) (1975) 618–620.
[38] X. Deng, et al., Widespread occurrence of N6-methyladenosine in bacterial mRNA, Nucleic Acids Res. 43 (13) (2015) 6557–6567.
[39] Z. Bodi, J.D. Button, D. Grierson, R.G. Fray, Yeast targets for mRNA methylation, Nucleic Acids Res. (2010). pp. gkq266.
[40] S. Zhong, et al., MTA is an Arabidopsis messenger RNA adenosine methylase and interacts with a homolog of a sex-specific splicing factor, Plant Cell 20 (5) (2008) 1278–1288.
[41] J.L. Nichols, "Cap" structures in maize poly(A)-containing RNA, Biochim. Biophys. Acta 563 (2) (1979) 490–495.
[42] T.D. Kennedy, B.G. Lane, Wheat embryo ribonucleates. XIII. Methyl-substituted nucleoside constituents and 5′-terminal dinucleotide sequences in bulk poly(AR)-rich RNA from imbibing wheat embryos, Can. J. Biochem. 57 (6) (1979) 927–931.
[43] R.A. Haugland, M.G. Cline, Post-transcriptional modifications of oat coleoptile ribonucleic acids, Eur. J. Biochem. 104 (1) (1980) 271–277.

[44] C.G. Edmonds, et al., Posttranscriptional modification of tRNA in thermophilic archaea (Archaebacteria), J. Bacteriol. 173 (10) (1991) 3138–3148.
[45] T. Csepany, A. Lin, C.J. Baldick, K. Beemon, Sequence specificity of mRNA N6-adenosine methyltransferase, J. Biol. Chem. 265 (33) (1990) 20117–20122.
[46] P. Narayan, R.L. Ludwiczak, E.C. Goodwin, F.M. Rottman, Context effects on N6-adenosine methylation sites in prolactin mRNA, Nucleic Acids Res. 22 (3) (1994) 419–426.
[47] S.E. Kane, K. Beemon, Precise localization of m6A in Rous sarcoma virus RNA reveals clustering of methylation sites: implications for RNA processing, Mol. Cell. Biol. 5 (9) (1985) 2298–2306.
[48] G.-Z. Luo, et al., Unique features of the m6A methylome in Arabidopsis thaliana, Nat. Commun. 5 (2014) 5630.
[49] D. Dominissini, et al., Topology of the human and mouse m6A RNA methylomes revealed by m6A-seq, Nature 485 (7397) (2012) 201–206.
[50] S. Schwartz, et al., High-resolution mapping reveals a conserved, widespread, dynamic mRNA methylation program in yeast meiosis, Cell 155 (6) (2013) 1409–1421.
[51] N. Liu, et al., Probing N6-methyladenosine RNA modification status at single nucleotide resolution in mRNA and long noncoding RNA, RNA 19 (12) (2013) 1848–1856.
[52] D. Dominissini, S. Moshitch-Moshkovitz, M. Salmon-Divon, N. Amariglio, G. Rechavi, Transcriptome-wide mapping of N6-methyladenosine by m6A-seq based on immunocapturing and massively parallel sequencing, Nat. Protoc. 8 (1) (2013) 176–189.
[53] J.A. Bokar, M.E. Shambaugh, D. Polayes, A.G. Matera, F.M. Rottman, Purification and cDNA cloning of the AdoMet-binding subunit of the human mRNA (N6-adenosine)-methyltransferase, RNA 3 (11) (1997) 1233–1247.
[54] J.A. Bokar, M.E. Rath-Shambaugh, R. Ludwiczak, P. Narayan, F. Rottman, Characterization and partial purification of mRNA N6-adenosine methyltransferase from HeLa cell nuclei. Internal mRNA methylation requires a multisubunit complex, J. Biol. Chem. 269 (26) (1994) 17697–17704.
[55] J. Liu, et al., A METTL3–METTL14 complex mediates mammalian nuclear RNA N6-adenosine methylation, Nat. Chem. Biol. 10 (2) (2014) 93–95.
[56] X.-L. Ping, et al., Mammalian WTAP is a regulatory subunit of the RNA N6-methyladenosine methyltransferase, Cell Res. 24 (2) (2014) 177–189.
[57] S. Schwartz, et al., Perturbation of m6A writers reveals two distinct classes of mRNA methylation at internal and 5′ sites, Cell Rep. 8 (1) (2014) 284–296.
[58] Y. Yang, et al., Dynamic m6A modification and its emerging regulatory role in mRNA splicing, Sci. Bull. 60 (1) (2015) 21–32.
[59] S.M. Carroll, P. Narayan, F.M. Rottman, N6-methyladenosine residues in an intron-specific region of prolactin pre-mRNA, Mol. Cell. Biol. 10 (9) (1990) 4456–4465.
[60] L. Shen, et al., N6-methyladenosine RNA modification regulates shoot stem cell fate in Arabidopsis, Dev. Cell 38 (2) (2016) 186–200.
[61] S.D. Agarwala, H.G. Blitzblau, A. Hochwagen, G.R. Fink, RNA methylation by the MIS complex regulates a cell fate decision in yeast, PLoS Genet. 8 (6) (2012). e1002732.
[62] C.F. Hongay, T.L. Orr-Weaver, Drosophila inducer of MEiosis 4 (IME4) is required for Notch signaling during oogenesis, Proc. Natl. Acad. Sci. U.S.A. 108 (36) (2011) 14855–14860.
[63] P.J. Batista, et al., m6A RNA modification controls cell fate transition in mammalian embryonic stem cells, Cell Stem Cell 15 (6) (2014) 707–719.
[64] S. Geula, et al., m6A mRNA methylation facilitates resolution of naïve pluripotency toward differentiation, Science 347 (6225) (2015) 1002–1006.

[65] G. Jia, et al., N6-methyladenosine in nuclear RNA is a major substrate of the obesity-associated FTO, Nat. Chem. Biol. 7 (12) (2011) 885–887.
[66] Y. Fu, et al., FTO-mediated formation of N6-hydroxymethyladenosine and N6-formyladenosine in mammalian RNA, Nat. Commun. 4 (2013) 1798.
[67] C. Xu, et al., Structural basis for selective binding of m6A RNA by the YTHDC1 YTH domain, Nat. Chem. Biol. 10 (11) (2014) 927–929.
[68] S. Luo, L. Tong, Molecular basis for the recognition of methylated adenines in RNA by the eukaryotic YTH domain, Proc. Natl. Acad. Sci. U.S.A. 111 (38) (2014) 13834–13839.
[69] N. Liu, et al., N6-methyladenosine-dependent RNA structural switches regulate RNA-protein interactions, Nature 518 (7540) (2015) 560–564.
[70] C. Roost, et al., Structure and thermodynamics of N6-methyladenosine in RNA: a spring-loaded base modification, J. Am. Chem. Soc. 137 (5) (2015) 2107–2115.
[71] C.R. Alarcón, H. Lee, H. Goodarzi, N. Halberg, S.F. Tavazoie, N6-methyladenosine marks primary microRNAs for processing, Nature 519 (7544) (2015) 482–485.
[72] J. Zhou, et al., Dynamic m6A mRNA methylation directs translational control of heat shock response, Nature 526 (7574) (2015) 591–594.
[73] H. Zhou, et al., m1A and m1G disrupt A-RNA structure through the intrinsic instability of Hoogsteen base pairs, Nat. Struct. Mol. Biol. 23 (9) (2016) 803–810.
[74] M. Helm, R. Giegé, C. Florentz, A Watson – Crick base-pair-disrupting methyl group (m1A9) is sufficient for cloverleaf folding of human mitochondrial tRNALys, Biochemistry (Mosc.) 38 (40) (1999) 13338–13346.
[75] D. Dominissini, et al., The dynamic N1-methyladenosine methylome in eukaryotic messenger RNA, Nature 530 (7591) (2016) 441–446.
[76] X. Li, et al., Transcriptome-wide mapping reveals reversible and dynamic N1-methyladenosine methylome, Nat. Chem. Biol. 12 (5) (2016) 311–316.
[77] T. Chujo, T. Suzuki, Trmt61B is a methyltransferase responsible for 1-methyladenosine at position 58 of human mitochondrial tRNAs, RNA 18 (12) (2012) 2269–2276.
[78] B.E.H. Maden, The numerous modified nucleotides in eukaryotic ribosomal RNA, in: W.E. Cohn, K. Moldave (Eds.), Progress in Nucleic Acid Research and Molecular Biology, vol. 39, Academic Press, Philadelphia, PA, 1990, pp. 241–303.
[79] A.T. Yu, J. Ge, Y.-T. Yu, Pseudouridines in spliceosomal snRNAs, Protein Cell 2 (9) (2011) 712–725.
[80] G. Wu, A.T. Yu, A. Kantartzis, Y.-T. Yu, Functions and mechanisms of spliceosomal small nuclear RNA pseudouridylation, Wiley Interdiscip. Rev. RNA 2 (4) (2011) 571–581.
[81] J. Ge, Y.-T. Yu, RNA pseudouridylation: new insights into an old modification, Trends Biochem. Sci. 38 (4) (2013) 210–218.
[82] F.F. Davis, F.W. Allen, Ribonucleic acids from yeast which contain a fifth nucleotide, J. Biol. Chem. 227 (2) (1957) 907–915.
[83] J.G. Arnez, T.A. Steitz, Crystal structure of unmodified tRNAGln complexed with glutaminyl-tRNA synthetase and ATP suggests a possible role for pseudo-uridines in stabilization of RNA structure, Biochemistry (Mosc.) 33 (24) (1994) 7560–7567.
[84] M.I. Newby, N.L. Greenbaum, Investigation of overhauser effects between pseudouridine and water protons in RNA helices, Proc. Natl. Acad. Sci. U.S.A. 99 (20) (2002) 12697–12702.
[85] I.S. Fernández, et al., Unusual base pairing during the decoding of a stop codon by the ribosome, Nature 500 (7460) (2013) 107–110.
[86] J. Karijolich, Y.-T. Yu, Converting nonsense codons into sense codons by targeted pseudouridylation, Nature 474 (7351) (2011) 395–398.

[87] J. Ofengand, A. Bakin, Mapping to nucleotide resolution of pseudouridine residues in large subunit ribosomal RNAs from representative eukaryotes, prokaryotes, archaebacteria, mitochondria and chloroplasts, J. Mol. Biol. 266 (2) (1997) 246–268.
[88] P. Ganot, M.-L. Bortolin, T. Kiss, Site-specific pseudouridine formation in preribosomal RNA is guided by small nucleolar RNAs, Cell 89 (5) (1997) 799–809.
[89] J. Ni, A.L. Tien, M.J. Fournier, Small nucleolar RNAs direct site-specific synthesis of pseudouridine in ribosomal RNA, Cell 89 (4) (1997) 565–573.
[90] B.S. Sibert, J.R. Patton, Pseudouridine synthase 1: a site-specific synthase without strict sequence recognition requirements, Nucleic Acids Res. 40 (5) (2012) 2107–2118.
[91] X. Ma, X. Zhao, Y.-T. Yu, Pseudouridylation (Ψ) of U2 snRNA in S. cerevisiae is catalyzed by an RNA-independent mechanism, EMBO J. 22 (8) (2003) 1889–1897.
[92] A. Urban, I. Behm-Ansmant, C. Branlant, Y. Motorin, RNA sequence and two-dimensional structure features required for efficient substrate modification by the Saccharomyces cerevisiae RNA:{Psi}-synthase Pus7p, J. Biol. Chem. 284 (9) (2009) 5845–5858.
[93] P. Ganot, M. Caizergues-Ferrer, T. Kiss, The family of box ACA small nucleolar RNAs is defined by an evolutionarily conserved secondary structure and ubiquitous sequence elements essential for RNA accumulation, Genes Dev. 11 (7) (1997) 941–956.
[94] I. Behm-Ansmant, et al., The Saccharomyces cerevisiae U2 snRNA:pseudouridine-synthase Pus7p is a novel multisite-multisubstrate RNA:Psi-synthase also acting on tRNAs, RNA 9 (11) (2003) 1371–1382.
[95] W.A. Decatur, M.N. Schnare, Different mechanisms for pseudouridine formation in yeast 5S and 5.8S rRNAs, Mol. Cell. Biol. 28 (10) (2008) 3089–3100.
[96] S. Schwartz, et al., Transcriptome-wide mapping reveals widespread dynamic-regulated pseudouridylation of ncRNA and mRNA, Cell 159 (1) (2014) 148–162.
[97] T.M. Carlile, et al., Pseudouridine profiling reveals regulated mRNA pseudouridylation in yeast and human cells, Nature 515 (7525) (2014) 143–146.
[98] X. Li, et al., Chemical pulldown reveals dynamic pseudouridylation of the mammalian transcriptome, Nat. Chem. Biol. 11 (8) (2015) 592–597.
[99] F.C. Courtes, et al., 28S rRNA is inducibly pseudouridylated by the mTOR pathway translational control in CHO cell cultures, J. Biotechnol. 174 (2014) 16–21.
[100] G. Wu, M. Xiao, C. Yang, Y.-T. Yu, U2 snRNA is inducibly pseudouridylated at novel sites by Pus7p and snR81 RNP, EMBO J. 30 (1) (2011) 79–89.
[101] Y. Motorin, H. Grosjean, Multisite-specific tRNA:m5C-methyltransferase (Trm4) in yeast Saccharomyces cerevisiae: identification of the gene and substrate specificity of the enzyme, RNA 5 (8) (1999) 1105–1118.
[102] D.T. Dubin, V. Stollar, C.-C. Hsuchen, K. Timko, G.M. Guild, Sindbis virus messenger RNA: the 5′-termini and methylated residues of 26 and 42 S RNA, Virology 77 (2) (1977) 457–470.
[103] S. Hussain, J. Aleksic, S. Blanco, S. Dietmann, M. Frye, Characterizing 5-methylcytosine in the mammalian epitranscriptome, Genome Biol. 14 (2013) 215.
[104] J.E. Squires, et al., Widespread occurrence of 5-methylcytosine in human coding and non-coding RNA, Nucleic Acids Res. 40 (11) (2012) 5023–5033.
[105] V. Khoddami, B.R. Cairns, Identification of direct targets and modified bases of RNA cytosine methyltransferases, Nat. Biotechnol. 31 (5) (2013) 458–464.
[106] B. Brzezicha, et al., Identification of human tRNA:m5C methyltransferase catalysing intron-dependent m5C formation in the first position of the anticodon of the pre-tRNA Leu (CAA), Nucleic Acids Res. 34 (20) (2006) 6034–6043.
[107] M.G. Goll, et al., Methylation of tRNAAsp by the DNA methyltransferase homolog Dnmt2, Science 311 (5759) (2006) 395–398.

[108] M. Schaefer, et al., RNA methylation by Dnmt2 protects transfer RNAs against stress-induced cleavage, Genes Dev. 24 (15) (2010) 1590–1595.
[109] S. Hussain, et al., NSun2-mediated cytosine-5 methylation of vault noncoding RNA determines its processing into regulatory small RNAs, Cell Rep. 4 (2) (2013) 255–261.
[110] L. Abbasi-Moheb, et al., Mutations in NSUN2 cause autosomal-recessive intellectual disability, Am. J. Hum. Genet. 90 (5) (2012) 847–855.
[111] S. Hussain, et al., The mouse cytosine-5 RNA methyltransferase NSun2 is a component of the chromatoid body and required for testis differentiation, Mol. Cell. Biol. 33 (8) (2013) 1561–1570.
[112] B. Delatte, et al., Transcriptome-wide distribution and function of RNA hydroxymethylcytosine, Science 351 (6270) (2016) 282–285.
[113] L. Fu, et al., Tet-mediated formation of 5-hydroxymethylcytosine in RNA, J. Am. Chem. Soc. 136 (33) (2014) 11582–11585.
[114] H.-Y. Zhang, J. Xiong, B.-L. Qi, Y.-Q. Feng, B.-F. Yuan, The existence of 5-hydroxymethylcytosine and 5-formylcytosine in both DNA and RNA in mammals, Chem. Commun. 52 (4) (2016) 737–740.
[115] J. Karijolich, C. Yi, Y.-T. Yu, Transcriptome-wide dynamics of RNA pseudouridylation, Nat. Rev. Mol. Cell Biol. 16 (10) (2015) 581–585.
[116] W. Park, J. Li, R. Song, J. Messing, X. Chen, CARPEL FACTORY, a dicer homolog, and HEN1, a novel protein, act in microRNA metabolism in Arabidopsis thaliana, Curr. Biol. 12 (17) (2002) 1484–1495.
[117] J. Li, Z. Yang, B. Yu, J. Liu, X. Chen, Methylation protects miRNAs and siRNAs from a 3′-end uridylation activity in Arabidopsis, Curr. Biol. 15 (16) (2005) 1501–1507.
[118] H.Y. Yi-Brunozzi, L.M. Easterwood, G.M. Kamilar, P.A. Beal, Synthetic substrate analogs for the RNA-editing adenosine deaminase ADAR-2, Nucleic Acids Res. 27 (14) (1999) 2912–2917.
[119] Z.-W. Dong, et al., RTL-P: a sensitive approach for detecting sites of 2′-O-methylation in RNA molecules, Nucleic Acids Res. 40 (20) (2012). e157–e157.
[120] U. Birkedal, et al., Profiling of ribose methylations in RNA by high-throughput sequencing, Angew. Chem. 127 (2) (2015) 461–465.
[121] V. Marchand, F. Blanloeil-Oillo, M. Helm, Y. Motorin, Illumina-based RiboMethSeq approach for mapping of 2′-O-Me residues in RNA, Nucleic Acids Res. 44 (16) (2016) e135.
[122] A.Y. Golovina, et al., Method for site-specific detection of m6A nucleoside presence in RNA based on high-resolution melting (HRM) analysis, Nucleic Acids Res. 42 (4) (2014) e27.
[123] C. Wetzel, P.A. Limbach, Mass spectrometry of modified RNAs: recent developments, Analyst 141 (1) (2016) 16–23.
[124] Z. Meng, P.A. Limbach, Mass spectrometry of RNA: linking the genome to the proteome, Brief. Funct. Genomic. Proteomic. 5 (1) (2006) 87–95.
[125] K.W. Gaston, P.A. Limbach, The identification and characterization of non-coding and coding RNAs and their modified nucleosides by mass spectrometry, RNA Biol. 11 (12) (2014) 1568–1585.
[126] C.M. Castleberry, C.-W. Chou, P.A. Limbach, UNIT 10.1 Matrix-assisted laser desorption/ionization time-of-flight mass spectrometry of oligonucleotides, Current Protocols in Nucleic Acid Chemistry, John Wiley & Sons, Inc., Hoboken, NJ, 2001.
[127] J. Mengel-Jørgensen, F. Kirpekar, Detection of pseudouridine and other modifications in tRNA by cyanoethylation and MALDI mass spectrometry, Nucleic Acids Res. 30 (23) (2002) e135.

[128] C.E. Dumelin, Y. Chen, A.M. Leconte, Y.G. Chen, D.R. Liu, Discovery and biological characterization of geranylated RNA in bacteria, Nat. Chem. Biol. 8 (11) (2012) 913–919.
[129] K. Miyauchi, S. Kimura, T. Suzuki, A cyclic form of N6-threonylcarbamoyladenosine as a widely distributed tRNA hypermodification, Nat. Chem. Biol. 9 (2) (2013) 105–111.
[130] T.-Y. Huang, J. Liu, S.A. McLuckey, Top-down tandem mass spectrometry of tRNA via ion trap collision-induced dissociation, J. Am. Soc. Mass Spectrom. 21 (6) (2010) 890–898.
[131] C.T.Y. Chan, et al., A quantitative systems approach reveals dynamic control of tRNA modifications during cellular stress, PLoS Genet. 6 (12) (2010). e1001247.
[132] P. Puri, et al., Systematic identification of tRNAome and its dynamics in Lactococcus lactis, Mol. Microbiol. 93 (5) (2014) 944–956.
[133] C. Wetzel, S. Li, P.A. Limbach, Metabolic de-isotoping for improved LC-MS characterization of modified RNAs, J. Am. Soc. Mass Spectrom. 25 (7) (2014) 1114–1123.
[134] Y. Motorin, S. Muller, I. Behm-Ansmant, C. Branlant, Identification of modified residues in RNAs by reverse transcription-based methods, Methods in Enzymology, vol. 425, Academic Press, Philadelphia, PA, 2007, pp. 21–53.
[135] G.G. Brownlee, E.M. Cartwright, Rapid gel sequencing of RNA by primed synthesis with reverse transcriptase, J. Mol. Biol. 114 (1) (1977) 93–117.
[136] D.J. Lane, et al., Rapid determination of 16S ribosomal RNA sequences for phylogenetic analyses, Proc. Natl. Acad. Sci. U.S.A. 82 (20) (1985) 6955–6959.
[137] A. Bakin, J. Ofengand, Four newly located pseudouridylate residues in Escherichia coli 23S ribosomal RNA are all at the peptidyltransferase center: Analysis by the application of a new sequencing technique, Biochemistry (Mosc.) 32 (37) (1993) 9754–9762.
[138] A.F. Lovejoy, D.P. Riordan, P.O. Brown, Transcriptome-wide mapping of pseudouridines: pseudouridine synthases modify specific mRNAs in S. cerevisiae, PLoS One 9 (10) (2014) e110799.
[139] C.T. Wittwer, G.H. Reed, C.N. Gundry, J.G. Vandersteen, R.J. Pryor, High-resolution genotyping by amplicon melting analysis using LCGreen, Clin. Chem. 49 (6) (2003) 853–860.
[140] C.N. Gundry, et al., Amplicon melting analysis with labeled primers: a closed-tube method for differentiating homozygotes and heterozygotes, Clin. Chem. 49 (3) (2003) 396–406.
[141] R.H.A.M. Vossen, E. Aten, A. Roos, J.T. den Dunnen, High-resolution melting analysis (HRMA)—more than just sequence variant screening, Hum. Mutat. 30 (6) (2009) 860–866.
[142] T.K. Wojdacz, A. Dobrovic, Methylation-sensitive high resolution melting (MS-HRM): a new approach for sensitive and high-throughput assessment of methylation, Nucleic Acids Res. 35 (6) (2007). e41.
[143] J. Eid, et al., Real-time DNA sequencing from single polymerase molecules, Science 323 (5910) (2009) 133–138.
[144] H.S. Rhee, B.F. Pugh, UNIT 21.24 ChIP-exo method for identifying genomic location of DNA-binding proteins with near-single-nucleotide accuracy, Current Protocols in Molecular Biology (100:21.24:21.24.1–21.24.14), John Wiley & Sons, Inc., Hoboken, NJ, 2001.
[145] B. Linder, et al., Single-nucleotide-resolution mapping of m6A and m6Am throughout the transcriptome, Nat. Methods 12 (8) (2015) 767–772.
[146] S. Kishore, et al., A quantitative analysis of CLIP methods for identifying binding sites of RNA-binding proteins, Nat. Methods 8 (7) (2011) 559–564.
[147] M. Zaringhalam, F.N. Papavasiliou, Pseudouridylation meets next-generation sequencing, Methods 107 (2016) 63–72.

[148] H.A. Ebhardt, et al., Meta-analysis of small RNA-sequencing errors reveals ubiquitous post-transcriptional RNA modifications, Nucleic Acids Res. 37 (8) (2009) 2461–2470.
[149] B.D. Gregory, et al., A link between RNA metabolism and silencing affecting Arabidopsis development, Dev. Cell 14 (6) (2008) 854–866.
[150] M.R. Willmann, N.D. Berkowitz, B.D. Gregory, Improved genome-wide mapping of uncapped and cleaved transcripts in eukaryotes—GMUCT 2.0, Methods 67 (1) (2014) 64–73.
[151] A. Rhoads, K.F. Au, PacBio sequencing and its applications, Genomics Proteomics Bioinformatics 13 (5) (2015) 278–289.
[152] B.A. Flusberg, et al., Direct detection of DNA methylation during single-molecule, real-time sequencing, Nat. Methods 7 (6) (2010) 461–465.
[153] C.-X. Song, et al., Sensitive and specific single-molecule sequencing of 5-hydroxymethylcytosine, Nat. Methods 9 (1) (2012) 75–77.
[154] M. Wanunu, et al., Discrimination of methylcytosine from hydroxymethylcytosine in DNA molecules, J. Am. Chem. Soc. 133 (3) (2011) 486–492.
[155] I.D. Vilfan, et al., Analysis of RNA base modification and structural rearrangement by single-molecule real-time detection of reverse transcription, J. Nanobiotechnology 11 (1) (2013) 8.
[156] A.M. Smith, R. Abu-Shumays, M. Akeson, D.L. Bernick, Capture, unfolding, and detection of individual tRNA molecules using a nanopore device, Front. Bioeng. Biotechnol. 3 (2015) 1–9.
[157] D. Aird, et al., Analyzing and minimizing PCR amplification bias in Illumina sequencing libraries, Genome Biol. 12 (2011) R18.
[158] D.I. Lou, et al., High-throughput DNA sequencing errors are reduced by orders of magnitude using circle sequencing, Proc. Natl. Acad. Sci. U.S.A. 110 (49) (2013) 19872–19877.
[159] M.J. Levesque, P. Ginart, Y. Wei, A. Raj, Visualizing SNVs to quantify allele-specific expression in single cells, Nat. Methods 10 (9) (2013) 865–867.

CHAPTER NINE

Impact of RNA Modifications and RNA-Modifying Enzymes on Eukaryotic Ribonucleases

Guillaume F. Chanfreau[1]
University of California, Los Angeles, CA, United States
[1]Corresponding author: e-mail address: guillom@chem.ucla.edu

Contents

Abstract

Constitutive and regulated turnover of RNAs is necessary to eliminate aberrant RNA molecules and control the level of specific mRNAs to maintain homeostasis or to respond to signals in living cells. Modifications of nucleosides in specific RNAs are important in modulating the functions of these transcripts, but they can also dramatically impact their fate and turnover. This chapter will review how RNA modifications impact the activities of ribonucleases that target these RNAs for degradation or cleavage, focusing more particularly on tRNAs and mRNAs in eukaryotic cells. Many nucleoside modifications are important to promote proper folding of tRNAs, and the absence of specific modifications makes them susceptible to degradation by quality control pathways that

The Enzymes, Volume 41
ISSN 1874-6047
http://dx.doi.org/10.1016/bs.enz.2017.03.008

eliminate improperly folded species. Modifications in tRNAs can also modulate their cleavage during stress or by fungal toxins that target modified nucleosides. Modifications of the cap structure found at the 5′-end of eukaryotic mRNAs are essential to control the degradation of these mRNAs. In addition, internal modifications of eukaryotic mRNAs can change their secondary structures or provide binding sites for reader proteins, which can dramatically impact their stability. Recent examples show that mRNA modifications play important roles in regulating mRNA stability during development, cellular differentiation and physiological responses. Finally, many modifications can impact microRNA- and siRNA-mediated gene regulation by direct or indirect effects. With the growing number of genomic techniques able to identify modifications genome wide, it is anticipated that novel chemical modifications or new modification sites will be identified, which will play additional regulatory functions for RNA turnover.

1. INTRODUCTION: RIBONUCLEASES THAT MEDIATE RNA DEGRADATION IN EUKARYOTIC CELLS

RNA degradation in eukaryotic cells is a key modulator of gene expression, as it counterbalances the rate of transcription and also serves to rapidly eliminate transcripts when they are no longer needed. In addition, RNA degradation fulfills a role of quality control, by eliminating RNA molecules that are deemed nonfunctional or aberrant because they lack specific sequences or modifications needed for their functions. In eukaryotic cells, RNA degradation relies on a set of enzymes that carry specific ribonuclease activities (reviewed in Refs. [1,2]). In general, RNAs are degraded by exoribonucleolytic reactions in the 5′–3′ or 3′–5′directions (Fig. 1). The main 5′–3′ exoribonucleases in eukaryotic cells are Xrn1 and Xrn2 (named Rat1 in yeast), which are localized in the cytoplasm and nucleus, respectively. These two activities are functionally interchangeable, as targeting of Xrn1 to the nucleus can functionally complement Rat1/Xrn2 defects [3]. In eukaryotic cells, the main 3′–5′ exonuclease activities are contained in a complex called the exosome, which is present in both nuclear and cytoplasmic compartments [4–6]. The nuclear and cytoplasmic exosomes contain subunits specific to each subcellular complex. In general, exoribonucleases need a free RNA 5′-end or 3′-end with specific termini (5′/3′phosphates or 5′/3′-OH) to catalyze degradation of their RNA substrates; these free ends with the proper chemical groups can be generated by removal of structures or sequences that protect the 5′- or 3′-ends (Fig. 1A), such as the cap structure or of the poly(A) tail, which is bound by poly(A)-binding proteins. Alternatively, RNA endonucleases can cleave phosphodiester bonds

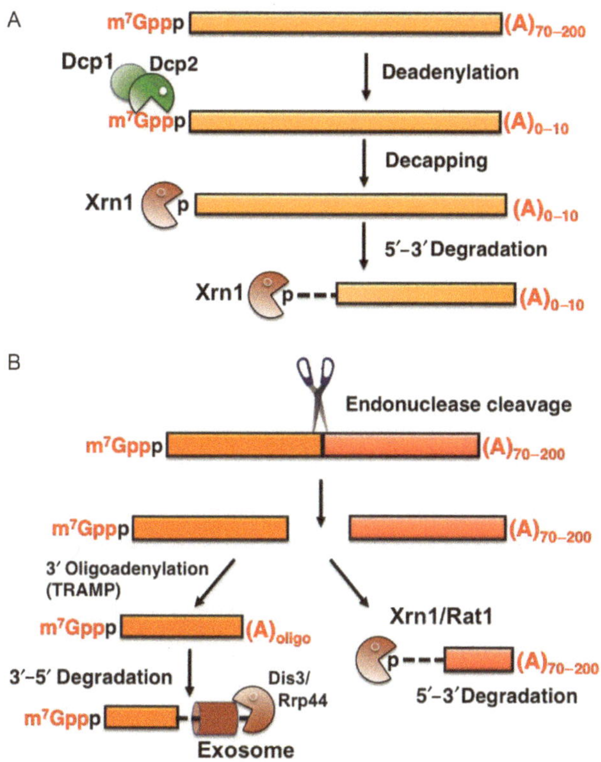

Fig. 1 Schematic representation of RNA degradation pathways in eukaryotic cells. (A) The major mRNA degradation pathway in eukaryotes involves removal of the bulk of the poly(A) tail (deadenylation) followed by decapping by the Dcp complex (Dcp1/Dcp2) and 5′–3′ degradation by the Xrn1 exonuclease or related enzymes. Following deadenylation, the RNA can also be degraded in the 3′–5′ direction by the exosome (not shown). (B) Specific RNAs can be degraded by endonucleolytic cleavage (*scissor*), followed by degradation of the cleaved fragments by Xrn1 or Rat1 (5′–3′ direction) or by the exosome (3′–5′direction). Dis3/Rrp44 is the catalytic subunit of the exosome.

internally, which will then expose free 5′ and 3′-ends for decay by exonucleases (Fig. 1B). Endoribonucleases are typically protein enzymes that bind specific RNA substrates, as described later for some enzymes that target transfer RNAs (tRNAs) or other molecules, but endoribonucleolytic cleavage can also be generated through small RNAs and RNA interference pathways. Within this general framework, we will consider the impact of RNA modifications on the stability of RNAs in eukaryotic cells and on the function of ribonucleases that target these RNAs. In general, RNA modifications will have three major effects on RNA degradation. First, they can chemically protect the RNA by preventing ribonucleases from binding or hydrolyzing their substrates because the RNA does not have the proper chemical groups needed for these activities. Second, modifications can also promote changes in the structure of the RNAs that are modified, which directly affect the activities of ribonucleases that target these RNAs. Finally, these RNA modifications can also serve as direct binding/recognition sites for proteins that mediate RNA turnover or promote stability. Because the impact of RNA modifications on the function of ribonucleases has been extensively documented for the enzymes that target tRNAs and mRNAs, I will focus this review primarily on these two types of transcripts.

2. IMPACT OF MODIFICATIONS ON NONCODING RNA TURNOVER

Noncoding RNAs are extensively modified, and some of the most thoroughly characterized RNA modification pathways and functions have been described for ribosomal RNAs (rRNAs) and tRNAs. The role and synthesis pathways of these modifications will not be covered here, as these are described extensively in other chapters of this volume. Rather, we will focus this section on describing the impact of these modifications on the activities of ribonucleases that target these RNAs, and how degradation pathways controlled by these modifications can function as quality control or signaling mechanisms. The bulk of this section will be devoted to tRNAs, as these RNAs have provided the most extensive examples of the impact of RNA modifications on turnover and cleavage.

2.1 Degradation of Methylation-Deficient pre-tRNA$_i^{Met}$ by the Nuclear Exosome

The first pathway that showed a link between tRNA modifications and RNA quality control in eukaryotes identified hypomodified initiator tRNA tRNA$_i^{Met}$ as a substrate of a quality control pathway mediated by the nuclear exosome (Fig. 2). Lack of N^1 methylation at position A_{58} (N^1-methyladenosine 58, m^1A_{58}) due to inactivation of the tRNA methyltransferase Trm6 was shown to destabilize the precursor and mature species of the initiator tRNA$_i^{Met}$ [7]. This phenotype is specific to the tRNA$_i^{Met}$ initiator, as elongator tRNAs do not exhibit such instability in mutant strains lacking m^1A methyltransferase activity. A genetic screen performed by Anderson and coworkers identified mutants which restored the growth phenotype of *trm6*-deficient strains and also restored tRNA$_i^{Met}$ levels [8]. This

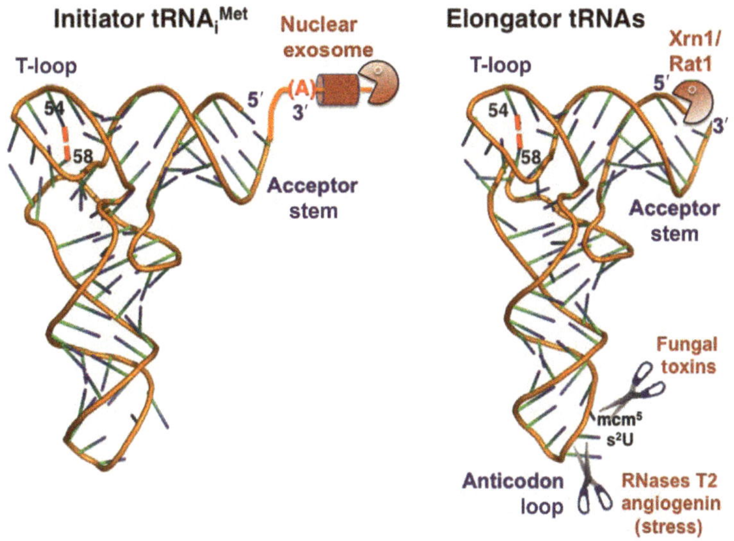

Fig. 2 tRNA degradation or cleavage pathways affected by RNA modifications. Shown are the tertiary structures of $tRNA_i^{Met}$ (*left*) and of an elongator $tRNA^{Phe}$ (*right*) and a schematic representation of the cleavage and degradative enzymes that target these tRNAs. In the case of $tRNA_i^{Met}$ (*left*), the precursor species are polyadenylated by TRAMP (not shown) and degraded from the 3′-end by the nuclear exosome. The *orange* 3′-extension with (A) represents the polyadenylated sequence of the precursor species. In the case of elongator tRNAs, Xrn1/Rat1 can degrade tRNA species from the 5′-end in the context of the RTD pathway, or if tRNA precursors are not capped (see main text). Location of the cleavage in the anticodon loop by fungal toxins or during stress is indicated by *scissors* and is not positioned accurately. Structures were generated using PyMol using PDB entries 1YFG ($tRNA_i^{Met}$) and 1EHZ ($tRNA^{Phe}$).

study was instrumental in discovering the mechanism of degradation mediated by the nuclear exosome, as one of the suppressor genes was mapped to the *RRP44* gene (also known as *DIS3*) encoding the catalytic subunit of the exosome, while another suppressor mapped to the gene encoding Trf4, a noncanonical poly (A) polymerase. Mutations inactivating either Rrp44 or Trf4 resulted in the suppression of the growth defects of the *trm6* mutants and also in the recovery of the levels of mature and precursor species of $tRNA_i^{Met}$. Further characterization of the species accumulating in various mutants showed that the precursor species of $tRNA_i^{Met}$ lacking m^1A_{58} are oligoadenylated at the 3′-end by Trf4p, which activates $3' \rightarrow 5'$ degradation of the tRNA precursors by the nuclear exosome (Fig. 2). Thus, degradation of hypomodified precursor species by the exosome competes with the normal processing pathway, leading to depletion of the mature species. This degradation pathway, proposed by genetic studies, was further supported by biochemical studies using purified components, which showed that the poly(A) polymerase activity of Trf4 contained in the TRAMP complex (Fig. 1) is specific to the hypomodified form of the $tRNA_i^{Met}$ lacking m^1A_{58} and that exosome fractions or the Rrp44 subunit can mediate degradation of oligoadenylated, hypomodified $tRNA_i^{Met}$ in vitro [9].

The Trf4/TRAMP-nuclear exosome-mediated degradation pathway of hypomodified $tRNA_i^{Met}$ is unique to this particular tRNA, despite many other tRNAs having the same m^1A modification at this position. This begs the question of why the lack of m^1A modification affects only the stability of $tRNA_i^{Met}$. One hypothesis proposed by Anderson and coworkers is that this is due to the fact that the T-loop structure of $tRNA_i^{Met}$ is different from that found in elongator tRNAs, and that m^1A_{58} is essential to promote the stabilization of this structure (Fig. 2). In $tRNA_i^{Met}$ this nucleotide forms a non-Watson–Crick interaction with an adenine base, while in other tRNAs it interacts with a 5-methyl uracil (m^5U or T) albeit also in a non-Watson–Crick manner (Fig. 3). The absence of the methyl group is proposed to affect the m^1A:A interaction more substantially than the m^1A:m^5U base pair, although the precise mechanism by which this occurs is not entirely clear based on the base pairing patterns (Fig. 3). In the absence of m^1A_{58} methyl group, disruption of the tertiary structure of the tRNA would promote escape of the $tRNA_i^{Met}$ precursor from the biogenesis pathway and its subsequent polyadenylation and decay. Interestingly, a recent study suggested that this mechanism might be conserved in mammalian cells. The ALKBH1 protein was recently demonstrated to catalyze RNA demethylation specific for m^1A, and the same study showed that the levels of the $tRNA_i^{Met}$ were elevated in an ALKBH1 knockout model [10]. Although this study did not demonstrate that the increase in $tRNA_i^{Met}$ levels were due to a decrease in exosome-mediated degradation of the hypomodified $tRNA_i^{Met}$ precursor as in yeast, it is tempting to speculate that a similar quality control pathway is in place in mammalian cells to ensure the integrity of the folding of $tRNA_i^{Met}$ when m^1A methylation is deficient or downregulated.

2.2 Degradation of Modification-Deficient tRNAs by the Rapid tRNA Degradation Pathway

While the previous section described how the nuclear exosome degrades hypomodified $tRNA_i^{Met}$ initiator tRNA, another degradation pathway

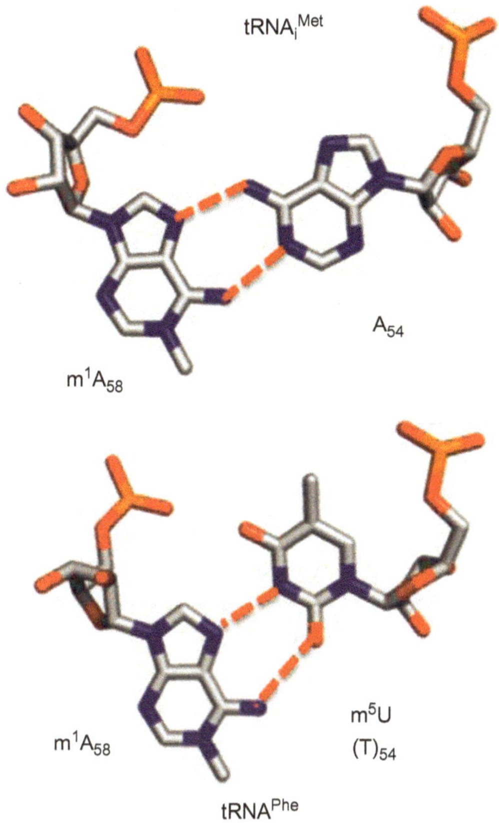

Fig. 3 Tertiary interactions involving m^{1}A58 in the T-loop for $tRNA_i^{Met}$ and a classical elongator tRNA ($tRNA^{Phe}$). Figures of the base pairs were generated using PyMol using PDB entries 1YFG ($tRNA_i^{Met}$) and 1EHZ ($tRNA^{Phe}$).

known as RTD (rapid tRNA degradation) is widely used to degrade improperly folded tRNAs, for instance because they lack specific modifications or because of other structural defects. This pathway was discovered when Phizicky and colleagues found that $tRNA^{Val}_{(AAC)}$ lacking methylated nucleosides following inactivation of the methyltransferases Trm4 and Trm8 was highly unstable. This increased turnover of hypomethylated $tRNA^{Val}_{(AAC)}$ is mediated by a pathway independent from the aforementioned nuclear exosome-mediated degradation of hypomodified $tRNA_i^{Met}$ [11]. This pathway was named RTD, and the enzymes responsible for this rapid degradation were later identified through genetics as Xrn1p and Rat1p [12], the two major

5′–3′ exonucleases located in the cytoplasm and nucleus, respectively. Inactivation of Met22p, which results in accumulation of the pAp metabolite that inhibits Xrn1p and Rat1p [13], also resulted in the same phenotype—rescue of the growth phenotypes of the *trm4 trm8* mutant and of the levels of hypomodified $tRNA^{Val}_{(AAC)}$. This pathway is not restricted to the degradation of hypomodified $tRNA^{Val}_{(AAC)}$ due to the absence of Trm4/Trm8-mediated methylations; for instance, inactivation of the RTD also rescued the growth defects of *trm8 pus7*, *trm8 dus3*, and *trm8 tan1* double mutants, which lack not only methyl groups added by Trm8 but also various other modifications [12]. More generally, the RTD pathway was shown to degrade tRNAs lacking a variety of modifications, and it was shown through genetic studies that RTD competes with the translation elongation factor EF1A for binding hypomodified tRNAs [14]. The primary mechanistic model emerging from these studies was that the lack of modifications induces a partial unfolding of the acceptor stem, which normally bridges the 5′ and 3′ ends of the tRNA (Fig. 2). Exposure of the 5′-end normally sequestered by the acceptor stem renders the hypomodified tRNAs susceptible to decay by 5′–3′ exonucleases [15]. This model is supported by genetic experiments showing that mutations that disrupt the integrity of the acceptor stem result in the targeting of tRNAs to RTD, even if these tRNAs are fully modified [15]. Moreover, a recent high-throughput screen used to identify tRNA mutations that trigger RTD showed that while some of these mutations promote the unfolding of the acceptor stem and exposure of the 5′-end to 5′–3′ exonucleases, other mutations also affect the entirety of the tRNA structure, including the anticodon stem [16]. Thus, the RTD pathway is a major quality control mechanism that eliminates tRNAs that are improperly folded, for instance when they lack specific modifications. Because both nuclear and cytoplasmic exonucleases are involved in RTD, it remains to be established where the major site(s) of degradation is located. Since Rat1p and Xrn1p have been shown to be functionally interchangeable [3], it is possible that the subcellular localization of each species of hypomodified tRNAs makes them more prone to attack by Xrn1p or Rat1p. In addition, it would be of the highest interest to determine whether this quality control pathway is also conserved in higher eukaryotes.

2.3 Precursors of tRNAs Are Stabilized by 5′-Cap Structures

The role of the cap structure in promoting the stability of eukaryotic mRNAs has been extensively described (see later). Recently, however, this modified structure has been shown to enhance the stability of tRNA precursors as well. tRNAs typically do not carry any modified 5′-ends or cap structure, as the mature 5′-end of these molecules is generated through cleavage by RNase P. However, the primary precursors of tRNAs synthesized by RNA polymerase III possess a 5′-triphosphate end due to incorporation of the first nucleotide as a triphosphate-nucleotide species by RNA polymerase. Until recently, this 5′ terminal nucleotide was thought to be unmodified. However, a recent study using mass spectrometry or immunoprecipitation showed that 7-methyl guanosine residue (m^7G) caps as well as dimethylated and trimethylated cap structures could be detected at the 5′-end of precursors of various tRNAs [17], in both yeast and human cells. While these modified 5′-ends were initially detected upon inactivation of RNase P, 5′-cap structures can also be found in wild-type cells, suggesting that these are not a consequence of the inactivation of the normal 5′-end processing pathway. The abundance of the cap varies from 5% to 22% depending on pre-tRNAs. Capping of these pre-tRNAs is mediated by the guanylyltransferase enzyme that also caps mRNAs, while trimethylation is catalyzed by the same enzyme that trimethylate snRNAs [17]. However, in contrast to mRNA capping, pre-tRNA capping does not seem to be coupled to transcription since 5′ caps are more abundantly detected in spliced pre-tRNAs, suggesting that they are generated posttranscriptionally. Inactivation of the 5′–3′ exonuclease Xrn1 increases the level of pre-tRNAs, showing that pre-tRNAs are targeted by this 5′–3′ degradation pathway. Inactivating or overexpressing the guanylyltransferase Ceg1 results in decreased and increased pre-tRNA levels, respectively, demonstrating the impact of the capping process on pre-tRNA stability. Collectively these observations suggest that the capping of pre-tRNA precursors has a protective effect against Xrn1-mediated 5′–3′ degradation (Fig. 2). This could help stabilize the tRNA precursors during their processing and prevent premature degradation of these capped species. However, because the fraction of pre-tRNAs containing a 5′-cap is relatively low and varies quite extensively between different tRNA species, the biological significance of pre-tRNA capping in tRNA biogenesis and tRNA processing pathways remains to be fully determined.

2.4 RNA Modifications and tRNA Cleavage

2.4.1 tRNA Cleavage, Stress, and Modifications

tRNA cleavage in the anticodon loop (Fig. 2) has emerged as a conserved cellular response to stress in many organisms including yeast, plants, mammalian cells [18], and protozoans such as *Tetrahymena* [19]. The function of this cleavage is not to deplete the pool of mature tRNA, as the steady-state level of tRNAs does not typically decrease in these conditions. Instead, cleaved tRNA fragments will accumulate to some extent and bind the translational machinery to inhibit translation in response to stress [20,21]. In yeast and protozoans, the enzymes responsible for these stress-mediated cleavage events belong to the T2 family of endonucleases such as the endonuclease Rny1 in the yeast *Saccharomyces cerevisiae* [22,23] and several T2-related enzymes in *Tetrahymena* [24]. In mammalian cells, tRNA cleavage is performed by angiogenin which is unrelated to the T2 family but shares homology with the RNase A family [25]. Angiogenin cleaves tRNAs during various stress conditions in mammalian cells, resulting in translational repression [21,20]. The modification status of cleaved tRNAs during stress has not been monitored carefully. However, it has been shown that inactivating tRNA methylases has a major impact on tRNA cleavage and cellular response to stress. In *Drosophila*, mutants of the tRNA methyltransferase Dnmt2 show an increase in tRNA cleavage, and these mutants also exhibit reduced viability in stress conditions [26]. This result suggests that Dnmt2-mediated tRNA methylation protects the tRNA against cleavage in *Drosophila.* A similar result was described in mammalian cells, as inactivation of the m^5C methyltransferase Nsun2 results in an increase of angiogenin-mediated cleavage of tRNAs and subsequent translation inhibition due to the accumulation of tRNA fragments [27,28]. Increased tRNA cleavage in the absence of modification is proposed to be due to a more efficient binding of angiogenin to tRNAs lacking the m^5C [27]. Because inactivation of Nsun2 results in microcephaly and other neurological disorders, it was suggested that increased tRNA cleavage is the cause of these neurological phenotypes [27,28]. In all these studies, angiogenin-mediated tRNA cleavage typically does not affect mature tRNA levels but instead modulates the amount of tRNA fragments to inhibit cellular translation. However, a recent study showed that mutation of human mitochondrial $tRNA^{Glu}$ at position 55 which results in the loss of pseudouridylation at this position decreases the levels of this tRNA in vivo and increases cleavage of this tRNA by angiogenin in vitro [29]. Reduction of the levels of this tRNA impacts normal mitochondrial function and is associated with maternally inherited diabetes and deafness [29]. Whether or not angiogenin is directly responsible for the reduction in the levels of this tRNA in vivo remains to be established, as it is possible that misfolding of the mutant, hypomodified tRNA may trigger quality control mechanisms similar to the ones summarized earlier. Nonetheless, this study provides another example of the importance of modifications in controlling tRNA stability and highlights how deficiencies in one modification can result in severe pathologies.

2.4.2 Ribonuclease Specific for Modified tRNAs

As described earlier, tRNA modifications generally have a protective role against RNA degradation, either because modified nucleosides are typically less efficient substrates for ribonucleases or because the lack of modification induces folding defects which make these tRNAs better substrates to specific ribonucleases. However, some ribonucleases have evolved to selectively target tRNAs containing modified nucleotides. This is the case for the multiprotein toxins zymocin from *Kluyveromyces lactis* and PaT from *Pichia acaciae*, which are expressed by these two yeast species to repress the growth of other yeast in natural environments, thus allowing them to outcompete other yeast species for natural resources. In both cases, specific subunits of the toxins will contain an RNA endonuclease activity that selectively cleaves the anticodon stem loop of tRNAs, when these tRNAs are modified. In the case of the zymocin toxin, the γ subunit cleaves $tRNA^{Glu}$ at a single site located 3′ from a modified nucleoside mcm^5s^2U located on the wobble position of the UUC anticodon of this tRNA [30,31] (Fig. 2). Other tRNAs having a mcm^5s^2U modification at the wobble position such as $tRNA^{Gln}$ and $tRNA^{Lys}$ can also be cleaved but at lower efficiency. The PaOrf2 subunit of the PaT toxin from *P. acaciae* performs a similar cleavage reaction on a similar set of substrates [32]. However, this endonuclease performs a secondary cleavage two nucleotides upstream from the first site; this double cleavage prevents the cleaved tRNAs from being sealed by the endogenous tRNA ligase [33]. This mechanism increases toxicity since the cleaved tRNAs can no longer be repaired by the endogenous cellular machinery of the host. As mentioned earlier, both of these endonucleases are highly specific for anticodons containing modified nucleosides. Accordingly, mutations in the machinery that catalyze formation of mcm^5s^2U confer resistance to the γ toxin [30,32,34]. Another unrelated example of a ribonuclease specific for modified nucleosides is human endonuclease V, which cleaves the anticodon loops of tRNAs that contain inosine at the wobble position in vitro [35]. However, it is currently unclear whether this activity has any biological relevance. Overall, the examples described earlier highlight how ribonucleases have evolved to specifically recognize tRNA modifications and have been selected to induce cellular toxicity in competing species through cleavage of fully modified functional tRNA molecules.

2.5 Modifications of Eukaryotic rRNAs and Stability

While the previous sections describe multiple roles for RNA modifications on influencing the turnover and cleavage of tRNAs, much less is known about the impact of these modifications on the degradation of other noncoding RNAs, even those for which extensive modifications have been identified. The major example identified so far showed that several pseudouridines residues within a specific region of the 25S rRNA are important to promote stability of rRNAs. Using genetic approaches, Fournier and colleagues showed that eliminating five pseudouridine modifications in helix 69 of the *S. cerevisiae* 25S rRNA destabilizes this rRNA, as shown by pulse-chase experiments [36]. Helix 69 interacts with both A and P site tRNAs and is involved in an intersubunit bridge. In agreement with this function, accelerated turnover of the mutant 25S seems to occur within the context of the entire ribosome, as the 18S rRNA—which is not directly affected by the loss of modifications—also exhibits a faster turnover in the mutant strain [36]. However, to the best of our knowledge, the mechanism responsible for this increased turnover has not been identified. It has been shown that nonfunctional rRNAs carrying point mutations in functionally important regions are subject to posttranscriptional quality control degradation through the so-called NRD pathway [37,38]. It remains to be established whether the NRD pathway is also responsible for the increased turnover of ribosomes lacking pseudouridines in helix 69 of the 25S rRNA, or whether this task is carried about by another, as-yet uncharacterized degradation pathway.

Besides this specific example, there is little evidence for a major contribution of specific modifications to rRNA stability. Several studies have shown that global reduction of 2′-O-methylation [39] or pseudouridylation [40] does not impact rRNA stability. Similarly, the 5′-cap structure is not required for U2 snRNA stability as its function can be fulfilled by an adenosine cap [41]. Several nucleoside modifications near the 5′-end of U2 are required for its function but not for its stability [42]. In addition, inactivation of the enzyme that promotes trimethylation of the cap structure of snRNAs has no effect on the stability of these snRNAs [43]. Thus, while most nucleoside modifications play important roles in the function of noncoding RNAs, they do not seem to impact their stability to the extent described earlier for tRNAs. A notable exception is the inosine modification of microRNA precursors, which will be discussed later.

3. IMPACT OF RNA MODIFICATIONS ON mRNA STABILITY

3.1 RNA Modifications in the 5′-Cap Structure and Their Impact on mRNA Turnover

The 5′-end of eukaryotic mRNAs is subject to modifications, which profoundly impact the fate and stability of these mRNAs. Early studies showed that polyadenylated RNAs from mammalian cells carry a m^{7}G [44,45] linked by a 5′-ppp-5′ linkage to the first nucleoside generated by transcription initiation (Fig. 4). In addition, the ribose moieties of the first two nucleotides incorporated by RNA polymerase can be methylated at the 2′ position (Fig. 4) [44,45]. In *Kinetoplastida* eukaryotic parasites, the first four nucleosides that follow the m^{7}G cap of all mRNAs are heavily modified [32,33]. They share an AACU sequence with all four nucleosides carrying a 2′-*O*-methyl group. In addition, the first adenine is modified as a dimethyladenine (m^{6,2}A), and the fourth base is modified by methylation at position 3 to form m^{3}U [32,33]. These modifications are not introduced into the mRNA directly, but rather generated in a spliced leader SL RNA mini exon

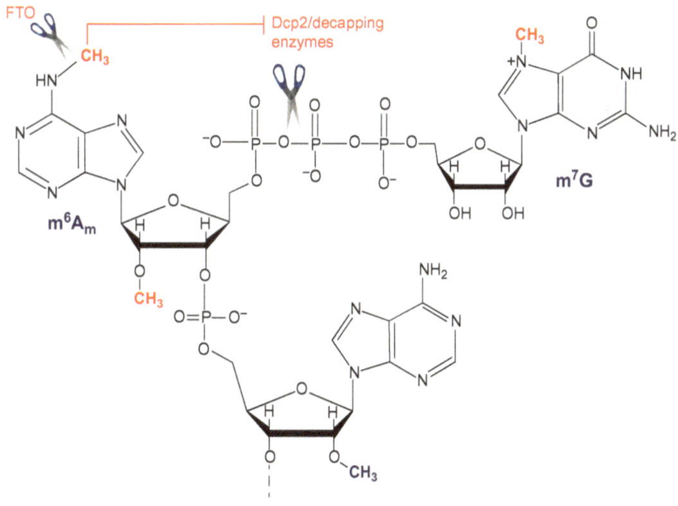

Fig. 4 Cap structure modifications and summary of the enzymes that modulate the structure or removal of the cap. The *scissors* symbolize the demethylase activity of FTO on m^{6}A and decapping activity of Dcp2. Not shown are the extensive modifications of the first four nucleosides of the transcript identified in *Kinetoplastida* or the 5′-NAD$^+$ caps (see text).

sequence, which is then *trans*-spliced onto all mRNAs [46–48]. In addition to the cap structure, it has been shown that the first position of mRNAs can contain a variety of noncanonical nucleotides, including NAD^+ [49,50]. These noncanonical "caps" are typically not generated through RNA modification per se, but rather through incorporation by the RNA polymerase when it uses one of these metabolites to initiate transcription. However, a recent study suggested that NAD^+ "caps" can be added posttranscriptionally to 5′-processed ends, such as those of small nucleolar RNAs [51]. This means that 5′-NAD^+ may correspond to a *bona fide* RNA modification. The role of NAD^+ caps in promoting turnover will be discussed later.

The cap structure plays multiple roles in gene expression in eukaryotes, but for the purpose of this section, we will focus on its impact on RNA degradation. The major degradation pathway of eukaryotic mRNAs involves a first step of deadenylation, followed by removal of the cap structure (also known as decapping) and 5′–3′ degradation [1,52] (Fig. 1). This degradation pathway was initially characterized in yeast and is conserved in mammalian cells [53]. The cap structure plays an important role in controlling mRNA decay because capped RNAs are inefficient substrates for 5′–3′ exonucleases, necessitating the removal of the cap in order for the mRNA to be degraded by these exonucleases. Enzymes that remove the cap structure from the 5′-end of mRNAs (decapping enzymes) have been identified in many eukaryotes, including the major decapping enzyme Dcp2 [54–56]. Other proteins interacting with Dcp2 in decapping complexes activate or modulate the activity of Dcp2 [57,58]. In addition to Dcp2, several additional decapping enzymes are present in mammalian cells, which target specific mRNAs in various degradation pathways [59–62]. Decapping enzymes belonging to the Nudix family of hydrolases were shown to cleave the cap structure between the α and β phosphate groups of the 5′-ppp-5′ linkage (Fig. 4), releasing m^7GDP and a 5′-monophosphate mRNA which is the proper substrate for 5′–3′ exonucleases [63]. Finally, the cap structure can also be a target of quality control mechanisms, which will degrade mRNAs with improperly methylated caps. Kiledjian and colleagues found that the fungal Rai1 protein has pyrophosphohydrolase activity specific for unmethylated 5′Gppp5′-RNA substrates, as opposed to the normal 5′m^7Gppp-5′cap structure [64]. This role is important to remove incomplete cap structures and expose incorrectly capped mRNAs to 5′–3′ degradation, as such improperly capped mRNAs are found more abundantly in stress conditions [64]. The yeast protein Dxo1 and its mammalian homologue DXO exhibit both decapping and 5′–3′ exonuclease activities and fulfill a function similar to

Rai1 in ensuring proper quality control of the cap structure [65,66]. Interestingly DXO/Rai1 have also recently been identified as enzymes capable of removing the NAD^+ "caps" from mRNA or snoRNA transcripts [51]. The presence of 5′-NAD^+ groups promotes NAD^+-decapping by DXO or Rai1, leading to accelerated turnover [51]. Therefore, as opposed to the m^7G cap, which acts as a stabilizing feature in eukaryotic mRNAs, NAD^+ caps lead to degradation. It remains to be identified if NAD^+-mediated decapping and degradation serves any regulatory role or is confined to quality control for transcripts that have initiated with an incorrect nucleoside such as NAD^+.

In addition to the terminal N^7-methyl guanosine cap, the 2′-hydroxyls of the two nucleosides immediately following the 5′-terminal N^7-methyl guanosine are methylated as well [44,45]. Typically, the nucleoside immediately following the terminal m^7G is always methylated, while methylation of the second one can vary. The functions of these methyl groups remained unknown until 2010–11, when it was shown that the presence of 2′-*O*-methyl groups at these positions distinguishes cellular mRNAs from viral mRNAs [67,68]. Viral mRNAs lacking 2′-*O*-methyl groups of the cap structure are attenuated through interferon-induced proteins with tetratricopeptide repeats which will negatively regulate translation of RNAs lacking these modifications [69]. However, because these modifications seem to act primarily by controlling the interferon response or translation initiation, their function will not be elaborated upon further.

Early in vivo labeling studies showed that in addition to the m^7G and 2′-*O*-methyl groups, an additional methyl group could be detected at the 5′-end of mammalian mRNAs [70]. This modification was recently identified as a N^6-methyl adenosine (m^6A) at the nucleoside immediately following the m^7G, which also carries a 2′-*O*-methyl group [71]. The presence of an m^6A at this position (hereby called 5′-m^6A_m; Fig. 4) stabilizes mRNAs, as demonstrated by the longer half-life of electroporated mRNAs containing an m^7Gppp-m^6A_m cap compared to mRNAs containing a standard m^7Gppp-A_m cap. This modification is dynamic and can be reversed by the FTO (fat mass and obesity-associated protein) demethylase. FTO demethylates m^6A_m adjacent to the m^7G cap in vitro and also controls the levels of 5′-m^6A_m in vivo [71]. Reducing the levels of 5′-m^6A_m by promoting the export of FTO out of the nucleus resulted in reduced mRNA stability, providing further evidence for the direct function the 5′-m^6A_m in conferring stability. The N^6-methyl group in the cap structure increases stability by conferring resistance to removal of the cap structure by the Dcp2 decapping enzyme (Fig. 4). Therefore, the 5′-m^6A_m directly inhibits 5′–3′ degradation by limiting the decapping of the transcripts (Fig. 4), and the subsequent exonucleolytic function of exonucleases such as Xrn1 or Xrn2/Rat1. In addition, genome wide analyses showed that mRNAs carrying a 5′-m^6A_m are less susceptible to microRNA-mediated RNA degradation [71]. In conclusion, this modification appears to play a major role in promoting RNA stability by antagonizing the decapping and/or microRNA-mediated degradation pathways. As this study was published only very recently, it is likely that additional roles will be established for this modification in mRNA turnover, and mRNA metabolism in general.

3.2 Internal Methylation of Bases and mRNA Stability

3.2.1 N^6-Methyl Adenosine (m^6A)

Eukaryotic mRNAs can contain a variety of methylated bases, mainly at positions 1 and 6 of adenosines (m^1A and m^6A, respectively), and at position 5 of cytosines (m^5C) [72–74] (Fig. 5). The presence of m^6A in the body of mRNAs has been known for decades, based on early in vivo labeling studies [75,76]. However, the functions of this modification in mRNA metabolism have only emerged in recent years, with the discovery of the enzymes that catalyze m^6A methylation ("writers"), its removal (the "erasers"), and the proteins that specifically bind to this modification (the "readers"). In general, m^6A is found in over 25% of human mRNAs, and this modification is typically more abundant in long exons, 3′ UTRs and in the vicinity of stop codons [77,78]. Many functions have been assigned for m^6A in mRNA metabolism (reviewed in Refs. [72–74]). In this section I will focus on the impact of m^6A on mRNA degradation and attempt to separate the direct effects of m^6A on turnover from other effects on mRNA metabolism that might impact degradation in an indirect manner.

Several studies have shown that the presence of m^6A can directly impact mRNA degradation [80,81]. A mechanistic link between m^6A and mRNA degradation was provided by the discovery that the YTH domain family protein YTHDF2 is an m^6A reader that can regulate degradation of m^6A-containing mRNAs [80] (Fig. 5). YTHDF2 was found to bind specific mRNAs, and YTHDF2 knockdown resulted in a stabilization of its mRNA targets. Interestingly, the impact of YTHDF2 on mRNA decay was correlated with the number of m^6A-binding sites [80], providing further support for a function for YTHDF2 in accelerating decay. YTHDF2 was later found to interact with the CNOT1 component of the CCR4–NOT complex, which promotes deadenylation of mRNAs [82] and mRNA decapping

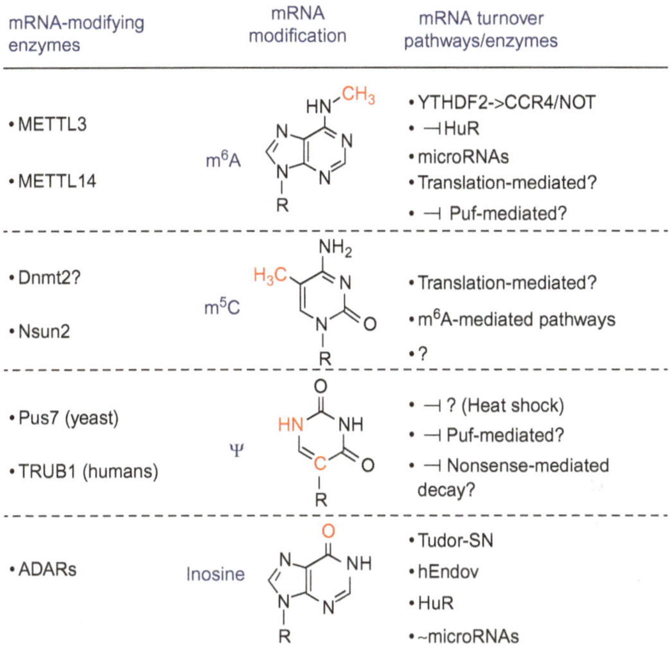

Fig. 5 Major internal base modifications found in mRNAs which are known to influence turnover. Shown are the chemical structures of major base modifications, the known enzymes that catalyze these modifications in eukaryotic mRNAs, and the enzymes or RNA degradation pathways that are modulated by each modification. Pathways for which the modification has not been shown to regulate the turnover of mRNAs in a physiological manner (e.g., based solely on in vitro studies) are labeled with a *question mark*. Unknown degradation pathways are indicated with a *question mark*. In the case of Dnmt2, its involvement in mRNA modification has not been directly demonstrated, thus the *question mark*. The ~ indicates that inosine can change the specificity of mRNA/ microRNA interactions. See text for more detail.

[83]. Therefore, recruitment of the CCR4–NOT complex by YTHDF2 accelerates the turnover of m^6A-containing mRNAs through the deadenylation–decapping turnover pathway. The localization of YTHDF2 in cytoplasmic P-bodies [80] where components of the mRNA degradation pathway are concentrated provided further circumstantial evidence for the involvement of YTHDF2 in mRNA decay. In addition to promoting deadenylation, m^6A was proposed to block the binding of the RNA-binding protein HuR (which promotes mRNA stability) and to increase microRNA binding (Fig. 5), leading to increased RNA turnover [81]. Therefore, the role of m^6A in controlling mRNA turnover is mechanistically complex and can involve a variety of degradation pathways.

The involvement of m^6A in promoting mRNA decay is essential to control several key transitions in the development of multicellular organisms. For instance, inactivation of Ythdf2 in zebrafish stabilizes maternally inherited transcripts that contain m^6A, which in turn results in developmental delays [84]. These results show that the decay of maternal transcripts must be tightly controlled through the m^6A pathway to ensure the proper timing of the maternal to zygotic transition. m^6A destabilizes mRNAs coding for developmental regulatory factors in embryonic stem cells [81], and YTHDF2 was found to regulate the turnover of key naïve pluripotency-promoting transcripts that contain m^6A such as Klf4, Nanog, Sox2, and Zfp42 [85], demonstrating the importance of m^6A-mediated degradation in the switch from pluripotency to differentiation. m^6A modification is also important for the processing or degradation of mRNAs that code for proteins which control the circadian clock [86], and an increase of m^6A modification in response to UV damage was recently reported [87], although it is not clear whether this increase impacts the degradation of m^6A-modified transcripts in these conditions. Nevertheless, these studies show a ubiquitous role for the m^6A modification in controlling key physiological processes.

In addition to this major function in promoting mRNA turnover through YTHDF2, the m^6A modification could also influence mRNA stability by modulating the binding of other factors that accelerate mRNA decay. For instance, m^6A was found to decrease the binding of the Puf family protein Pumilio2 in vitro [88]. Because Puf proteins can promote the decay of their mRNA targets [89], it is possible that m^6A might regulate Puf proteins-mediated mRNA decay (Fig. 5), although no physiological evidence was provided in this study. It is unclear whether the function for m^6A in promoting mRNA decay is conserved in lower eukaryotes. Representatives of the YTH domain family m^6A readers have been identified in *S. cerevisiae* [90] and *Schizosaccharomyces pombe*. These proteins have been shown to control the decay of meiotic transcripts [91], unspliced mRNAs [92], or of mRNAs involved in phosphate metabolism [90]. However, the mode of RNA binding of the *S. pombe* YTH domain protein Mmi1 seems to be different from that which typically recognizes m^6A, and the Mmi1-predicted m^6A-binding pocket does not seem capable of binding its purported substrate [93]. Therefore, the function of these YTH domain proteins in promoting decay seems to be independent from their ability to recognize m^6A, and based on current evidence, it is unlikely that this modification plays a major role in promoting mRNA turnover in the fungal species examined thus far.

Besides a direct role for m^6A in promoting decay through the recruitment of YTH domain proteins or other factors, m^6A can also indirectly affect mRNA turnover through its impact on various aspects of mRNA metabolism [74]. Because m^6A can affect alternative splicing [78], its presence could result in different mRNA isoforms that each have their own stabilities due to the inclusion or exclusion of various regulatory elements. m^6A also promotes mRNA export [86,94]. Therefore, the level of methylation of specific mRNAs will impact their cellular localization and thus their exposure to nuclear vs cytoplasmic degradation pathways. In addition, the YTHDF1 m^6A reader recruits translation initiation factors to promote translation initiation of m^6A-containing mRNAs [95]. Therefore, any RNA degradation pathway mechanistically connected to translation will be indirectly impacted by this modification as well. Overall because m^6A impacts so many RNA processing pathways, it may directly or indirectly influence the turnover of mRNAs that are subject to this modification in many ways.

3.2.2 Other Internal Methylation Events

Besides m^6A, methylation of internal nucleotides at other positions could also influence mRNA turnover. For instance, methylation at position 1 of adenosines (m^1A) was recently found to be more abundant near translation initiation codons upstream from the first splice site, and in highly structured RNA regions near translation initiation sites [96]. The presence of m^1A is correlated with an increase in protein expression [96], although the precise mechanism responsible for this effect has not been identified. To date, there is no direct evidence that links the presence of this modification to mRNA turnover. However, given the impact of RNA modifications on translation and the mechanistic interconnections between translation and mRNA decay [97–99], it is possible that m^1A will emerge in the next few years as one of the determinants of mRNA degradation. Finally, m^5C is also present in eukaryotic mRNAs [72]. A recent study showed that it can directly enhance translation and also increase the efficiency of m^6A formation, thereby indirectly impacting translation and possibly RNA turnover through that modification (Fig. 5) [100].

3.3 Impact of Pseudouridylation on mRNAs Turnover

Recent genomic studies have detected the presence of pseudouridines (Ψ) in mRNAs from yeast and human cells [101–103]. In addition, the presence of these modifications is dynamically regulated, as Ψ levels in mRNAs vary depending on the growth culture conditions [102] and increases dramatically during heat shock [103] in yeast. The presence of Ψ in mammalian mRNAs can also be regulated during serum starvation [102]. Most Ψ in mRNAs seem to be synthesized by protein-only pseudouridines synthases as opposed to box H/ACA-snoRNA-mediated pseudouridylation events. A direct example of the impact of Ψ on mRNA stability was provided by the demonstration that in yeast, heat shock induces over 200 Pus7-mediated pseudouridylation events [103]. Furthermore, inactivation of Pus7 results in a reduction of the level of transcripts for which pseudouridylation is induced during heat shock, suggesting that these Ψ play a role in conferring stability to these specific mRNAs during heat shock in yeast (Fig. 5). However, the precise mechanism by which Ψ increases the stability of these mRNAs remains to be established. In mammalian cells, TRUB1 is the major pseudouridine synthase for mRNAs (Fig. 5), and in contrast to the results observed in yeast, the level of mRNAs that are pseudouridylated by TRUB1 did not change upon TRUB1 knockdown [104]. This result suggested that the function of Ψ in stabilizing mRNAs might not be conserved among eukaryotes. However, this study did not analyze the impact of TRUB1 knockdown on pseudouridylation during heat shock or stress, as was done previously in yeast for Pus7. Thus, the function of Ψ in mammalian mRNAs stability might not be apparent unless cells are challenged in stress conditions.

Because Ψ impact the secondary structure of RNA by stabilizing base pairs, the presence of these modified nucleosides can in principle modulate any degradation pathway that involves the formation of specific RNA secondary or tertiary structures containing this modification. This would be the case for microRNA-mediated gene regulation, or for proteins that promote mRNA decay or stability by binding to specific secondary or tertiary structures in which pseudouridines are found. In addition, the presence of Ψ could also directly affect protein binding to single-stranded RNA. A recent study showed that introduction of Ψ into a simple RNA-binding motif for the Puf protein Pumilio2 reduced the binding affinity of this protein for its RNA motif in vitro [88]. If regulated pseudouridylation can be demonstrated in vivo at the binding sites for some of the Puf proteins, it may provide evidence for a role for pseudouridylation in regulating RNA turnover mediated by this class of proteins (Fig. 5). It is likely that additional examples of the impact of this modification on mRNA turnover will emerge with a more extensive and dynamic characterization of Ψ sites in eukaryotic mRNAs.

A final aspect of the potential impact of pseudouridylation on mRNA stability was provided by the demonstration that UAA, UAG, or UGA stop codons can be artificially pseudouridylated by expression of synthetic guide RNAs [105]. These synthetic small RNAs promote the conversion of the uridine of these stop codons to Ψ through the box H/ACA snoRNP pathway [105]. Engineered pseudouridylation of these nonsense codons resulted in the translational machinery reading these nonsense codons as specific amino acids, as ΨAA and ΨAG are now read as serine and threonine, and ΨGA as tyrosine and phenylalanine. While this recoding does not directly impact mRNA turnover, pseudouridylation of stop codons may provide a mechanism to influence the turnover of mRNAs that are subject to degradation by the nonsense-mediated decay (NMD) pathway. This degradation system affects mRNA containing premature stop codons; however, if these premature stop codons can be artificially pseudouridylated, they would then be read as normal codons and would potentially escape degradation by NMD (Fig. 5). Therefore, inducing this modification at sequences of premature stop codons may provide a mechanism to limit the impact of NMD on specific transcripts, for instance in genetic diseases when mRNAs are produced from genes that contain mutations causing premature translation termination [106,107].

3.4 Inosine Modification and mRNA Turnover

Adenosine to inosine (A–I) modification of mRNAs is catalyzed by adenosine deaminases that act on RNA enzymes (ADARs). These enzymes have a preference for double-stranded RNA regions, and the conversion of A–I (also described as editing in this section) will change the base pairing properties of the modified bases, as they will now base pair similarly to guanine. As was done previously, we will separate the direct effects of inosine on mRNA stability from those that arise from indirect effects. Introduction of inosine into double-stranded regions of RNA can trigger degradation through the recruitment of ribonucleases specific for inosine-containing RNAs. The first enzyme proposed to mediate this effect is the Tudor-staphylococcal nuclease (SN), which was shown to promote cleavage of hyperedited double-stranded RNAs in extracts [108]. Most of the results described in this study demonstrate a biochemical activity for Tudor-SN on hyperedited substrates in vitro. However, it is unclear from these studies what fraction of mRNAs containing inosine are subject to cleavage by Tudor-SN in vivo, and whether or not Tudor-SN was the direct endonuclease. Tudor-SN was found to be localized in cytoplasmic stress granules [109], so it is possible that some of its target transcripts may be preferentially regulated during stress conditions. The second nuclease known to be specific for inosine-containing RNAs is human endonuclease V (hEndoV), which can cleave a variety of inosine-containing RNA substrates in vitro [35,110]. One study suggested that hEndoV is the actual inosine endonuclease, with Tudor-SN providing a stimulatory activity to hEndoV [110]. Interestingly, hEndoV also localizes to cytoplasmic stress granules [111], suggesting a function similar to Tudor-SN in potentially regulating the levels of inosine-containing mRNAs during stress. Overall while it is clear that inosine-specific nuclease activities exist in eukaryotic cells (Fig. 5), their contribution to mRNA cleavage and turnover remains poorly understood. Therefore, the physiological impact of the cleavage of hyperedited dsRNA by Tudor-SN and/or hEndoV needs to be explored more thoroughly.

ADAR enzymes can also play a major role in mRNA stability by regulating the binding of RNA-binding proteins that influence mRNA decay. The influence of ADARs on the level of their target mRNAs was measured globally, and in general their impact on mRNA levels does not correlate well with their capacity to promote inosine formation in these mRNAs [112]. Rather, it appears that ADARs' major impact on mRNA stability is through the recruitment of the RNA-binding protein HuR (ELAVL1), which is a major regulator of mRNA decay. ADAR1 interacts with HuR in an RNA-dependent manner, and most mRNAs downregulated in the absence of ADAR1 contain HuR-binding sites [112]. Thus, ADAR1 seems to influence the stability of its mRNA targets through recruitment of HuR to target sites located nearby. ADARs can also regulate the levels of several mRNAs and ncRNAs by preventing the binding of the decay factor PARN, which is a poly(A)-specific nuclease [113]. This function also seems independent from editing, as a catalytically inactive ADAR mutant is able to fulfill the functions of the enzyme in controlling decay. Thus, the impact of ADARs on the stability of their mRNA targets seems to be mostly independent from their ability to promote the formation of inosine.

A major consequence of inosine modification on mRNA turnover is due to the impact of inosine on microRNA-mediated gene regulation. The presence of inosine can influence two major steps of this pathway: (i) the biogenesis of microRNAs and (ii) the specificity of targeting of mRNAs by microRNAs. Regarding the impact of inosine on microRNAs biogenesis, it was shown that primary precursors of microRNAs can be edited by ADAR1 or ADAR2 [114,115]. The presence of inosine in these precursors has two detrimental effects on the production of mature microRNAs

[114,115]. First, inosine reduces the efficiency of processing of the primary precursors by Drosha; second, inosine-containing precursors now become a target for inosine-specific ribonucleases such as Tudor-SN and/or hEndoV. These combined effects of inosine modification on the primary precursors of microRNAs result in a decrease of the levels of microRNAs produced from these precursors, with a concomitant increase in the target mRNAs. This process can be regulated, as precursors of the miR-151 microRNA were shown to be edited during early development, which results in precursor degradation by Tudor-SN [116]. More generally, precursors of microRNAs containing inosine are degraded during postzygotic stages in mice [116]. These results show that A–I editing of microRNAs precursors can regulate their biogenesis during development, which in turn directly affects the expression of their mRNA targets. Yet, the effect of ADARs on microRNA biogenesis is complex because ADAR1 was also shown to heterodimerize with the RNase III enzyme Dicer to *increase* the efficiency of microRNA processing [117], thus playing a positive role in microRNA biogenesis. The complexity of the direct and indirect effects of ADARs on small RNA biogenesis was also demonstrated genome-wide in *Caenorhabditis elegans* [118]. Finally, ADAR1 can also bind siRNAs independently from RNA editing in mammalian cells, which limits the efficacy of siRNA treatment [119]. Therefore, ADARs appear to be a double-edged sword for microRNA- and small RNA-mediated gene silencing, as the suppressive effect of inosine modification on microRNA precursor processing and stability can be counterbalanced by ADARs' ability to promote microRNA processing through their association with Dicer, and their effect on siRNA availability. Nevertheless, these enzymes play key functions in controlling cellular differentiation through editing of microRNA precursors. For instance, hyperediting of the Let-7 microRNA precursor by ADAR1 in leukemia was shown to promote leukemia stem cells renewal [115], highlighting the importance of fine tuning microRNA precursors editing during cell differentiation.

Inosine also influences microRNA-mediated mRNA regulation by affecting RNA duplex formation, which is critical in determining the specificity of microRNAs–mRNA interactions (Fig. 5). This was first demonstrated by showing that introduction of inosine into the miR-376 microRNA resulted in the repression of a different group of mRNAs relative to those repressed by the unmodified miR-376 [120]. If the edited site resides in the seed sequence of the microRNA, the modification results in a change of specificity for the microRNAs [121] because inosine preferentially base pairs with C. Reciprocally, A–I editing in 3′-UTR sequences of mRNAs will change the potential targeting of these UTRs by microRNAs and can create new microRNA-binding sites [122]. Regulated editing was shown to be potentially important in the regulation of oncogenes and tumor suppressor genes through microRNAs during cellular transformation [123].

Finally, inosine can influence mRNA stability through indirect effects. Inosine-containing mRNAs have been found to be restricted to the nucleus in LPS-stimulated immune cells [124], preventing their exposure to the cytoplasmic mRNA decay machinery. Nuclear retention of these mRNAs, however, would subject them to preferential degradation by nuclear degradation pathways such as the nuclear exosome. This may result in an increase or decrease in stability, depending on the relative efficiency of each of these degradation pathways on specific mRNAs. Inosine also impacts alternative splicing, which in turn could result in mRNA isoforms with different stabilities. This is especially true if the alternatively spliced species contain binding sites for distinct RNA-binding proteins that modulate their stability. Therefore, inosine and ADARs can affect mRNA stability and expression in many ways through a combination of both direct and indirect effects.

4. CONCLUDING REMARKS

In this chapter, I have attempted to summarize most of the current knowledge of the impact of RNA modifications and RNA-modifying enzymes on the activities of ribonucleases in eukaryotes. Some modifying enzymes can influence RNA stability independently from their role in modifying RNA, and the effect of modifications on turnover can be complex due to the integration of direct and indirect effects. While most of the examples described have focused on tRNAs, microRNAs, and mRNAs, it is possible that in the next few years, new studies will highlight a role for these modifications in controlling the stability of other classes of eukaryotic transcripts. It is also likely that the genomic repertoire of RNA modifications will expand—both in the scope of the sites where these modifications are found, as well as in the chemical diversity of these modifications. Thus, it is expected that some of novel sites or types of modifications will emerge as new regulatory signals for RNA degradation.

ACKNOWLEDGMENTS

I sincerely apologize to authors and colleagues whose work could not be cited due to space constraints. I would like to thank Edouard Bertrand, Remy Bordonné, and Anthony Henras for helpful discussions, and Juan Alfonzo, Peter Beal, Shahriyar Jahanbakhsh, Jason Gabunilas, Mike Kiledjian, and Charles Wang for critical reading and suggestions on the manuscript. Work in my laboratory is supported by Grant GM61518 from NIGMS.

REFERENCES

[1] R. Parker, RNA degradation in Saccharomyces cerevisae, Genetics 191 (2012) 671–702.

[2] G. Chanfreau, F. Tamanoi, Eukaryotic RNases and Their Partners in RNA Degradation and Biogenesis Part A, Enzymes 30 (2012) x–xii.

[3] A.W. Johnson, Rat1p and Xrn1p are functionally interchangeable exoribonucleases that are restricted to and required in the nucleus and cytoplasm, respectively, Mol. Cell. Biol. 17 (1997) 6122–6130.

[4] S. Lykke-Andersen, D.E. Brodersen, T.H. Jensen, Origins and activities of the eukaryotic exosome, J. Cell Sci. 122 (2009) 1487–1494.

[5] D.L. Makino, F. Halbach, E. Conti, The RNA exosome and proteasome: common principles of degradation control, Nat. Rev. Mol. Cell Biol. 14 (2013) 654–660.

[6] K. Januszyk, C.D. Lima, The eukaryotic RNA exosome, Curr. Opin. Struct. Biol. 24 (2014) 132–140.

[7] J. Anderson, L. Phan, R. Cuesta, B.A. Carlson, M. Pak, K. Asano, G.R. Bjork, M. Tamame, A.G. Hinnebusch, The essential Gcd10p-Gcd14p nuclear complex is required for 1-methyladenosine modification and maturation of initiator methionyl-tRNA, Genes Dev. 12 (1998) 3650–3662.

[8] S. Kadaba, A. Krueger, T. Trice, A.M. Krecic, A.G. Hinnebusch, J. Anderson, Nuclear surveillance and degradation of hypomodified initiator tRNAMet in S. cerevisiae, Genes Dev. 18 (2004) 1227–1240.

[9] S. Vanacova, J. Wolf, G. Martin, D. Blank, S. Dettwiler, A. Friedlein, H. Langen, G. Keith, W. Keller, A new yeast poly(A) polymerase complex involved in RNA quality control, PLoS Biol. 3 (2005) e189.

[10] F. Liu, W. Clark, G. Luo, X. Wang, Y. Fu, J. Wei, X. Wang, Z. Hao, Q. Dai, G. Zheng, et al., ALKBH1-mediated tRNA demethylation regulates translation, Cell 167 (2016) 816–828.e16.

[11] A. Alexandrov, I. Chernyakov, W. Gu, S.L. Hiley, T.R. Hughes, E.J. Grayhack, E.M. Phizicky, Rapid tRNA decay can result from lack of nonessential modifications, Mol. Cell 21 (2006) 87–96.

[12] I. Chernyakov, J.M. Whipple, L. Kotelawala, E.J. Grayhack, E.M. Phizicky, Degradation of several hypomodified mature tRNA species in Saccharomyces cerevisiae is mediated by Met22 and the 5′-3′ exonucleases Rat1 and Xrn1, Genes Dev. 22 (2008) 1369–1380.

[13] B. Dichtl, A. Stevens, D. Tollervey, Lithium toxicity in yeast is due to the inhibition of RNA processing enzymes, EMBO J. 16 (1997) 7184–7195.

[14] J.M. Dewe, J.M. Whipple, I. Chernyakov, L.N. Jaramillo, E.M. Phizicky, The yeast rapid tRNA decay pathway competes with elongation factor 1A for substrate tRNAs and acts on tRNAs lacking one or more of several modifications, RNA 18 (2012) 1886–1896.

[15] J.M. Whipple, E.A. Lane, I. Chernyakov, S. D'Silva, E.M. Phizicky, The yeast rapid tRNA decay pathway primarily monitors the structural integrity of the acceptor and T-stems of mature tRNA, Genes Dev. 25 (2011) 1173–1184.

[16] M.P. Guy, D.L. Young, M.J. Payea, X. Zhang, Y. Kon, K.M. Dean, E.J. Grayhack, D.H. Mathews, S. Fields, E.M. Phizicky, Identification of the determinants of tRNA function and susceptibility to rapid tRNA decay by high-throughput in vivo analysis, Genes Dev. 28 (2014) 1721–1732.
[17] T. Ohira, T. Suzuki, Precursors of tRNAs are stabilized by methylguanosine cap structures, Nat. Chem. Biol. 12 (2016) 648–655.
[18] D.M. Thompson, C. Lu, P.J. Green, R. Parker, tRNA cleavage is a conserved response to oxidative stress in eukaryotes, RNA 14 (2008) 2095–2103.
[19] S.R. Lee, K. Collins, Starvation-induced cleavage of the tRNA anticodon loop in Tetrahymena thermophila, J. Biol. Chem. 280 (2005) 42744–42749.
[20] S. Yamasaki, P. Ivanov, G.F. Hu, P. Anderson, Angiogenin cleaves tRNA and promotes stress-induced translational repression, J. Cell Biol. 185 (2009) 35–42.
[21] P. Ivanov, M.M. Emara, J. Villen, S.P. Gygi, P. Anderson, Angiogenin-induced tRNA fragments inhibit translation initiation, Mol. Cell 43 (2011) 613–623.
[22] N. Luhtala, R. Parker, Structure-function analysis of Rny1 in tRNA cleavage and growth inhibition. PLoS One 7 (7) (2012) e41111, http://dx.doi.org/10.1371/journal.pone.0041111. Epub 2012 Jul 19.
[23] D.M. Thompson, R. Parker, The RNase Rny1p cleaves tRNAs and promotes cell death during oxidative stress in Saccharomyces cerevisiae, J. Cell Biol. 185 (2009) 43–50.
[24] K.L. Andersen, K. Collins, Several RNase T2 enzymes function in induced tRNA and rRNA turnover in the ciliate Tetrahymena, Mol. Biol. Cell 23 (2012) 36–44.
[25] S.K. Saxena, S.M. Rybak, R.T. Davey, R.J. Youle, E.J. Ackerman, Angiogenin is a cytotoxic, tRNA-specific ribonuclease in the RNase a superfamily, J. Biol. Chem. 267 (1992) 21982–21986.
[26] M. Schaefer, T. Pollex, K. Hanna, F. Tuorto, M. Meusburger, M. Helm, F. Lyko, RNA methylation by Dnmt2 protects transfer RNAs against stress-induced cleavage, Genes Dev. 24 (2010) 1590–1595.
[27] S. Blanco, S. Dietmann, J.V. Flores, S. Hussain, C. Kutter, P. Humphreys, M. Lukk, P. Lombard, L. Treps, M. Popis, et al., Aberrant methylation of tRNAs links cellular stress to neuro-developmental disorders, EMBO J. 33 (2014) 2020–2039.
[28] J.V. Flores, L. Cordero-Espinoza, F. Oeztuerk-Winder, A. Andersson-Rolf, T. Selmi, S. Blanco, J. Tailor, S. Dietmann, M. Frye, Cytosine-5 RNA methylation regulates neural stem cell differentiation and motility, Stem Cell Reports 8 (2017) 112–124.
[29] M. Wang, H. Liu, J. Zheng, B. Chen, M. Zhou, W. Fan, H. Wang, X. Liang, X. Zhou, G. Eriani, et al., A deafness- and diabetes-associated tRNA mutation causes deficient pseudouridinylation at position 55 in $tRNA^{Glu}$ and mitochondrial dysfunction, J. Biol. Chem. 291 (2016) 21029–21041.
[30] J. Lu, B. Huang, A. Esberg, M.J.O. Johansson, A.S. Byström, The Kluyveromyces lactis γ-toxin targets tRNA anticodons, RNA 11 (2005) 1648–1654.
[31] J. Lu, A. Esberg, B. Huang, A.S. Byström, Kluyveromyces lactis γ-toxin, a ribonuclease that recognizes the anticodon stem loop of tRNA, Nucleic Acids Res. 36 (2008) 1072–1080.
[32] R. Klassen, J.P. Paluszynski, S. Wemhoff, A. Pfeiffer, J. Fricke, F. Meinhardt, The primary target of the killer toxin from Pichia acaciae is tRNA Gln, Mol. Microbiol. 69 (2008) 681–697.
[33] B. Meineke, A. Kast, B. Schwer, F. Meinhardt, S. Shuman, R. Klassen, A fungal anticodon nuclease ribotoxin exploits a secondary cleavage site to evade tRNA repair, RNA 18 (2012) 1716–1724.
[34] F. Frohloff, L. Fichtner, D. Jablonowski, K.D. Breunig, R. Schaffrath, Saccharomyces cerevisiae elongator mutations confer resistance to the Kluyveromyces lactis zymocin, EMBO J. 20 (2001) 1993–2003.
[35] E. Sebastian Vik, M. Sameen Nawaz, P. Strøm Andersen, C. Fladeby, M. Bjørås, B. Dalhus, I. Alseth, Endonuclease V cleaves at inosines in RNA, Nat. Commun. 4 (2013) 2271.

[36] X.H. Liang, Q. Liu, M.J. Fournier, rRNA modifications in an intersubunit bridge of the ribosome strongly affect both ribosome biogenesis and activity, Mol. Cell 28 (2007) 965–977.
[37] F.J. LaRiviere, S.E. Cole, D.J. Ferullo, M.J. Moore, A late-acting quality control process for mature eukaryotic rRNAs, Mol. Cell 24 (2006) 619–626.
[38] S.E. Cole, F.J. LaRiviere, C.N. Merrikh, M.J. Moore, A convergence of rRNA and mRNA quality control pathways revealed by mechanistic analysis of nonfunctional rRNA decay, Mol. Cell 34 (2009) 440–450.
[39] D. Tollervey, H. Lehtonen, R. Jansen, H. Kern, E.C. Hurt, Temperature-sensitive mutations demonstrate roles for yeast fibrillarin in pre-rRNA processing, pre-rRNA methylation, and ribosome assembly, Cell 72 (1993) 443–457.
[40] Y. Zebarjadian, T. King, M.J. Fournier, L. Clarke, J. Carbon, Point mutations in yeast CBF5 can abolish in vivo pseudouridylation of rRNA, Mol. Cell. Biol. 19 (1999) 7461–7472.
[41] A.M. Kleinschmidt, T. Pederson, RNA processing and ribonucleoprotein assembly studied in vivo by RNA transfection, Proc. Natl. Acad. Sci. U.S.A. 87 (1990) 1283–1287.
[42] G. Dönmez, K. Hartmuth, R. Lührmann, Modified nucleotides at the 5′ end of human U2 snRNA are required for spliceosomal E-complex formation, RNA 10 (2004) 1925–1933.
[43] J. Mouaikel, C. Verheggen, E. Bertrand, J. Tazi, R. Bordonne, Hypermethylation of the cap structure of both yeast snRNAs and sno RNAs requires a conserved methyltransferase that is localized to the nucleolus, Mol. Cell 9 (2002) 891–901.
[44] J.M. Adams, S. Cory, Modified nucleosides and bizarre 5′-termini in mouse myeloma mRNA, Nature 255 (1975) 28–33.
[45] C.M. Wei, A. Gershowitz, B. Moss, Methylated nucleotides block 5′ terminus of HeLa cell messenger RNA, Cell 4 (1975) 379–386.
[46] K.L. Perry, K.P. Watkins, N. Agabian, Trypanosome mRNAs have unusual 'cap 4' structures acquired by addition of a spliced leader, Proc. Natl. Acad. Sci. U.S.A. 84 (1987) 8190–8194.
[47] J.D. Bangs, P.F. Crain, T. Hashizume, J.A. McCloskey, J.C. Boothroyd, Mass spectrometry of mRNA cap 4 from trypanosomatids reveals two novel nucleosides, J. Biol. Chem. 267 (1992) 9805–9815.
[48] M. Parsons, R.G. Nelson, K.P. Watkins, N. Agabian, Trypanosome mRNAs share a common 5??? spliced leader sequence, Cell 38 (1984) 309–316.
[49] J.G. Bird, Y. Zhang, Y. Tian, N. Panova, I. Barvík, L. Greene, M. Liu, B. Buckley, L. Krásný, J.K. Lee, et al., The mechanism of RNA 5′ capping with NAD(+), NADH and desphospho-CoA. Nature 535 (2016) 444–447, http://dx.doi.org/10.1038/nature18622.
[50] R.W. Walters, T. Matheny, L.S. Mizoue, B.S. Rao, D. Muhlrad, R. Parker, Identification of NAD^+ capped mRNAs in *Saccharomyces cerevisiae*, Proc. Natl. Acad. Sci. U.S.A. 114 (2017) 480–485.
[51] X. Jiao, S.K. Doamekpor, J.G. Bird, B.E. Nickels, L. Tong, R.P. Hart, M. Kiledjian, 5′ End nicotinamide adenine dinucleotide cap in human cells promotes RNA decay through DXO-mediated deNADding, Cell 168 (2017) 1015–1027.e10.
[52] C.J. Decker, R. Parker, A turnover pathway for both stable and unstable mRNAs in yeast: evidence for a requirement for deadenylation, Genes Dev. 7 (1993) 1632–1643.
[53] P. Couttet, M. Fromont-Racine, D. Steel, R. Pictet, T. Grange, Messenger RNA deadenylylation precedes decapping in mammalian cells, Proc. Natl. Acad. Sci. U.S.A. 94 (1997) 5628–5633.
[54] T. Dunckley, R. Parker, The DCP2 protein is required for mRNA decapping in Saccharomyces cerevisiae and contains a functional MutT motif, EMBO J. 18 (1999) 5411–5422.

[55] E. van Dijk, N. Cougot, S. Meyer, S. Babajko, E. Wahle, B. Séraphin, Human Dcp2: a catalytically active mRNA decapping enzyme located in specific cytoplasmic structures, EMBO J. 21 (2002) 6915–6924.
[56] Z. Wang, X. Jiao, A. Carr-Schmid, M. Kiledjian, The hDcp2 protein is a mammalian mRNA decapping enzyme, Proc. Natl. Acad. Sci. U.S.A. 99 (2002) 12663–12668.
[57] N.G. D'Lima, J. Ma, L. Winkler, Q. Chu, K.H. Loh, E.O. Corpuz, B.A. Budnik, J. Lykke-Andersen, A. Saghatelian, S.A. Slavoff, A human microprotein that interacts with the mRNA decapping complex, Nat. Chem. Biol. 13 (2) (2017) 174–180, http://dx.doi.org/10.1038/nchembio.2249. Epub 2016 Dec 5.
[58] M. Arribas-Layton, D. Wu, J. Lykke-Andersen, H. Song, Structural and functional control of the eukaryotic mRNA decapping machinery, Biochim. Biophys. Acta, Gene Regul. Mech. 1829 (2013) 580–589.
[59] M.G. Song, Y. Li, M. Kiledjian, Multiple mRNA decapping enzymes in mammalian cells, Mol. Cell 40 (2010) 423–432.
[60] Y. Li, M. Song, M. Kiledjian, Differential utilization of decapping enzymes in mammalian mRNA decay pathways, RNA 17 (2011) 419–428.
[61] M.-G. Song, S. Bail, M. Kiledjian, Multiple Nudix family proteins possess mRNA decapping activity, RNA 19 (2013) 390–399.
[62] E. Grudzien-Nogalska, X. Jiao, M.-G. Song, R.P. Hart, M. Kiledjian, Nudt 3 is an mRNA decapping enzyme that modulates cell migration, RNA 22 (2016) 773–781, http://dx.doi.org/10.1261/rna.055699.115.
[63] A. Stevens, T.L. Poole, 5′-exonuclease-2 of Saccharomyces cerevisiae. Purification and features of ribonuclease activity with comparison to 5′-exonuclease-1, J. Biol. Chem. 270 (1995) 16063–16069.
[64] X. Jiao, S. Xiang, C. Oh, C.E. Martin, L. Tong, M. Kiledjian, Identification of a quality-control mechanism for mRNA 5′-end capping, Nature 467 (2010) 608–611.
[65] J.H. Chang, X. Jiao, K. Chiba, C. Oh, C.E. Martin, M. Kiledjian, L. Tong, Dxo1 is a new type of eukaryotic enzyme with both decapping and 5′-3′ exoribonuclease activity, Nat. Struct. Mol. Biol. 19 (2012) 1011–1017.
[66] X. Jiao, J.H. Chang, T. Kilic, L. Tong, M. Kiledjian, A mammalian pre-mRNA 5′ end capping quality control mechanism and an unexpected link of capping to pre-mRNA processing, Mol. Cell 50 (2013) 104–115.
[67] S. Daffis, K.J. Szretter, J. Schriewer, J. Li, S. Youn, J. Errett, T.-Y. Lin, S. Schneller, R. Zust, H. Dong, et al., 2′-O methylation of the viral mRNA cap evades host restriction by IFIT family members, Nature 468 (2010) 452–456.
[68] R. Zust, L. Cervantes-Barragan, M. Habjan, R. Maier, B.W. Neuman, J. Ziebuhr, K.J. Szretter, S.C. Baker, W. Barchet, M.S. Diamond, et al., Ribose 2′-O-methylation provides a molecular signature for the distinction of self and non-self mRNA dependent on the RNA sensor Mda5, Nat. Immunol. 12 (2011) 137–143.
[69] P. Kumar, T.R. Sweeney, M.A. Skabkin, O.V. Skabkina, C.U.T. Hellen, T.V. Pestova, Inhibition of translation by IFIT family members is determined by their ability to interact selectively with the 5′-terminal regions of cap 0-, cap1- and 5'ppp-mRNAs, Nucleic Acids Res. 42 (2014) 3228–3245.
[70] W. Cha-Mer, A. Gershowitz, B. Moss, N6, O2′-dimethyladenosine a novel methylated ribonucleoside next to the 5′ terminal of animal cell and virus mRNAs, Nature 18 (1975) 251–253.
[71] J. Mauer, X. Luo, A. Blanjoie, X. Jiao, A.V. Grozhik, D.P. Patil, B. Linder, B.F. Pickering, J.-J. Vasseur, Q. Chen, et al., Reversible methylation of m6Am in the 5′ cap controls mRNA stability, Nature 541 (2016) 371–375.
[72] B.S. Zhao, I.A. Roundtree, C. He, Post-transcriptional gene regulation by mRNA modifications, Nat. Rev. Mol. Cell Biol. 18 (1) (2017) 31–42, http://dx.doi.org/10.1038/nrm.2016.132. Epub 2016 Nov 3.

[73] W.V. Gilbert, T.A. Bell, C. Schaening, F.F. Davis, F.W. Allen, M.A. Machnicka, K. Milanowska, O.O. Oglou, E. Purta, M. Kurkowska, et al., Messenger RNA modifications: form, distribution, and function, Science 352 (2016) 1408–1412.
[74] C.J.T. Lewis, T. Pan, A. Kalsotra, RNA modifications and structures cooperate to guide RNA-protein interactions, Nat. Rev. Mol. Cell Biol. 18 (2017) 202–210.
[75] R. Desrosiers, K. Friderici, F. Rottman, Identification of methylated nucleosides in messenger RNA from Novikoff hepatoma cells, Proc. Natl. Acad. Sci. U.S.A. 71 (1974) 3971–3975.
[76] R.P. Perry, D.E. Kelley, Existence of methylated messenger RNA in mouse L cells, Cell 1 (1974) 37–42.
[77] K.D. Meyer, Y. Saletore, P. Zumbo, O. Elemento, C.E. Mason, S.R. Jaffrey, Comprehensive analysis of mRNA methylation reveals enrichment in 3′ UTRs and near stop codons, Cell 149 (2012) 1635–1646.
[78] D. Dominissini, S. Moshitch-Moshkovitz, S. Schwartz, M. Salmon-Divon, L. Ungar, S. Osenberg, K. Cesarkas, J. Jacob-Hirsch, N. Amariglio, M. Kupiec, et al., Topology of the human and mouse m6A RNA methylomes revealed by m6A-seq, Nature 485 (2012) 201–206.
[79] Deleted in review.
[80] X. Wang, Z. Lu, A. Gomez, G.C. Hon, Y. Yue, D. Han, Y. Fu, M. Parisien, Q. Dai, G. Jia, B. Ren, T. Pan, C. He, N6-methyladenosine-dependent regulation of messenger RNA stability, Nature 505 (2014) 1–20.
[81] Y. Wang, Y. Li, J.I. Toth, M.D. Petroski, Z. Zhang, J.C. Zhao, N6-methyladenosine modification destabilizes developmental regulators in embryonic stem cells, Nat. Cell Biol. 16 (2014) 191–198.
[82] C. Temme, S. Zaessinger, S. Meyer, M. Simonelig, E. Wahle, A complex containing the CCR4 and CAF1 proteins is involved in mRNA deadenylation in Drosophila, EMBO J. 23 (2004) 2862–2871.
[83] N. Alhusaini, J. Coller, The deadenylase components Not2p, Not3p, and Not5p promote mRNA decapping, RNA 22 (2016) 709–721, http://dx.doi.org/10.1261/rna.054742.115.
[84] B.S. Zhao, X. Wang, A.V. Beadell, Z. Lu, H. Shi, A. Kuuspalu, R.K. Ho, C. He, m6A-dependent maternal mRNA clearance facilitates zebrafish maternal-to-zygotic transition, Nature 542 (2017) 475–478.
[85] S. Geula, S. Moshitch-Moshkovitz, D. Dominissini, A.A. Mansour, N. Kol, M. Salmon-Divon, V. Hershkovitz, E. Peer, N. Mor, Y.S. Manor, et al., Stem cells. m6A mRNA methylation facilitates resolution of naïve pluripotency toward differentiation, Science 347 (2015) 1002–1006.
[86] J.-M. Fustin, M. Doi, Y. Yamaguchi, H. Hida, S. Nishimura, M. Yoshida, T. Isagawa, M.S. Morioka, H. Kakeya, I. Manabe, et al., RNA-methylation-dependent RNA processing controls the speed of the circadian clock, Cell 155 (2013) 793–806.
[87] Y. Xiang, B. Laurent, C.-H. Hsu, S. Nachtergaele, Z. Lu, W. Sheng, C. Xu, H. Chen, J. Ouyang, S. Wang, et al., RNA m6A methylation regulates the ultraviolet-induced DNA damage response, Nature 543 (2017) 573–576, http://dx.doi.org/10.1038/nature21671. Epub 2017 Mar 15.
[88] P.P. Vaidyanathan, I. AlSadhan, D.K. Merriman, H. Al-Hashimi, D. Herschlag, Pseudouridine and N-6 methyladenosine modifications weaken PUF protein/RNA interactions, RNA (2017), pii: rna.060053.116. http://dx.doi.org/10.1261/rna.060053.116. [Epub ahead of print].
[89] M.A. Miller, W.M. Olivas, Roles of Puf proteins in mRNA degradation and translation, Wiley Interdiscip. Rev. RNA 2 (2011) 471–492.
[90] H.-J. Kang, S.-J. Jeong, K.-N. Kim, I.-J. Baek, M. Chang, C.-M. Kang, Y.-S. Park, C.-W. Yun, A novel protein, Pho92, has a conserved YTH domain and regulates

phosphate metabolism by decreasing the mRNA stability of PHO4 in Saccharomyces cerevisiae, Biochem. J. 457 (2014) 391–400.
[91] Y. Harigaya, H. Tanaka, S. Yamanaka, K. Tanaka, Y. Watanabe, C. Tsutsumi, Y. Chikashige, Y. Hiraoka, A. Yamashita, M. Yamamoto, Selective elimination of messenger RNA prevents an incidence of untimely meiosis, Nature 442 (2006) 45–50.
[92] C. Kilchert, S. Wittmann, M. Passoni, S. Shah, S. Granneman, L. Vasiljeva, Regulation of mRNA levels by decay-promoting introns that recruit the exosome specificity factor Mmi1, Cell Rep. 13 (2015) 2504–2515.
[93] C. Wang, Y. Zhu, H. Bao, Y. Jiang, C. Xu, J. Wu, Y. Shi, A novel RNA-binding mode of the YTH domain reveals the mechanism for recognition of determinant of selective removal by Mmi1, Nucleic Acids Res. 44 (2016) 969–982.
[94] G. Zheng, J.A. Dahl, Y. Niu, P. Fedorcsak, C.M. Huang, C.J. Li, C.B. Vågbø, Y. Shi, W.L. Wang, S.H. Song, et al., ALKBH5 is a mammalian RNA demethylase that impacts RNA metabolism and mouse fertility, Mol. Cell 49 (2013) 18–29.
[95] X. Wang, B.S. Zhao, I.A. Roundtree, Z. Lu, D. Han, H. Ma, X. Weng, K. Chen, H. Shi, C. He, N^6-methyladenosine modulates messenger RNA translation efficiency, Cell 161 (2015) 1388–1399.
[96] D. Dominissini, S. Nachtergaele, S. Moshitch-Moshkovitz, E. Peer, N. Kol, M.S. Ben-Haim, Q. Dai, A. Di Segni, M. Salmon-Divon, W.C. Clark, et al., The dynamic N1-methyladenosine methylome in eukaryotic messenger RNA, Nature 530 (2016) 441–446.
[97] W. Hu, T.J. Sweet, S. Chamnongpol, K.E. Baker, J. Coller, Co-translational mRNA decay in Saccharomyces cerevisiae, Nature 461 (2009) 225–229.
[98] W. Hu, C. Petzold, J. Coller, K.E. Baker, Nonsense-mediated mRNA decapping occurs on polyribosomes in Saccharomyces cerevisiae, Nat. Struct. Mol. Biol. 17 (2010) 244–247.
[99] V. Presnyak, N. Alhusaini, Y.H. Chen, S. Martin, N. Morris, N. Kline, S. Olson, D. Weinberg, K.E. Baker, B.R. Graveley, et al., Codon optimality is a major determinant of mRNA stability, Cell 160 (2015) 1111–1124.
[100] Q. Li, X. Li, H. Tang, B. Jiang, Y. Dou, M. Gorospe, W. Wang, NSUN2-mediated m5C methylation and METTL3/METTL14-mediated m6A methylation cooperatively enhance p21 translation, J. Cell. Biochem. (2017), http://dx.doi.org/10.1002/jcb.25957.
[101] X. Li, P. Zhu, S. Ma, J. Song, J. Bai, F. Sun, C. Yi, Chemical pulldown reveals dynamic pseudouridylation of the mammalian transcriptome, Nat. Chem. Biol. 11 (2015) 592–597.
[102] T.M. Carlile, M.F. Rojas-Duran, B. Zinshteyn, H. Shin, K.M. Bartoli, W.V. Gilbert, Pseudouridine profiling reveals regulated mRNA pseudouridylation in yeast and human cells, Nature 515 (2014) 143–146.
[103] S. Schwartz, D.A. Bernstein, M.R. Mumbach, M. Jovanovic, R.H. Herbst, B.X. León-Ricardo, J.M. Engreitz, M. Guttman, R. Satija, E.S. Lander, et al., Transcriptome-wide mapping reveals widespread dynamic-regulated pseudouridylation of ncRNA and mRNA, Cell 159 (2014) 148–162.
[104] M. Safra, R. Nir, D. Farouq, I.V. Slutzkin, S. Schwartz, TRUB1 is the predominant pseudouridine synthase acting on mammalian mRNA via a predictable and conserved code, Genome Res. 27 (2017) 393–406.
[105] J. Karijolich, Y.-T. Yu, Modifying the genetic code: converting nonsense codons into sense codons by targeted pseudouridylation, Nature 474 (2012) 395–398.
[106] P.A. Frischmeyer, H.C. Dietz, Nonsense-mediated mRNA decay in health and disease, Hum. Mol. Genet. 8 (1999) 1893–1900.
[107] J. Karijolich, Y.-T. Yu, Therapeutic suppression of premature termination codons: mechanisms and clinical considerations (review), Int. J. Mol. Med. 34 (2014) 355–362.
[108] A.D.J. Scadden, The RISC subunit Tudor-SN binds to hyper-edited double-stranded RNA and promotes its cleavage, Nat. Struct. Mol. Biol. 12 (2005) 489–496.

[109] R. Weissbach, A.D.J. Scadden, Tudor-SN and ADAR1 are components of cytoplasmic stress granules, RNA 18 (2012) 462–471.
[110] Y. Morita, T. Shibutani, N. Nakanishi, K. Nishikura, S. Iwai, I. Kuraoka, Human endonuclease V is a ribonuclease specific for inosine-containing RNA, Nat. Commun. 4 (2013) 2273.
[111] M.S. Nawaz, E.S. Vik, N. Berges, C. Fladeby, M. Bjørås, B. Dalhus, I. Alseth, Regulation of human endonuclease V activity and relocalization to cytoplasmic stress granules, J. Biol. Chem. (2016), http://dx.doi.org/10.1074/jbc.M116.730911.
[112] I.X. Wang, E. So, J.L. Devlin, Y. Zhao, M. Wu, V.G. Cheung, ADAR regulates RNA editing, transcript stability, and gene expression, Cell Rep. 5 (2013) 849–860.
[113] A. Anantharaman, V. Tripathi, A. Khan, J.-H. Yoon, D.K. Singh, O. Gholamalamdari, S. Guang, J. Ohlson, H. Wahlstedt, M. Öhman, et al., ADAR2 regulates RNA stability by modifying access of decay-promoting RNA-binding proteins, Nucleic Acids Res, (2017), pii: gkw1304. http://dx.doi.org/10.1093/nar/gkw1304. [Epub ahead of print].
[114] W. Yang, T.P. Chendrimada, Q. Wang, M. Higuchi, P.H. Seeburg, R. Shiekhattar, K. Nishikura, Modulation of microRNA processing and expression through RNA editing by ADAR deaminases, Nat. Struct. Mol. Biol. 13 (2006) 13–21.
[115] M.A. Zipeto, A.C. Court, A. Sadarangani, N.P. Delos Santos, L. Balaian, H.-J. Chun, G. Pineda, S.R. Morris, C.N. Mason, I. Geron, et al., ADAR1 activation drives leukemia stem cell self-renewal by impairing let-7 biogenesis, Cell Stem Cell 19 (2016) 177–191.
[116] J. García-López, J.D.D. Hourcade, J. Del Mazo, Reprogramming of microRNAs by adenosine-to-inosine editing and the selective elimination of edited microRNA precursors in mouse oocytes and preimplantation embryos, Nucleic Acids Res. 41 (2013) 5483–5493.
[117] H. Ota, M. Sakurai, R. Gupta, L. Valente, B.E. Wulff, K. Ariyoshi, H. Iizasa, R.V. Davuluri, K. Nishikura, ADAR1 forms a complex with dicer to promote MicroRNA processing and RNA-induced gene silencing, Cell 153 (2013) 575–589.
[118] M.B. Warf, B.A. Shepherd, W.E. Johnson, B.L. Bass, Effects of ADARs on small RNA processing pathways in C. elegans, Genome Res. 22 (2012) 1488–1498.
[119] W. Yang, Q. Wang, K.L. Howell, J.T. Lee, D.-S.C. Cho, J.M. Murray, K. Nishikura, ADAR1 RNA deaminase limits short interfering RNA efficacy in mammalian cells, J. Biol. Chem. 280 (2005) 3946–3953.
[120] Y. Kawahara, B. Zinshteyn, P. Sethupathy, H. Iizasa, A.G. Hatzigeorgiou, K. Nishikura, Redirection of silencing targets by adenosine-to-inosine editing of miRNAs, Science 315 (80) (2007) 1137–1140.
[121] H. Kume, K. Hino, J. Galipon, K. Ui-Tei, A-to-I editing in the miRNA seed region regulates target mRNA selection and silencing efficiency, Nucleic Acids Res. 42 (2014) 10050–10060.
[122] G.M. Borchert, B.I. Gilmore, R.M. Spengler, Y. Xing, W. Lanier, D. Bhattacharya, B.L. Davidson, Adenosine deamination in human transcripts generates novel microRNA binding sites, Hum. Mol. Genet. 18 (2009) 4801–4807.
[123] L. Zhang, C.-S. Yang, X. Varelas, S. Monti, Altered RNA editing in 3′ UTR perturbs microRNA-mediated regulation of oncogenes and tumor-suppressors, Sci. Rep. 6 (2016) 23226.
[124] M.G. Blango, B.L. Bass, Identification of the long, edited dsRNAome of LPS-stimulated immune cells, Genome Res. 26 (2016) 852–862.

Kolibri 360 GmbH
Ferdinand-Jühlke-Straße 7
99095 Erfurt, Deutschland
produktsicherheit@kolibri360.de